S0-CNH-581

Engineering Biosensors: Kinetics and Design Applications

This book is dedicated to my parents,
Dr. J. C. Sadana and Mrs. Jinder Sadana, to whom I owe more than
they will ever know.

CONTENTS

PREFACE

Biosensors are becoming increasingly important bioanalytical tools in the pharmaceutical, biotechnology, food, and other consumer-oriented industries. Although well developed in Europe, this technology has only recently begun to generate interest in the United States and is developing slowly. Much research is now being directed toward the development of biosensors that are versatile, economical, and simple to use.

There is a critical need to provide a better understanding of the mode of operation of biosensors with the goal being to improve its stability, specificity, response time, regenerability, and robustness. Diffusional limitations are invariably present in biosensors because of their construction and principle of operation. A better knowledge of the kinetics involved in the binding and dissociation assays of the biosensors will provide valuable physical insights into the nature of the biomolecular interactions sensed by the biosensors. In addition to these kinetics, knowledge regarding the nature of the sensor surface is an important consideration in the design. However, this aspect is sadly overlooked in many texts and publications dealing with biosensors. The main aim of this book is to address the kinetics involved in analyte-receptor binding using a novel mathematical approach called *fractals*. We will attempt to model the binding and dissociation of an analyte and a receptor using examples obtained from literature using fractal analysis. In doing so, we wish to delineate the role of the biosensor surface and diffusional limitations on the binding and dissociation reactions involved.

In the introductory chapter, we have given a background for the need for biosensors and the different types of immunoassays. Traditional kinetics are described under the influence of diffusion on antigen–antibody binding kinetics in biosensors in Chapter 2. Lateral interactions are included in Chapter 3.

In our opinion, Chapter 4 is one of the most important chapters in the book as there we first introduce the concept of fractals, fractal kinetics, and fractal dimensions. We also give a background of the factors that contribute toward heterogeneity on a biosensor surface and how it can be explained using fractal kinetics. There are a host of other parameters—such as analyte/ligand concentration, regeneration conditions, etc.—that affect biosensor performance characteristics. In Chapter 5, we try to explain the influence of these parameters on the surface and consequently on the fractal dimension values.

Havlin (1989) developed an equation for relating the rate of complex formation on the surface to the existing fractal dimension in electrochemical reactions. We have extended this idea to relate the binding rate coefficient and fractal dimension for an analyte-receptor reaction on a biosensor surface. A detailed explanation of Havlin's equation and how it can be made amenable to suit our needs can be found in Chapter 6. Just as the association between the analyte and the receptor is important, the reverse (dissociation) is equally important, perhaps more so from the viewpoint of reusability of the biosensor. Recognizing its importance, we have treated the dissociation separately in Chapter 7, where we present equations that we feel can adequately describe and model the dissociation kinetics involved. We have extended Havlin's ideas and applied them successfully, with slight modifications and reasonable justifications to model the dissociation kinetics. We feel that the analysis of binding and dissociation kinetics is our contribution in the application of fractal modeling techniques to model analyte-receptor systems.

There is a very slight shift in focus in Chapter 8 as we go back to the traditional kinetic models described in Chapters 3 and 4 to describe the problem of nonspecific binding in biosensors and how design considerations may have to be altered to account for this phenomenon. We also analyze this problem using fractals in Chapter 9.

In Chapter 10, we analyze examples from literature wherein DNA hybridization reactions have been studied using biosensors. In Chapter 11, we look at cell analyte-receptor examples, and in Chapter 12 we present examples of biomolecular interactions analyzed using the surface plasmon resonance (SPR) biosensor. The SPR biosensor is finding increasing application as an analytical technique in industrial and research laboratories. We have developed expressions for relating the fractal dimensions and binding rate coefficients, fractal dimensions/binding rate coefficients and analyte concentration, and so on.

We conclude with what in our opinion is the highlight of this book: a chapter on the biosensor market economics. What makes this chapter special is the effort that has gone into compiling it from hard-to-obtain industry and market sales figures over the last several years. Although some of the projection figures may be outdated, the chapter does give the reader a feel for the costs involved, and the realistic returns on the investment involved, and the potential for growth and improvement. Just to emphasize the point and to make it easier to understand, we have presented a 5-year economic analysis of a leading biosensor company, BIACORE AB.

We have targeted this book for graduate students, senior undergraduate students, and researchers in academia and industry. The book should be particularly interesting for researchers in the fields of biophysics, biochemical engineering, biotechnology, immunology, and applied mathematics. It can also serve as a handy reference for people directly involved in the design and manufacture of biosensors. We hope that this book will foster better interactions, facilitate a better appreciation of all perspectives, and help in advancing biosensor design and technology.

Ajit Sadana

CHAPTER 1

Introduction

1.1. BACKGROUND, DEFINITION, AND THE NEED FOR BIOSENSORS

A biosensor is a device that uses a combination of two steps: a recognition step and a transducer step. The recognition step involves a biological sensing element, or receptor, on the surface that can recognize biological or chemical analytes in solution or in the atmosphere. The receptor may be an antibody, enzyme, or a cell. This receptor is in close contact with a transducing element that converts the analyte-receptor reaction into a quantitative electrical or optical signal. The signal may be transduced by optical, thermal, electrical, or electronic elements. Lowe (1985) emphasizes that a transducer should be highly specific for the analyte of interest. Also, it should be able to respond in the appropriate concentration range and have a moderately fast response time (1–60 sec). The transducer also should be reliable, able to be miniaturized, and suitably designed for practical application. Figure 1.1 shows the principle of operation of a typical biosensor (Byfield and Abuknesha, 1994).

As early as 1985, Lowe (1985) indicated that most of the major developments in biosensor technology will come from advances in the health care field. Efficient patient care is based on frequent measurement of many analytes, such as blood cations, gases, and metabolites. Emphasizing that, for inpatient and outpatient care, key metabolites need to be monitored on tissue fluids such as blood, sweat, saliva, and urine, Lowe indicated that implantable biosensors could, for example, provide real-time data to direct drug release by

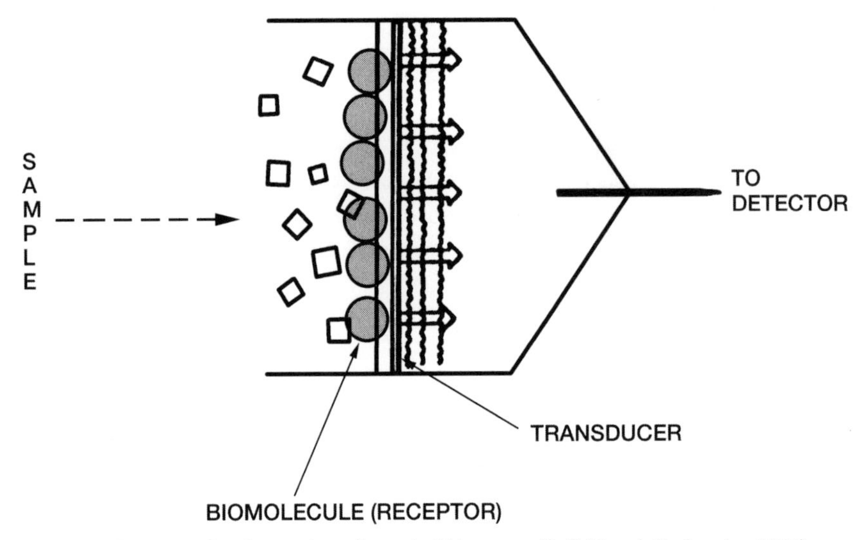

FIGURE 1.1 Principle of operation of a typical biosensor (Byfield and Abuknesha, 1994). Reprinted from *Biosensors and Bioelectronics*, vol 9, M. P. Byfield and R. A. Abuknesha, pp. 373–400, 1994, with permission of Elsevier Science.

a drug dispenser. Such a feedback system could be incorporated into, for example, an artificial pancreas. Thus, two types of biosensors could be used: *in vitro* and *in vivo*.

Byfield and Abuknesha (1994) have clearly identified the biological components that may be used as receptors. These include cofactors, antibodies, receptors, enzymes, enzyme systems, membranes, organelles, cells tissues, and organisms. As the number of biosensor applications increases (which it will, due to the need for rapid, reliable, reproducible, accurate, and sensitive analyses) the types of receptors that may be used in biosensors will increase. These same authors have also identified the different types of transducers that may be used. These include optical (fluorescence, absorbance), electrochemical (amperometric, potentiometric, conductimetric), piezoelectric, calorimetric, acoustic, and mechanical.

Let us illustrate the flexibility of a typical biosensor—for example, an enzyme biosensor—by using three different transduction elements. The glucose–glucose oxidase (analyte-receptor) biosensor is a good example of an electrochemical transduction wherein the analyte is converted to an electroactive product (Byfield and Abuknesha, 1994). Similarly, the lactate–lactate monoxygenase biosensor is a good example of an optical transduction wherein the optical properties of the enzyme are changed upon reacting with the analyte. Finally, the glucose–glucose oxidase biosensor may also have a calorimetric transduction wherein the analyte reacting with the enzyme gives off heat energy.

Among the various properties of a biosensor to be considered in the design—such as specificity, sensitivity, reproducibility, stability, regenerability, and response time—the two most important are specificity and sensitivity. Alvarez-Icaza and Bilitewski (1993) emphasize that due to the specificity of the biosensor it may be used in complex media such as blood, serum, urine, fermentation broths, and food, often with minimum sample treatment.

A simple example would be of assistance here. Lowe (1985) indicates that the concentration of certain proteins in blood serum may be as low as a few μg/L or less as compared to a total protein concentration of 70 g/L. This requires a discrimination ratio of 10^7–10^8 to specifically estimate the desired protein. Since other chemicals will also be present in the blood, an even higher discrimination ratio will be required. Lowe defines sensitivity as the ability of a biosensor to discriminate between the desired analyte and a host of potential contaminants. There are two ways to further delineate sensitivity: the smallest concentration of analyte that a biosensor can detect or the degree of discrimination between measurements at any level.

Lukosz (1991) indicates that optics provide the high sensitivity in, for example, integrated optical (IO) and surface plasmon resonance (SPR) biosensors. However, it is the biochemistry that provides the specificity of biosensors. Lukosz adds that the chemoreceptive coating on IO or SP sensors makes them biosensors. In this text we will cover different types of biosensors but we will emphasize and illustrate most of the concepts using antigen–antibody binding. By focusing on one type of example, we can develop it in detail with the hope that similar development is possible for other types of biosensors. Such further development is essential since, in spite of the vast literature available on biosensors and the increasing funding in this area, there are but a handful of commercially available biosensors (Paddle, 1996). This underscores the inherent difficulties present in these types of systems. However, due to their increasing number of applications and their potential of providing a rapid and accurate analysis of different analytes, worldwide research in this area is bound to continue at an accelerated pace.

The transduction of the biochemical signal to the electrical signal is often a critical step wherein a large fraction of the signal loss (for example, fluorescence by quenching) may occur (Sadana and Vo-Dinh, 1995). This leads to deleterious effects on the sensitivity and the selectivity of the biosensor, besides decreasing the quality of the reproducibility of the biosensor. The sensing principles for biosensors may be extended to a chemical sensor (Alarie and Vo-Dinh, 1991). Here the analyte molecule to be detected is sequestered in β-cyclodextrin molecules immobilized on the distal

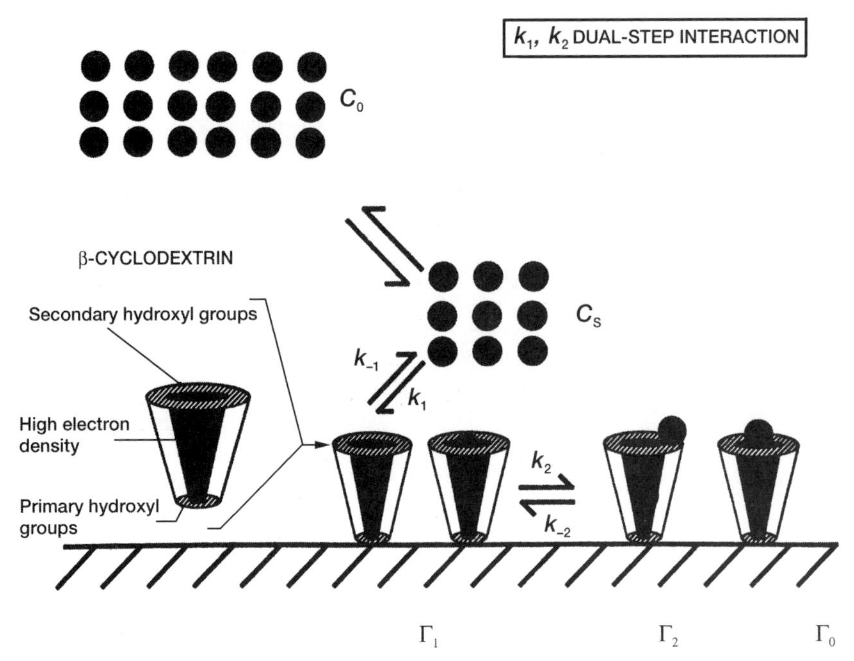

FIGURE 1.2 Truncated-cone structure of β-cyclodextrins exhibiting a hydrophobic cavity (Sadana and Vo-Dinh, 1995).
Reprinted from *Talanta*, vol 42, A. Sadana and T. Vo-Dinh, pp. 1567–1574, 1995, with permission from Elsevier Science.

end of a fiber-optic chemical sensor. Cyclodextrins are sugar molecules that possess the structure of a truncated cone with a hydrophobic cavity (Fig. 1.2). It is in this hydrophobic cavity that the analyte is complexed and placed in a hydrophobic environment. This leads to fluorescence quenching protection (for example, from water) that can lead to a fluorescence enhancement effect (Alak *et al.*, 1984).

Antibodies frequently have been used for the detection of various analytes due to their high specificity. They have been immobilized on various supports for application in immuno-diagnostic assays. Antibodies may be immobilized on, for example, optical fibers, electrodes, or semiconductor chips (Ogert *et al.*, 1992; Rosen and Rishpon, 1989; Jimbo and Saiti, 1988). Lu *et al.* (1996) indicate that different immobilization chemistries and strategies have been utilized. Linkages to solid surfaces are frequently made by glutaraldehyde, carbodiimide, and other reagents such as succinimide ester, maleinimide, and periodate. However, one has to be very careful during the immobilization procedure as quite a bit of the activity may be destroyed or become unavailable. Some of the factors that decrease the specificity of biosensors include the cross-reactivity of enzymes, nonspecific binding (analyte binding

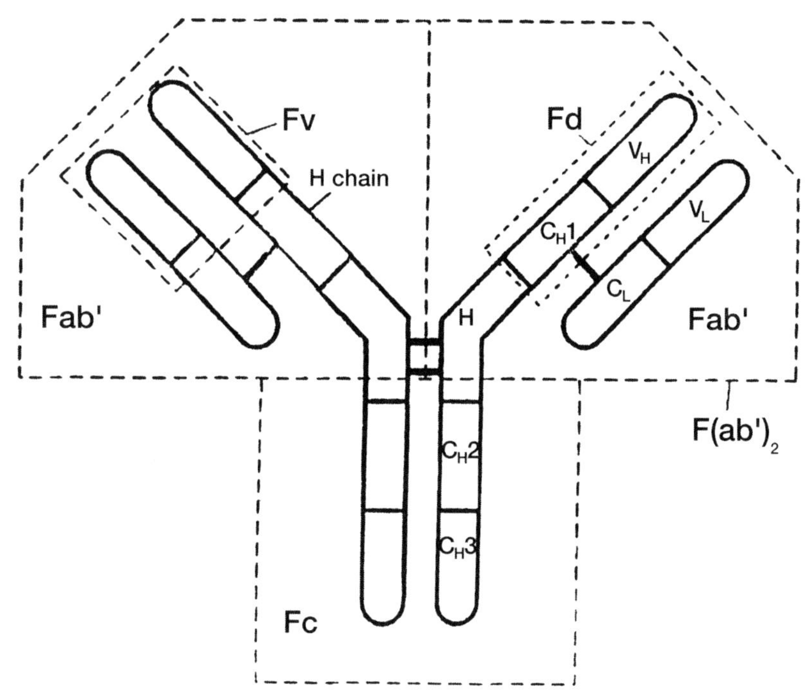

FIGURE 1.3 Schematic diagram of an antibody molecule and its two fragments (Lu *et al.*, 1996). Lu, B., Smyth, M. R., and O'Kennedy, R., *Analyst*, 121, 29–32R (1996). Reproduced by permission of the The Royal Society of Chemistry.

occurs at places where it should not), and interferences at the transducer (Scheller *et al.*, 1991).

Figure 1.3 shows the schematic diagram of an antibody molecule with its fragments (Lu *et al.*, 1996). The Fc fragment comprises the effector functions, such as complement activation, cell membrane receptor interaction, and transplacental transfer (Staines *et al.*, 1993). The $F(ab')_2$ contains two identical Fab′ fragments, which are held together by disulfide linkages in the hinge (H) region. The Fab′ fragments contain the antigen-binding site. The V_H and V_L are the variable heavy and light chains, respectively. The C_H1 and the C_H2 are the constant chains. Note that as the antibody is immobilized on a support, it generally loses some of its activity, as noted in Figure 1.4 wherein some orientations (or conformations) inhibit the formation of the antigen–antibody complex. Lu *et al.* (1996) emphasize that if the immobilization occurs through the antigen-binding site, then the ability of the antibody on the surface to bind to the antigen in solution may be lost completely, or at least to a high degree. Therefore, one needs to be careful during the immobilization procedure to preserve most, if not all, of the inherent antibody

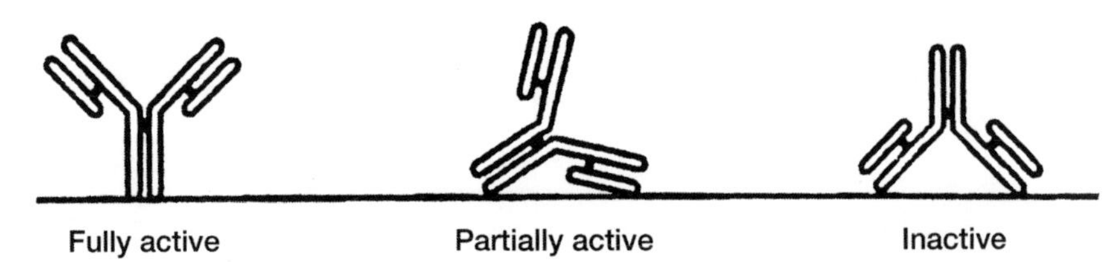

FIGURE 1.4 IgG antibody configurations on a surface during a random coupling procedure (Lu *et al.*, 1996).
Lu, B., Smyth, M. R., and O'Kennedy, R., *Analyst*, 121, 29–322 (1996). Reproduced by permission of The Royal Society of Chemistry.

activity. Thus, several approaches have been developed to obtain an appropriate orientation of the antibody during the immobilization process. Figure 1.5 shows some of these approaches (Lu *et al.*, 1996).

The antigen–antibody reaction in a typical biosensor has a much higher specificity than the average chemical sensor, and this advantage needs to be

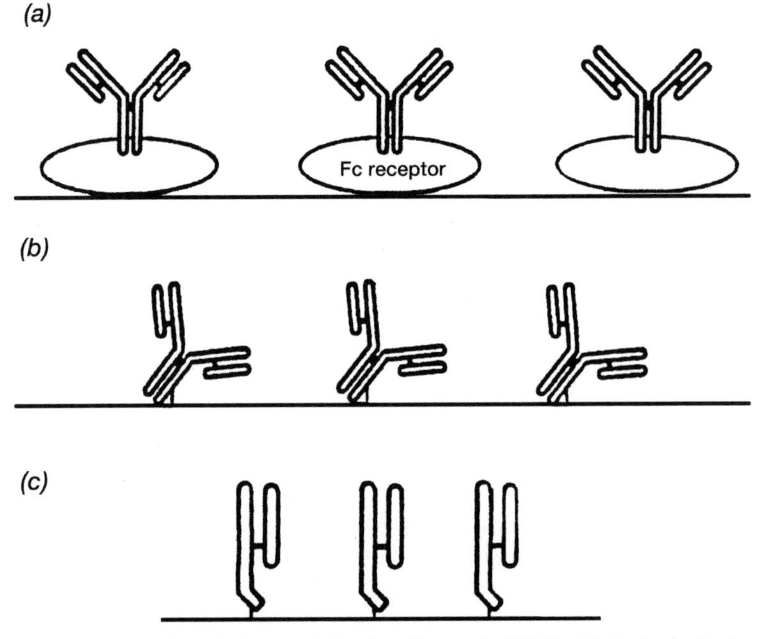

FIGURE 1.5 Oriented antibody immobilization (Lu *et al.*, 1996). (a) Antibody binds to Fc receptors on surface; (b) antibody is bound to a solid support through an oxidized carbohydrate moiety on its C_H2 domain of the Fc fragment; (c) monovalent Fab′ fragment bound to insoluble support through a sulfhydryl group in the C-terminal region.
Lu, B., Smyth, M. R., and O'Kennedy, R., *Analyst*, 121, 29–322 (1996). Reproduced by permission of The Royal Society of Chemistry.

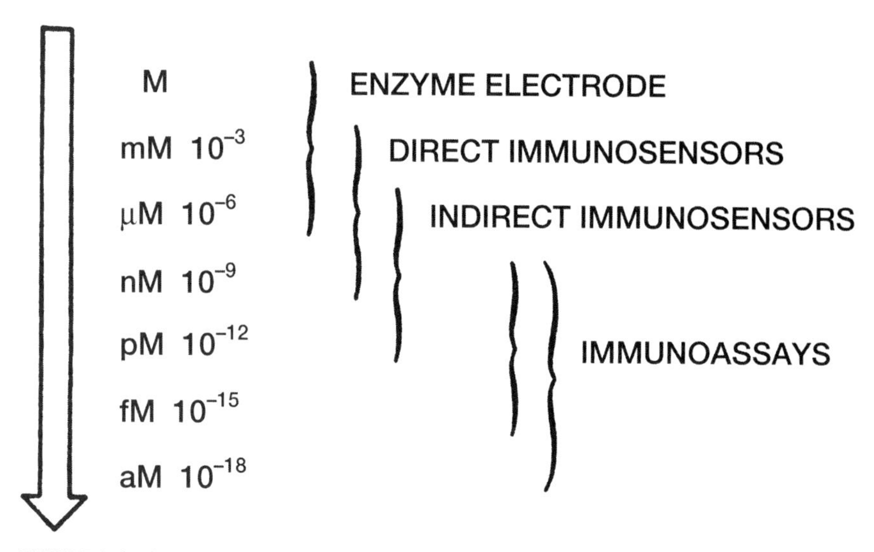

FIGURE 1.6 Concentration ranges measured by various types of biosensors and immunoassays (Byfield and Abuknesha, 1994).
Reprinted from *Biosensors and Bioelectronics*, vol 9, M. P. Byfield and R. A. Abuknesha, pp. 373–400, 1994, with permission from Elsevier Science.

exploited. The relative concentration ranges that may be measured using chemical and biological sensors are given in Figure 1.6, once again emphasizing the importance of antigen–antibody-based biosensors for the detection of analytes at lower concentration ranges (Byfield and Abuknesha, 1994). However, this exquisite specificity in antigen–antibody interactions does come at a price, in the sense that minor modifications may seriously affect the affinity of these interactions. Figure 1.7 shows the dramatic decrease in the cross-reactivity of antibodies raised against 17B-estradiol (an estrogen hormone) for different estrogens with a structure similar to 17B-estradiol. This explains the ever-increasing efforts in studying and analyzing antigen–antibody interactions.

Paddle (1996) indicates that it is the requirement of sensors to be highly selective for their specific analyte in a matrix of other chemical and biological components that makes their development so attractive. Paddle further delineates the handicaps involved in developing biosensors in that specific target molecules (such as those used in biological warfare) are (1) not available, (2) not identifiable, or (3) cannot be isolated in a stable form.

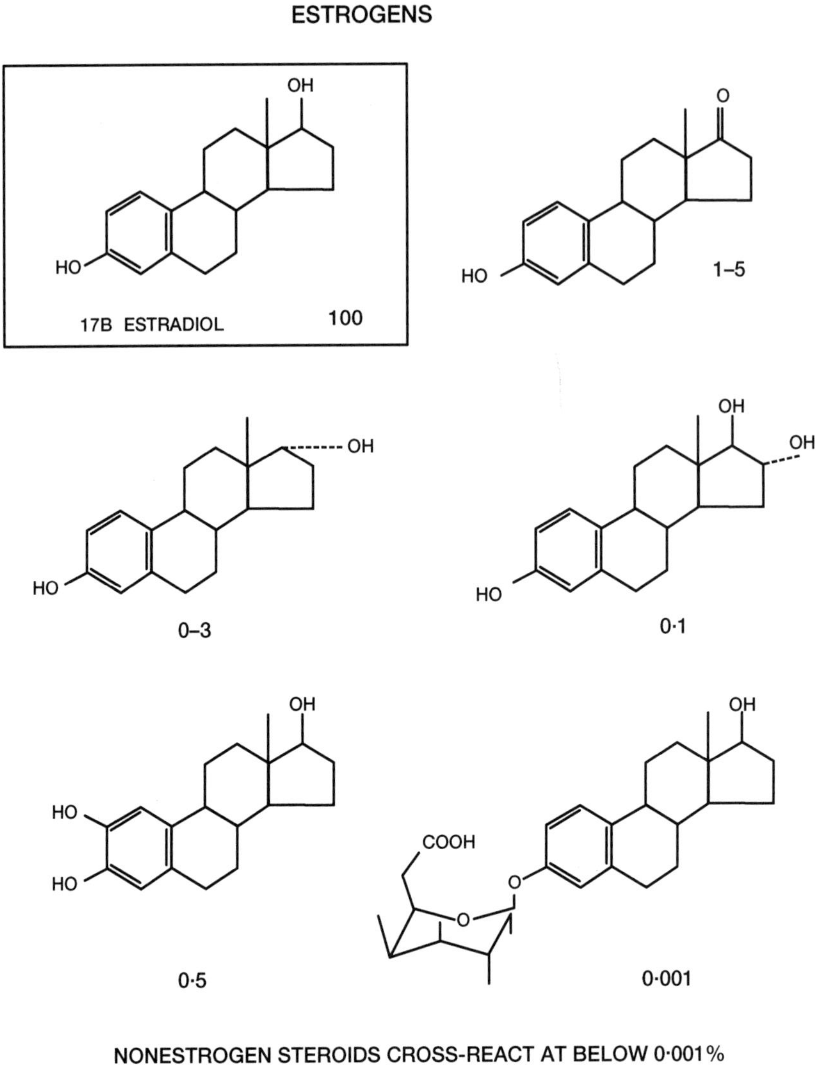

ESTROGENS

17B ESTRADIOL 100

1–5

0–3

0·1

0·5

0·001

NONESTROGEN STEROIDS CROSS-REACT AT BELOW 0·001%

FIGURE 1.7 Cross-reactivity of antibodies raised to 17B-estradiol for similarly structured estrogens. The affinity of 17B-estradiol is assumed to be 100% (Byfield and Abuknesha, 1994). Reprinted from *Biosensors and Bioelectronics*, vol 9, M. P. Byfield and R. A. Abuknesha, pp. 373–400, 1994, with permission from Elsevier Science.

Therefore, especially in terms of national defense, these factors and others may limit the development of biosensors.

One of the intentions of this text is to promote the commercialization of biosensors in spite of several limiting factors. One such limiting factor is the availability of suitable receptor molecules such as enzymes, antibodies, etc. In general, if an enzyme (that is to be used as a receptor for a biosensor) is readily available, then its corresponding analyte may be detected at low concentrations. Another limiting factor is that the stability of the biological receptor (such as proteins or enzymes) is subject to denaturation either during storage or in actual use. For example, Paddle (1996) indicates, that the average enzyme-based biosensor used for clinical applications has a lifetime of about a month. However, an enzyme-based potentiometric biosensor for cyanide was stable for six months on storage in buffer. A third limiting factor is that biosensors do require tedious preparation (Scheller *et al.*, 1991). Furthermore, these authors note that enzyme stability depends to a large extent on enzyme loading. They state that an enzyme reserve is built up by employing more active enzyme in front of the probe. This is greater than the minimum required to achieve final conversion.

In the design of biosensors one often has to optimize conditions to obtain the best possible value of the biosensor performance parameters such as stability, specificity, affinity, reliability, response time, and selectivity. For example, Paddle (1996) indicates that the high selectivity of antibodies for antigens is often achieved at the cost of high affinity. Thus, regeneration of the antibody becomes a problem and is achieved only at slightly "severe" conditions. Thus, if one wants to reuse the biosensor (with an expensive antibody as the receptor), then one often encounters progressive denaturation of the receptor (antibody, protein, etc.). Similarly, stability and high activity often are at cross-purposes. For example, immobilization of a protein on a biosensor surface leads to stability of the protein but this is generally achieved with a loss in activity. Another problem is the uniformity and geometry of the receptor on the biosensor surface, a very important aspect as far as reproducibility is concerned. The binding site of the receptor on the surface should be accessible by the analyte in solution with a minimum of hindrance. This uniformity and geometry is of particular interest to us since we will explore the effects of uniformity in great detail throughout the text. The nonuniformity of the receptors on the biosensor surface will be quantitatively characterized using a fractal dimension, with higher degrees of heterogeneity being represented by higher values of the fractal dimension. This degree of heterogeneity on the biosensor surface may be directly related to the binding rate coefficient.

To obtain a better perspective of biosensors, we now present some different assay formats.

1.2. ASSAY FORMATS

Figure 1.8 shows the four typical assay formats used: (a) direct binding, (b) sandwich assays, (c) displacement assays, and (d) replacement assays (Lukosz, 1991). The figure shows the different forms of immunoassays that may be used either with integrated optical (IO) or surface plasmon resonance (SPR) biosensors.

(a) *Direct binding* The analyte in solution binds directly to the receptor on the surface, and the analyte-receptor complex is monitored. For example, if an antigen in solution binds to the antibody immobilized on the biosensor surface, we are looking at an immunoassay format. If a DNA in solution binds directly to a complementary DNA immobilized on a biosensor surface, we looking at a hybridization assay.

(b) *Sandwich assay* When low molecular weight antigens (haptens, Ag) are to be analyzed, one often resorts to sandwich assays. This is because the

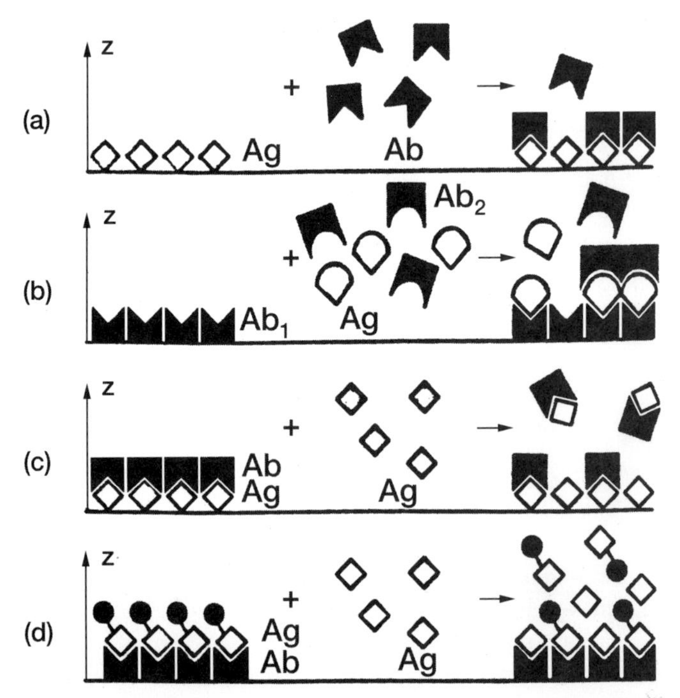

FIGURE 1.8 Different forms of immunoassays (Lukosz, 1991). Immobilized molecules shown at left; the addition of the analyte in the middle; resulting adlayer on the surface on the right. (a) Direct binding; (b) sandwich assay; (c) displacement assay; (d) replacement assay.
Reprinted from *Biosensors and Bioelectronics*, vol 6, W. Lukosz, pp. 215–225, 1991, with permission from Elsevier Science.

binding of the hapten to the primary antibody (Ab_1) in this case induces only insignificant or very small changes in the signal to be measured (such as the change in refractive index when using an evanescent fiber-optic biosensor). A secondary antibody (Ab_2) is simultaneously added to the analyte (antigen in our case) in solution, or soon thereafter. The complex formed ($Ab_2 \cdot Ab_1 \cdot Ag$) is then continuously monitored.

(c) *Displacement assay* Here an analogue of the analyte to be measured is immobilized on the biosensor surface. Then, the corresponding antibody is bound. On the addition of the analyte in solution to be measured, antibodies are displaced from the surface and bind to the free antigens. In this case, one observes and measures a negative change (for example, a negative change in the refractive index).

(d) *Replacement assay* In this case the surface is coated with the antibody molecules. An analogue of the antigen (conjugated to a larger molecule) is bound to the antibody immobilized on the sensor surface. This is the refractive index probe (Drake *et al*., 1988). On the addition of the analyte in solution to be measured to the biosensor, the analyte partially replaces the bound conjugated antigen and a negative index change (for example, refractive index occurs) and is measured.

The understanding of biological processes at the molecular level is becoming increasingly important, especially for medical purposes. Two basic approaches may be employed: structural and functional analysis. Ramakrishnan (2000) indicates that under ideal conditions these should complement each other and provide a complete and a comprehensive picture of the molecular process. Some of the structural techniques routinely employed are electron microscopy, sequence analysis, mass spectrometry, and x-ray, and electron diffraction studies. These techniques are quite effective and have provided information about the atomic organization of individual as well as interacting molecules, but have a major drawback in that they are static and frozen in time. Functional investigation techniques such as affinity chromatography and immunological and spectrometric techniques give valuable information on the conditions and the specificity of the interaction. However, these techniques are either unable to follow a process in time or are too slow to be rendered suitable for most biospecific interactions. Moreover, these techniques demand some kind of labeling of interactants, which is undesirable as it may interfere with the interaction and necessitate purification of the expensive interactants in large quantities. Another difficulty encountered while monitoring biomolecular interactions is that a number of experimental artifacts such as surface-imposed heterogeneity, mass transport, aggregation, avidity, crowding, matrix effects, and nonspecific binding complicate binding responses (Morton and Myszka, 1998).

Ekins and Chu (1994) have reviewed the techniques to develop multianalyte assays. In discussing the need for and importance of developing miniaturized assays, the authors emphasize the decrease in traumatic effects in taking blood (fingertip), especially for young children. Also, the cost of using miniaturized assays at the point of application (for example, in the medical doctor's laboratory itself) should be less than the cost of using large centralized analyzers in hospital laboratories. These authors state that the real advantage of miniaturization is that it permits multianalyte assays, which is especially helpful in medical diagnostics where the trend is to detect and determine more and more analytes to provide a better perspective for diagnosis. This is especially true for complex diseases where multiple criteria need to be satisfied to help discriminate between closely related diseases with similar symptoms.

1.3. DIFFICULTIES WITH BIOSENSOR APPLICATIONS

We now examine some of the difficulties for the application of biosensors on-line. A workshop on the use of biosensors for on-line analysis (Scheper *et al.*, 1994) emphasized that the biosensor should be considered as an integral part of the bioprocess as a whole. The workshop indicated that much work needs to be done before biosensors can be used on-line, especially with regard to the "harsh conditions" to which some of these biosensors may be subjected. The workshop also pointed out some of the drawbacks that arise when biosensors are used *in situ*: (1) the biosensors cannot be sterilized; (2) they function within a limited range of analyte concentrations; (3) if an enzyme is used in the detection process, its pH optimum is different from the pH optimum of the process. The workshop further emphasizes that since information from a wide variety of disciplines is required and the process itself is very time consuming, it may be advisable to design a biosensor in a modular fashion and then tailor it to meet specific analytical protocols. Also indicated was the need for the incorporation of "intelligent" software to run the analytical systems.

In a guest editorial, Weetall (1996) indicates that, in general, biosensors suffer from quite a few problems that need to be overcome before they can be used on a large scale. In essence, the biosensor should be inexpensive, user friendly, sufficiently sensitive and accurate, and easily manufactured with high select rates. Weetall emphasizes that the biosensor technique will be competing with the more established clinical technologies that are already inexpensive, rapid, automated, multianalytic, accurate, and sensitive. Furthermore, this author indicates that the true test for biosensor competitiveness is the cost per test; the more labor required, the greater the cost.

Weetall further indicates that electrochemical sensors can measure analytes in the 10^{-6} M concentration range, a range that is sufficient for measuring glucose, urea, cholesterol, drugs of abuse, and many large and small molecules. Weetall emphasizes that biosensors have difficulty measuring analytes in the 10^{-9} M range, the range required to measure hormones and other serum components. Thus sensitivity is an issue, as indicated earlier in this chapter, that still requires improvement.

Weetall cautions, with tongue in cheek, that many scientists developing biosensors do not consider the manufacturing aspects of the device they want to work with. But this aspect is essential and should be brought into consideration early in the development process. Finally, this author predicts the use of biosensors in the next 15 years in the following areas: single-molecule detection, nano-size sensors, multianalyte arrays, home imaging systems for wellness screening (using noninvasive biosensors), and interface between the human central nervous system and artificial intelligence (using implantable sensors).

1.4. NEWER APPLICATIONS FOR BIOSENSORS

The development of biosensors is particularly attractive in the sense that the detection of different (new) analytes in solution is made possible by using basic principles in immunoassays and either expanding the old concepts or using newer concepts, or employing a combination of concepts. Let us demonstrate with a few examples.

Suzuki *et al.* (2000) indicate that the sandwich enzyme-linked immunosorbent assay (ELISA) (Ueda *et al.*, 1999) is one of the most commonly used immunoassay formats because it exhibits high specificity and sensitivity, superior dynamic range, and a low background. However, the procedure is time consuming and tedious, plus monovalent antigens such as haptens are not measurable due to their small size. Suzuki *et al.* have applied the open-sandwich ELISA (OS-ELISA)—an immunoassay based on antigen stabilization of the antibody variable regions (V_H and V_L domains)—to make more quantitative hapten concentration in solution. They emphasize that they have already used this technique to analyze large proteins; thus according to them it is the first practical immunoassay approach capable of quantitating the analyte concentration in solution regardless of the size of the analyte.

Bouquet *et al.* (2000) have recently indicated the need for the development of assays for total and active recombinant antibodies in different expression systems. These authors have developed a new approach involving two different immunometric assays. Each immunometric assay makes more quantitative both the total and the active fragments of a recombinant antibody (single-chain fragment variable, or ScFv) as obtained in a crude extract from an *Escherichia coli* expression system. Total ScFv expression is determined using a sandwich assay. This involves the recognition of two different Tag sequences introduced at the N-terminal (Ha-Tag) and C-terminal (Myc-Tag) extremities of the recombinant protein. An assay was also developed in parallel for active ScFv in which capture is ensured by the immobilized antigen.

According to Souteyrand *et al.* (2000), affinity systems permit the specific and selective recognition of analytes or species without the consummation or transformation of species. However, there are difficulties with regard to the transduction of the signal since these types of interactions do not produce electronic transfers; thus no current is produced. Thus, presently the detection step involves the use of labeled targets (enzymatic, radioactive, etc.). These authors describe the use of semiconductors to transduce the biological recognition into an electrical signal that is easily measurable. They were able to develop a sensitive optical fiber immunosensor to detect anti-*S. Pneumoniae* antibodies.

In a study by Marks *et al.* (2000), *Streptococcus pneumoniae* (pneumococcus) (a bacterial pathogen) has been associated with life-threatening diseases such as meningitis, pneumonia, and sepsis. These authors indicate that this pathogen is the most common cause of bacterial pneumonia and especially affects children and the elderly. These authors emphasize that for the efficacy of a vaccine, one needs to monitor the antipneumococcal IgG levels in multiple blood samples. Since multiple blood sampling is traumatic, especially for small children, there is a need for noninvasive sampling. Thus, these authors developed a chemiluminescent-based optical fiber immunosensor for the detection of antipneumococcal antibodies. They developed a chemical procedure that used 3-aminopropyl trimethoxysilane and cyuranic chloride to conjugate pneumococcal cell wall polysaccharides to the optical fiber tips. This significantly improved the sensitivity of the detection system.

Cullum and Vo-Dinh (2000) have recently reviewed the development of optical nanosensors—sensors with dimensions on the nanometer scale—for biological measurement. These authors indicate that optical nanosensors, like larger sensors, can be used as either chemical or biological sensors, depending on the probe used. The optical nanosensors have been used to monitor various chemicals in microscopic environments and to detect different chemicals in single cells. These authors emphasize that these sensors offer significant enhancements over the employment of traditional biosensors since in this case very little or insignificant diffusion is involved. Diffusion, as expected, in most cases exacerbates the kinetics of binding and the overall biosensor performance. This is of particular interest to us since our main focus is to describe the influence of diffusion and heterogeneity on the biosensor surface on the kinetics of binding and other biosensor performance parameters.

Kleinjung *et al.* (1998) have recently used high-affinity RNA as a recognition element in a biosensor. According to these authors, the future development of biosensors is bound to be influenced by advancements in the understanding of nucleic acids. Some nucleic acids combine both the genotype (nucleotide sequence) and a phenotype (ligand binding or catalytic activity) into one molecule. There is a need for the development of newer techniques, since traditional antigen–antibody binding still does have some handicaps such as stability and regenerability. This is so even though antibodies are excellent diagnostic reagents. Furthermore, antibodies cannot be raised against all analytes of interest. Antibodies are most often raised *in vivo* and are subject to animal-to-animal variation. Also, the immunosystem prevents the production of antibodies against analytes present in the body such as amino-acids and nucleotides. Thus, Kleinjung *et al.* developed a biosensor for L-adenosine using high-affinity RNA as a binder. This high-affinity RNA was attached to an optical fiber using an avidin-biotin bridge. The authors indicate that competitive inhibition with L-adenosine permits the device they developed to detect L-adenosine in the submicromolar range.

Drolet *et al.* (1996) indicated that the large size (150 kD) and the complexity of the antibody often makes a typical antibody difficult to modify with enzymes or other labels. Also, these modifications may reduce the antibody's affinity. These authors add that combinatorial chemistry methodologies exhibit the potential to provide high-affinity ligands for use as therapeutic and diagnostic reagents (Gallop *et al.*, 1994; Gordon *et al.*, 1993; Abelson, 1990; Tuerk and Gold, 1990; Gold, 1995). The advantage of this technology is that it seems to overcome all the handicaps present when using antibodies. The diagnostic reagents provided by this technique are smaller in size and complexity than antibodies, are easy to manufacture and modify, and are stable during storage.

Drolet *et al.* (1996) indicate that the SELEX (systematic evolution of ligands by exponential enrichment) method is one such combinatorial chemistry process that permits the rapid identification of the few nucleotide sequences that bind to the desired target molecule with high affinity and specificity. This nucleotide sequence identification is done from a large random sequence pool. Using the SELEX method, Drolet *et al.* identified a ligand specific for human vascular endothelial growth factor (VEGF)/vascular permeability factor and were able to quantify the analyte in serum.

It is worthwhile to mention again the surface plasmon resonance (SPR) biosensor that has gained increasing popularity over the years. Silin and Plant

(1997) have indicated that two advantages of the SPR biosensor are that one obtains measurements of kinetic interactions in real time and that no labeling of molecules is required. Thus, these SPR biosensors have been used to detect a wide variety of analytes in biotechnological applications, immunological studies, protein–protein interaction studies, signal transduction and cell–cell interactions, screening of ligands, etc. Phizicicky and Fields (1995) state, "The development of a machine to monitor protein–protein and ligand–receptor interactions by using changes in surface plasmon resonance measured in real time spells the beginning of a minor revolution in biology." Silin and Plant (1997) emphasize that more recent SPR biosensors can detect 0.5 ng/cm^2 surface concentration, 0.1-nm film thickness, or 10^{-12} M analyte concentration. These authors emphasize that to operate the instrument at these detection levels all reagents, buffers, protein solutions, etc. should be very pure and properly prepared. The authors further indicate that the sensitivity to analyte concentration depends significantly on affinity constants, and the specificity of the immobilized biomolecules on the surface, as well as the biomolecules surface concentration, packing density, orientation, and denaturation level. Another particular advantage of the SPR technique is that it is a versatile technique and may be combined with other techniques, such as electrochemical measurements (Flatgen *et al.*, 1995), HPLC (high-pressure liquid chromatography) (Nice *et al.*, 1994), and atomic force microscopy (Chen *et al.*, 1996). Silin and Plant (1997) predict that the SPR should find considerable and increasing applications for clinical as well as scientific laboratories.

Jhaveri *et al.* (2000) have recently attempted to use aptamers in biosensor formats. Aptamers are selected nucleic acid binding species that can recognize molecules as simple as amino acids or as complex as red blood cell membranes. These authors have attempted to develop and combine the recognition properties of aptamers with the signal transduction. They indicate that as aptamers bind to their cognate ligands they undergo conformational changes. Furthermore, if a fluorophore is introduced into the aptamer as it undergoes a conformational change due to binding, changes in fluorescence intensity results.

Sole *et al.* (1998) have recently developed a flow injection immunoanalysis based on a magnetimmunosensor system. The authors indicate that one of the main problems affecting immunosensors is the reproducible regeneration of the sensing surface (Hall, 1990). They also express the need for the renewal of the sensing surface due to the high-affinity constants involved in the strong antigen–antibody reaction. Drastic procedures could be used to regenerate the surface, but that would lead to denaturation of the sensing surface. These authors further indicate that magnetic particles with immunoreagents may be used in manual immunoassay methods (Gascon *et al.*, 1995; Varlan *et al.*, 1995).

The new immunosensor developed by Sole *et al.* (1998) is integrated to a flow system. Magnetic immunoparticles are immobilized on a solid-state transducer using a magnetic field. This technique simplifies the renewal of the solid phase. Furthermore, it has been reported that biologically modified magnetic particles have been used in conjunction with analytical flow systems (Kindervater *et al.*, 1990; Gunther and Bilitewski, 1995; Pollema *et al.* 1992). Sole *et al.* emphasize that these biologically magnetic particles are easy to handle, which makes the regeneration procedures of the biological material unnecessary.

1.5. COMMERCIALLY AVAILABLE BIOSENSORS

Ward and Windzor (2000) have recently reviewed the relative merits of flow-cell and cuvette designs. They indicate that the interaction of a soluble ligate with an affinity ligand immobilized on a sensor surface can be monitored by changes in electrochemical parameters with the aid of potentiometric and amperometric electrodes, changes in mass using quartz crystals or surface acoustic wave oscillators, and changes in capacitance. They mention four commercially available (biosensor) instruments, and out of these three use optical systems. The four commercially available instruments are

1. BIACORE instruments (Pharmacia Biosensor, Uppsala, Sweden), which may be used to monitor the refractive index in the vicinity of the sensor surface.
2. The BIOS-1 Sensor (Artificial Sensing Instruments, Zurich, Switzerland), in which the refractive index change is due to the binding of the analyte in solution to the receptor on the surface. This change is followed by an optical grating coupler (Bernard and Bosshard, 1995).
3. The IAsys biosensor (Affinity Sensors, Cambridge, UK), which uses resonant mirror technology (Cush *et al.*, 1993) to monitor the surface refractive index changes.
4. The IBIS instrument (Intersens, Amersfoot, The Netherlands), which uses surface plasmon resonance to make the refractive index changes at the sensor surface quantitative (de Mol *et al.*, 2000).

1.6. BIOMEDICAL APPLICATIONS

Killard and Smyth (2000) have recently created a biosensor based on potentiometric and amperometric principles and used for the measurement of creatinine in solution. As indicated earlier in this chapter, there are still some problems with biosensor performance parameters, specifically, the balance between sensitivity and selectivity, sensor stability, and interference rejection. These authors indicate, however, that creatinine biosensors appear to be close to the standards required for widespread application.

Killard and Smyth note the importance of developing a reliable biosensor for creatinine, since according to Bakker *et al.* (1999) it is the most requested analyte in the clinical laboratory. Creatinine is a by-product of amino acid metabolism and is the energy source for muscle tissue. The normal clinical range for blood creatinine is 44–106 μM (Tietz, 1987). Under certain pathological conditions it may exceed 1000 μM (Sena *et al.*, 1988). Creatinine levels greater than 140 μM indicate the onset of a pathological condition and need to be investigated further (Madras and Buck, 1996). Creatinine levels greater than 530 μM indicate a severe renal disorder.

The impetus for designing a biosensor for creatinine is that present-day analysis is inconvenient and is subject to interference (by NH_4^+ in blood and tissue samples). Finally, Killard and Smyth (2000) indicate that amperometric biosensors are becoming increasing popular, with the Nova Biomedical (Rodermark, Germany) being the first to successfully commercialize such a sensor.

Ohlson *et al.* (2000) very recently indicated that noncovalent and weak (biological) interactions are becoming more and more important in biological applications, especially for diagnostic purposes. These have dissociation constants, K_d, in the range of 0.10 to 0.01 mM. These authors indicate that these types of bioaffinity reactions have been exploited in biosensor development (Nicholson *et al.*, 1998; Strandh *et al.*, 1998; Mann *et al.*, 1998), in addition to analytical applications such as chromatography, capillary electrophoresis, and electrochemiluminescence. The authors emphasize that in weak biological interactions the binding and release of one-to-one contact may occur in a fraction of a second.

Figure 1.9 shows a typical setup for continuous immunosensor monitoring. The subject may be, for example, a bioreactor, a laboratory animal, or a patient under medical surveillance. The analyte passes continuously over an immunosensor surface, which is interfaced with a transducer of the biosensor system that permits the quantitative measurement of the analyte. Ohlson *et al.* emphasize that the mass transport to the antibody sensor surface is significantly influenced by the cell geometry and the flow properties. The influence of diffusional limitation and heterogeneity on the surface on antigen–antibody (in general, analyte-receptor) binding kinetics is analyzed in detail in the chapters that follow in this text.

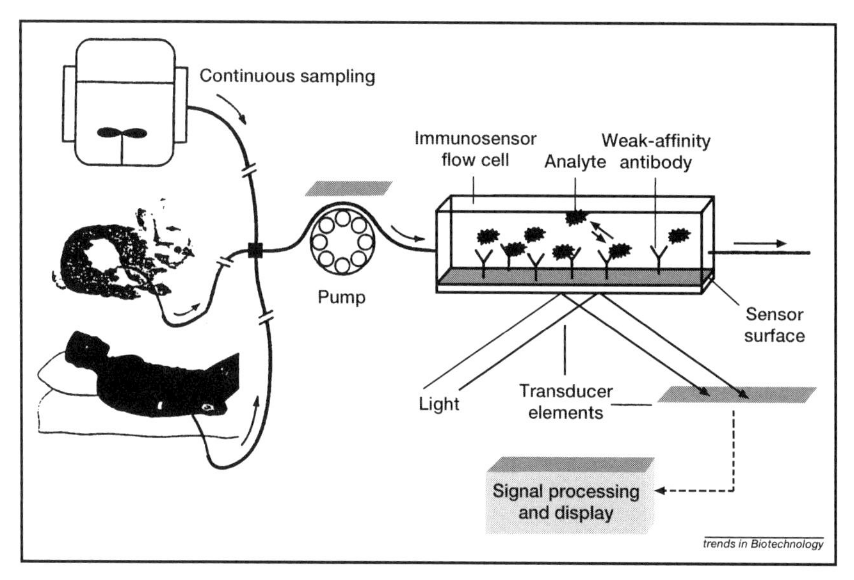

FIGURE 1.9 A schematic diagram showing the continuous immunosensor setup (Ohlson *et al.*, 2000).

1.7. OVERVIEW

We will now briefly describe the material that is to be presented in the remaining chapters. In Chapter 2, we look at the influence of diffusional limitations and reaction order on antigen–antibody binding kinetics. We analyze the influence of classical reaction order and diffusional limitations on antigen–antibody binding kinetics. We present the influence of heterogeneity on the surface on the binding kinetics through a temporal binding rate coefficient. In Chapter 3, we look at the influence of lateral interactions on the biosensor surface and diffusional limitations on antigen–antibody kinetics. We examine the influence of a variable binding rate coefficient on binding kinetics as well as the influence of multivalency of either the antigen or the antibody on antigen–antibody binding.

In Chapter 4, we introduce the terms *fractal* and *fractal kinetics*. Fractal kinetics typically exhibit noninteger kinetic reaction orders and may be used to make the degree of heterogeneity that exists on the surface more quantitative. An increase in the degree of heterogeneity on the surface may be described by a higher fractal dimension. In Chapter 5, we analyze the influence of different parameters on fractal dimension values and binding rate coefficients during antigen–antibody binding on different biosensor surfaces.

Examples are provided wherein the binding kinetics may be described by both single- and dual-fractal analysis. In Chapter 6, we analyze the influence of the fractal dimension on the surface on the binding rate coefficient. In the examples provided we generally note that an increase in the fractal dimension on the surface (increase in the heterogeneity on the surface) leads to an increase in the binding rate coefficient.

In Chapter 7, we analyze the influence of the fractal dimension on the binding as well as on the dissociation rate coefficient. Again, examples are provided wherein the binding and the dissociation may each be described either by single- or dual-fractal analysis. In Chapter 8, we analyze the influence of nonspecific binding on the rate and the amount of specific binding. We also present a classical kinetic analysis, where the specific binding may be described by a temporal binding rate coefficient. In Chapter 9, we again analyze the influence of nonspecific binding on the rate and the amount of specific binding but time we present a fractal analysis.

In Chapter 10, we analyze the influence of the fractal dimension, or the degree of heterogeneity on the surface, on the binding kinetics during hybridization. We once again give both single- and dual-fractal analysis examples. In Chapter 11, we extend our analysis to analyte-receptor binding in cells. Herein the binding is detected by biosensors. Surface plasmon resonance (SPR) biosensors are becoming more and more popular, so we cover quite a few of these examples. In Chapter 12, we analyze the influence of the fractal dimension on the binding and on the dissociation rate coefficient using the SPR biosensor. The degree of heterogeneity on the surface may be different for these two phases. Finally, in Chapter 13, we present some of the economics involved with biosensors. Understandably, economic data is very difficult to get in the open literature; thus it is a worthwhile exercise to present this type of data under one cover.

REFERENCES

Abelson, J., *Science* **249**, 488–489 (1990).

Alak, A. M., Eiwel, E., Hinze, W. L., Oh, K., and Armstrong, D. W., *J. Liq. Chromatogr.* 7, 1273 (1984).

Alarie, J. P. and Vo-Dinh, T., *Talanta* **38**(5), 525 (1991).

Alvarez-Icaza, M. and Bilitewski, U., *Anal. Chem.* **65**, 525A–533A (1993).

Bakker, E. *et al.*, *Anal. Chim. Acta* **393**, 11–18 (1999).

Bernard, A. and Bosshard, H. R., *Eur. J. Biochem.* **230**, 416 (1995).

Bouquet, D., Creminion, C., Clement, G., Frobert, Y., Nevers, M. C., Essono, S. and Grassi, J., *Anal. Biochem.* **284**, 221 (2000).

Byfield, M. P. and Abuknesha, R. A., *Biosen. & Bioelectr.* **9**, 373–400 (1994).

Chen, X., *et al.*, *J. Vac. Sci. Technol.* **B14**, 1582–1586 (1996).

Cullum, B. M. and Vo-Dinh, T., *TIBTECH* **18**, 388 (2000).

Cush, R., Cronin, J. M., Stewart, W. J., Maule, C. H., and Goddard, N. J., *Biosen. & Bioelectr.* **8**, 347–364 (1993).
Drake, R., Sawyers, C., and Robinson, G., Assay technique, EPA 883 00458.2 (1988).
Drolet, D. W., Moon-McDermott, L., and Ronig, T. S., *Nature Biotech.* **14**, 1021–1025 (1996).
de Mol, N. J., Plomp, E., Fischer, M. J. E., and Ruijtenbeek, R., *Anal. Biochem.* **279**, 61–70 (2000).
Ekins, R. P. and Chu, F., *TIBTECH* **12**, 89–94 (1994).
Flatgen, G., *et al. Science* **269**, 668–671 (1995).
Gallop, M. A., Barrett, R. W., Dower, W. J., Fodor, S. P. A., and Gordon, E. M., *J. Med. Chem.* **37**, 1233–1251 (1994).
Gascon, J., Martinez, E., and Barcelo, D., *Anal. Chim. Acta* **311**, 357 (1995).
Gold, L., *J. Biol. Chem.* **270**, 13,581–13,584 (1995).
Gordon, E. M., Barrett, R. W., Dower, W. J., Fodor, S. P. A., and Gallop, M. A., *J. Med. Chem.* **37**, 1385–1401 (1993).
Gunther, A. and Bilitewski, U., *Anal. Chim. Acta* **300**, 117–123 (1995).
Hall, E. A. H., *Biosensors*, 1990. Milton Keynes, Buckinghamshire, England: Open University Press, 1990.
Jhaveri, S., Rajendran, M., and Ellington, A. D., *Nature Biotech.* **18**, 1293 (2000).
Jimbo, Y. and Saiti, M., *J. Mol. Electro.* **4**, 111 (1988).
Killard, A. J. and Smyth, M. R., *TIBTECH* **18**, 432 (2000).
Kindervater, R., Kunneche, P., and Schmid, R. D., *Anal. Chim. Acta* **234**, 113 (1990).
Kleinjung, F., Klussman, S., Erdmann, V. A., Scheller, F. W., Furste, J. P., and Bier, F. F., *Anal. Chem.* **70**, 328–331 (1998).
Lowe, C. R., *Biosensors* **1**, 3–16 (1985).
Lu, B., Smyth, M. R., and O'Kennedy, R., *Analyst* **121**, 29R–32R (1996).
Lukosz, W., *Biosen. & Bioelectr.* **6**, 215–225 (1991).
Madras, M. B. and Buck, R. P., *Anal. Chem.* **68**, 3832–3839 (1996).
Mann, D. A. *et al.*, *J. Am. Chem. Soc.* **120**, 10575–10582 (1998).
Marks, R. S., Margalit, A., Bychenko, A., Bassis, E., Porat, N., and Dagar, R., *Applied Biochemistry and Biotechnology*, **89**, 117–125 (2000).
Morton, T. A. and Myszka, D. G., in *Methods in Enzymology*, Vol. 295, (G. K. Ackers and M. L. Johnson, eds.), San Diego: Academic Press, 1998.
Nice, E., Lackmann, M., Smyth, F., Fabri, L., and Burgess, A. W., *J. Chromatogr. A.* **66**, 169–185 (1994).
Nicholson, M. W., *et al.*, *J. Biol. Chem.* **273**, 763–770 (1998).
Ogert, R. A., Brown, J. E., Singh, B. R., Shriver-Lake, L. C., and Ligler, F. S., *Anal. Biochem.* **205**, 306 (1992).
Ohlson, S., Jungar, C., Strandh, M., and Mandenius, C.-F., *TIBTECH*, **18**, 49 (2000).
Paddle, B. M., *Biosen. & Bioelectr.*, **11**, 1079–1113 (1996).
Phizicicky, E. M. and Fields, S., *Microbiol, Rev.* **59**, 94–123 (1995).
Pollema, C. H., Ruzicka, J., Christian, G. D., and Lernmark, A., *Anal. Chem.* **64**, 1356 (1992).
Ramakrishnan, A., "Evaluation of Binding Kinetics in Surface Plasmon Resonance (SPR) Biosensors and Cellular Analyte-Receptor Systems: A Fractal Analysis," Master's thesis, University of Mississippi, May 2000.
Rosen, I. and Rishpon, J., *J. Electroanal. Chem.* **258**, 27 (1989).
Sadana, A. and Vo-Dinh, T., *Talanta* **42**, 1567–1574 (1995).
Scheller, F. W., Hintsche, R., Pfeiffer, D., Schubert, F., Riedel, K., and Kindervater, R., *Sensors & Actuators* **4**, 197–206 (1991).
Scheper, T., Plotz, F., Muller, C., and Hitzmann, B., *TIBTECH*, **12**, February, 42 (1994).
Sena, F. S., *et al.*, *Clin. Chem.* **34**, 594–595 (1988).
Silin, V. and Plant, A., *Trends in Biotechnol.* September, **15**, 353–359 (1997).

Sole, S., Alegret, S., Cespedes, F., Fabergas, E., and Diez-Caballero, T., *Anal. Chem.* **70**, 1462–1467 (1998).

Souteyrand, E., Chen, C., Cloarec, J. P., Nesme, X., Simonet, P., Navarro, I., and Martin, J. R., *Appl. Biochem. & Biotechnol.* **80**, 195 (2000).

Staines, N., Brostoff, J., and James, K., *Introducing Immunology*, Mosby, St. Louis (1993).

Strandh, M., *et al.*, *J. Mol. Recog.* **11**, 188–190 (1998).

Suzuki, C., Ueda, H., Mahoney, W., and Nagamune, T., *Anal. Biochem.* **286**, 238 (2000).

Tietz, N. W., *Fundamentals of Clinical Chemistry*, 3rd ed., W. B. Saunders, Philadelphia, PA, 1987.

Tietz, N. W., in *Textbook of Clinical Chemistry*, p. 1810, 1986.

Ueda, H., Kubota, K., Wang, Y., Tsumoto, K., Mahoney, W. C., Kumagai, I., and Nagamune, T., *Biotechniques* **27**, 738 (1999).

Varlan, A. R., Suls, J., Jacobs, P., and Sansen, W., *Biosen. & Bioelectr.* **10**(8), xv (1995).

Ward, L. D. and Winzor, D. J., *Anal. Biochem.* **285**, 179 (2000).

Weetall, H., *Biosen. & Bioelectr.* **11**(1,2) i–iv (1996).

CHAPTER 2

Influence of Diffusional Limitations and Reaction Order on Antigen–Antibody Binding Kinetics

2.1. INTRODUCTION

The solid-phase immunoassay technique provides a convenient means for the separation of reactants (for example, antigens) in a solution. Such a separation is possible due to the high specificity of the analyte for the immobilized antibody. However, in these types of assays, external diffusional limitations are generally present, and these must be taken into account. Giaever (1977) has indicated that these external diffusion limitations are present, and Nygren and coworkers (Stenberg *et al.*, 1982; Nygren and Stenberg, 1985a,b; Stenberg and Nygren, 1982) have analyzed the influence of external diffusional limitations on immunoassays.

Stenberg *et al.* (1986) analyzed the effect of external diffusion on solid-phase immunoassays in which the antigen is immobilized on a solid surface and the antibodies are in solution. These authors obtained an analytical solution for a first-order system. They noted that external diffusional limitations play a significant role when high concentrations of antigens (or binding sites) are immobilized on the surface. This analysis is a general description of diffusion limitation. The "reverse" system, wherein the antibody is immobilized on the surface and the antigen is in solution, is also of interest. Under different conditions, these systems would yield non-first-order systems, which would require nonanalytical methods of solution.

In general, real-life situations are not first-order, and solutions to these types of situations would be helpful. Under certain circumstances, such as by linearization, non-first-order systems may be reduced to first-order systems. Nevertheless, reaction rate expressions for non-first-order systems would be helpful, as would (presumably nonanalytical) methods for the solutions of the equations that result from such developments.

In protein adsorption systems, which exhibit behavior similar to that of antigen–antibody systems at solid–liquid interfaces (Stenberg and Nygren, 1988), the influence of the surface-dependent intrinsic adsorption and desorption rate constants on the amount of protein adsorption has been analyzed (Cuypers *et al.*, 1987; Nygren and Stenberg, 1990). Cuypers *et al.* (1987) basically analyzed the influence of a variable adsorption rate coefficient on protein adsorption. This influence may be due to nonidealities or heterogeneity on the surface. Nygren and Stenberg (1990), while studying the adsorption of ferritin from a water solution to a hydrophobic surface, noted that initially the adsorption rate coefficient of new ferritin molecules increased with time. In an earlier study, these same authors (Nygren and Stenberg, 1985a,b) had noted a decrease in the binding rate coefficient with time while studying the kinetics of antibody binding to surface-immobilized bovine serum albumin (antigen) by ellipsometry. They indicated that the decrease in binding rate with time is probably due to a saturation through steric hindrance at the surface.

In this chapter, we present a theoretical model for the binding of an antigen molecule in solution to a single binding site of an antibody on the surface. We analyze the effect of external diffusional limitations on the different reaction orders. We also examine the influence of a decreasing and an increasing adsorption rate coefficient on binding kinetics for different reaction orders. Finally, we generalize the model to include binding to bivalent antibodies as well as, briefly, to nonspecific binding.

2.2. THEORY

Figure 2.1 describes the steps involved in the binding of the antigen in solution to the antibody covalently attached to a surface. The rate of binding of a single antigen by an antibody is given by

$$\frac{d\Gamma_1}{dt} = k_1 c_s(\Gamma_0 - \Gamma_1) - k_{-1}\Gamma_1, \tag{2.1}$$

where Γ_0 is the total concentration of the antibody sites on the surface; Γ_1 is the surface concentration of antibodies that are bound by antigens at any time t; c_s is the concentration of the antigen close to the surface; k_1 is the forward

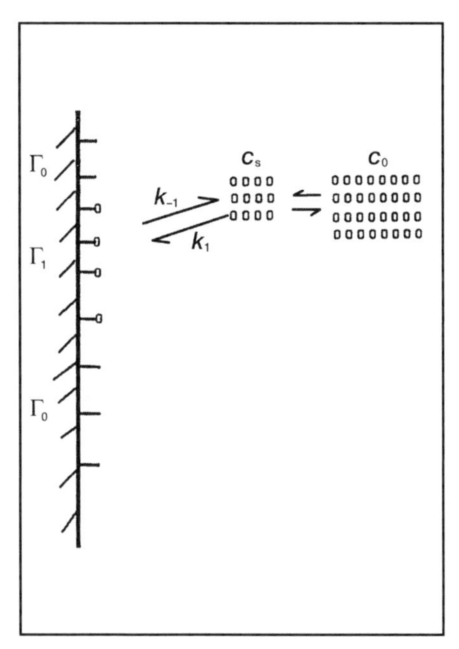

FIGURE 2.1 Elementary steps involved in the binding of the antigen to the antibody covalently attached to the surface. Each arm of the antibody molecule reacts with an antigen molecule independent of the other arm.
Sadana, A. and Sii, D., *J. Colloid & Interface Science*, 151(1), 166 (1992). Reprinted with permission from Academic Press.

reaction rate constant; and k_{-1} is the reverse reaction rate constant. Even though the antibody molecule has two binding sites, for all practical purposes we believe that an antigen molecule reacts with the antibody as if had only one binding site. The simplified reaction scheme, then, is

$$\mathrm{Ab} + \mathrm{Ag} \underset{k_r}{\overset{k_f}{\rightleftharpoons}} \mathrm{Ab{\cdot}Ag}, \tag{2.2}$$

where Ab is the antibody binding site; and Ag is the antigen; and $k_f = k_1$ and $k_r = k_{-1}$.

Since we are interested in initial binding kinetics, $\Gamma_1 < \Gamma_0$. Also, $k_1 > k_{-1}\Gamma_1$. Physically, even if there is some desorption, the antigen quickly readsorbs. These two conditions simplify Eq. (2.1) to

$$\frac{d\Gamma_1}{dt} = k_1 c_s \Gamma_0 = \frac{d\Gamma_{\mathrm{Ag}}}{dt}, \tag{2.3}$$

where Γ_{Ag} is the surface concentration of the bound reactant or antigen. The first-order dependence on antigen concentration close to the surface is expected if one antigen molecule in solution binds to a single binding site on the surface. For first-order reaction kinetics, note that k_f and k_r are equal to k_1 and k_{-1}, respectively.

The diffusion limitation of the reaction scheme can be determined for purely radial diffusion by using the equation

$$\frac{\partial c}{\partial t} = D\frac{1}{r}\frac{\partial}{\partial r}\left(r\frac{\partial c}{\partial r}\right), \tag{2.4a}$$

where D is the diffusion coefficient.

In cylindrical geometry (or coordinates), the preceding analysis is useful in understanding biosensor applications. In other coordinate systems (for example, spherical), Shoup and Szabo (1982) indicate that the diffusive motion of the reactants play an important role in determining the rates of biomolecular reactions, such as ligand binding to cell macromolecules or cell-bound receptors. In sensor applications, fiber-optic sensors are finding considerable applications. These sensors are cylindrical in nature and have diameters that are typically 400–600 μm. For all practical purposes, analysis that assumes single-dimension diffusion on a flat plate is appropriate considering the dimensions of the molecule, the diffusion coefficient of the reactant in solution, and the radius of the fiber-optic sensor. Place *et al.* (1991), in their review of immunoassay kinetics at continuous surfaces, have utilized diffusion coefficients in the range of 10^{-11} to $10^{-10}\,\mathrm{m^2/sec}$ to estimate the effect of the molecular weight of the diffusing species with a cell dimension of 1 mm) on the equilibrium time. Assuming a typical value of $4 \times 10^{-7}\,\mathrm{cm^2/sec}$ for the diffusion coefficient, D, and a reaction period of 100 sec yields $\sqrt{Dt} = 0.0063\,\mathrm{cm}$. Since $\sqrt{Dt}$ is much smaller than the diameter of the fiber-optic sensor (0.06 cm), the cylindrical surface may be approximated by a plane surface.

In this case, the diffusion limitation of the reaction scheme can be determined by the equation

$$\frac{\partial c}{\partial t} = D\nabla^2 c = D\frac{\partial^2 c}{\partial x^2}. \tag{2.4b}$$

Equation (2.4b) may be rewritten in dimensionless form as

$$\frac{\partial y}{\partial \theta} = \frac{\partial^2 y}{\partial z^2}, \tag{2.4c}$$

where $y = c/c_0$, $z = x/L$, where L is a characteristic length dimension (for example, the diameter of a fiber-optic biosensor), and $\theta = t/(L^2/D)$.

The boundary condition for Eq. (2.4c) is

$$\frac{d\Gamma_{\mathrm{Ag}}}{dt} = D\frac{\partial c}{\partial x}\bigg|_{x=0}. \tag{2.5a}$$

Here, $x = 0$ represents the origin of the Cartesian coordinate system and is physically the surface of, for example, the fiber to which the antibody is attached. Equation (2.5a) arises because of mass conservation, wherein the flow of antigens to the surface must be equal to the rate of antigens reacting with antibodies at the surface of the fiber.

From Eqs. (2.3) and (2.5a),

$$\frac{dc}{dx}\bigg|_{x=0} = \frac{k_1\Gamma_0 c(0,t)}{D}. \tag{2.5b}$$

Equation (2.5b) may be rewritten in dimensionless form as

$$\frac{\partial y}{\partial z}\bigg|_{z=0} = \mathrm{Da}u, \tag{2.5c}$$

where $y = c/c_0$, $z = x/L$, $u = c(0,t)/c_0$, and Da is the Damkohler number and is equal to $Lk_1(\Gamma_0/D)$. The Damkohler number is the ratio between the maximum reaction rate and the maximum rate of external diffusional mass transport.

Prior to solving Eq. (2.4a), it is instructive to estimate the Damkohler number for typical antibody–antigen systems. For fiber-optic biosensors, some typical values are as follows: L, the diameter of the fiber-optic biosensor = 0.06 cm; D, the forward association rate constant $= 10^9\ \mathrm{cm}^3/(\mathrm{mol\text{-}sec})$ (DeLisi, 1976); the concentration of the antibody attached to the fiber-optic surface $= 0.96\ \mathrm{ng/mm}^2$; and the molecular weight of the antibody = 160,000 (Bhatia *et al.*, 1989). Then, Γ_0 equals $6 \times 10^{-12}\ \mathrm{g\,mol/cm}^2$. Substituting these values into the Damkohler number yields Da = 900. This is a high value for the Da and should lead to significant external diffusional limitations.

Another parameter of interest that defines the diffusional mass transport in these systems is φ_t (Stenberg *et al.*, 1986), which is equal to $\sqrt{\pi}\Gamma_0/(2c_0Dt)$. Here, c_0 is the initial concentration of the antigen in solution. A typical value of c_0 is 50 μg/ml. Using the numbers presented yields a value of $\varphi_t = 2.69$. It is estimated that φ_t should be less than 0.5 for the system to be away from external diffusional limitations (Stenberg *et al.*, 1986). The high estimated values of Da and φ_t indicate the presence of external diffusional limitations.

The preceding estimates of Da and φ_t were made with a stagnant fluid model. One may be tempted to reduce diffusional limitations effects by increasing convection. This is impractical in these types of systems because one is usually dealing with small volumes (antigens and antibodies are expensive), which are difficult to stir. However, as the sample is introduced (for example, the immersion of the fiber-optic biosensor in a solution of antigen), there is always some motion of fluid in the droplets, which would enhance mass transfer significantly over the values obtained by the stagnant film model.

An estimate of the local stirring required to enhance the antibody–antigen interactions is instructive. The following argument is adapted from Berg and Purcell (1977). Transport by stirring in our case is given by some velocity, V_s, and by the distance of travel l. The characteristic time in this case is given by l/V_s. A good approximation for l is the diameter of the antigen molecule. Let $l = 100 \times 10^{-9}$ m (Humphrey, 1972). Movement of molecules over distance l by diffusion alone is characterized by l^2/D. Stirring is effective only if $L/V_s < L^2/D$. In this case $V_s > D/l = 4 \times 10^{-2}\ \mathrm{cm}^2/\mathrm{sec}$. Thus, stirring is effective for speeds on the order of 10^{-1} cm/sec.

The appropriate initial condition for Eq. (2.4a) is

$$\begin{aligned} c(x,0) &= c_0 \quad \text{for } x > 0, t = 0, \\ c(0,0) &= 0 \quad \text{for } x = 0, t = 0. \end{aligned} \tag{2.5d}$$

Condition (2.5d) is equivalent to the rapid immersion of a sensor into a solution with antigens.

The solution for Eqs. (2.4a), (2.5b), and (2.5d) may be obtained from Carslaw and Jaeger (1959), who describe a semi-infinite solid, initially at temperature zero, heated at $x = 0$ by radiation from a medium at a particular temperature. Our equations for the binding of the antigen to the antibody binding site and the heat transfer case correspond exactly. Transforming the solution in Carslaw and Jaeger to our notation yields

$$u = \frac{c(0,t)}{c_0} = 1 - e^{t/\tau}\mathrm{erfc}(\mathrm{Da}\gamma). \tag{2.6}$$

Here $\tau = D/(k_1^2\Gamma_0^2)$ and $\gamma = Dt/L$.

Starting with $\Gamma_{\mathrm{Ag}} = 0$ at time $t = 0$, integrating Eq. (2.3) yields

$$\Gamma_{\mathrm{Ag}}(t) = k_1\Gamma_0 \int_0^t c(x = 0, t')dt'. \tag{2.7a}$$

The solution of this integral can be obtained by integration of parts. The solution may be adapted from the solution given earlier (Carslaw and Jaeger,

1959). Then,

$$\Gamma_{\mathrm{Ag}}(\bar{t}) = c_0\sqrt{D\tau}\left[2\sqrt{\frac{\bar{t}}{\pi}} + \exp(\bar{t})\ \mathrm{erfc}(\sqrt{\bar{t}}) - 1\right], \qquad (2.7b)$$

where $\bar{t} = t/\tau$.

Equation (2.7b) may be utilized to model the concentration of the antigen bound to the antibodies that are, for example, covalently attached to an optical fiber. Equation (2.7b) may be rewritten as

$$\begin{aligned} v &= \frac{\Gamma_{\mathrm{Ag}}}{c_0\sqrt{D\tau}} \\ &= 2\sqrt{\frac{\bar{t}}{\pi}} + \exp(\bar{t})\ \mathrm{erfc}(\sqrt{\bar{t}}) - 1. \end{aligned} \qquad (2.7c)$$

The influence of the Damkohler number on c_s and on Γ_{Ag} for a first-order reaction is shown in Figs. 2.2a and 2.2b, respectively. Clearly, either an increase in the Damkohler number or an increase in the mass transfer limitations will substantially decrease the saturation level of c_s. Figure 2.2b shows that as the Damkohler number decreases (gradual elimination of external mass transfer limitations), the rate of increase of Γ_{Ag}/c_0 increases, as expected.

Figure 2.2 represents an idealized analysis. Surely, heterogeneity of adsorption is a more realistic picture of the actual situation and should be carefully examined to determine its influence on external mass transfer limitations on the ultimate analytical procedure. In general, a heterogeneity of the antibody (or antigen) immobilization on the solid surface should yield lower specific rates of binding, thereby alleviating the diffusional constraints to a certain extent. Heterogeneity in the covalent attachment of the antibody to the surface can probably be accounted for by considering an appropriate distribution of covalent energies for attachment. Although heterogeneity needs to be considered in the analysis, such an attempt is beyond the scope of this book.

Heterogeneity may arise due to different factors. The antibodies, especially polyclonal antibodies, possess an "inherent heterogeneity" in that the antibodies in a particular sample are not identical. Furthermore, different sites on the antibody may become covalently bound to the surface. As a result, especially in large antibodies, steric factors will play a significant role in determining the Ag/Ab ratio. It would be useful to make the influence of heterogeneity on the kinetics of antigen–antibody interactions more quantitative.

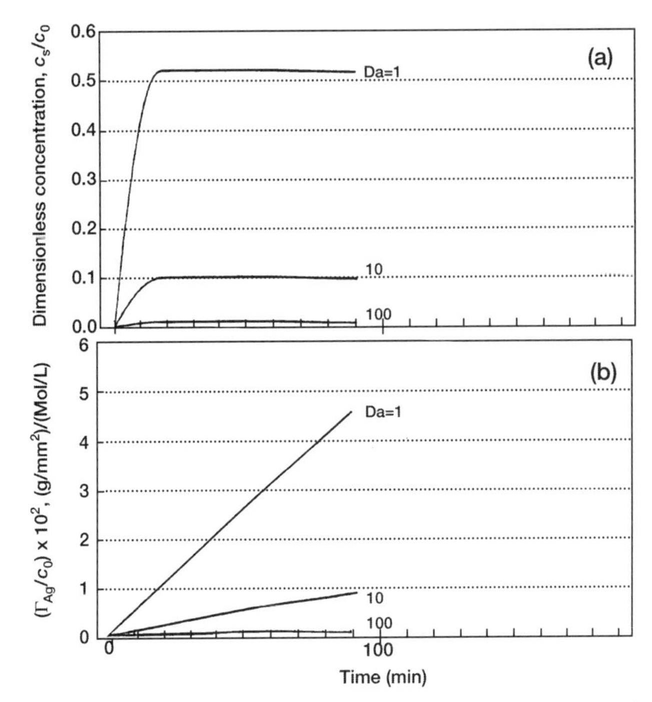

FIGURE 2.2 Influence of the Damkohler number, Da, on (a) c_s/c_0 and (b) Γ_{Ag}/c_0 for a first-order reaction.
Sadana, A. and Sii, D., *J. Colloid & Interface Science*, 151(1), 166 (1992). Reprinted with permission from Academic Press.

One way of introducing heterogeneity in the analysis is to consider a time-dependent adsorption rate coefficient. According to Kopleman (1988), classical reaction kinetics is sometimes unsatisfactory when the reactants are spatially constrained on the microscopic level by walls, phase boundaries, or force fields. These types of "heterogeneous" reactions (for example, bioenzymatic reactions) that occur at the interfaces of different phases exhibit fractal orders for elementary reactions and rate coefficients with temporal memories. In these types of reactions the rate coefficient exhibits a form given by

$$k = k't^{-b}, \quad 0 \leq b \leq 1 \quad (t \geq 1). \tag{2.8a}$$

In general, k depends on time, whereas $k' = k(t = 1)$ does not. Kopelman indicates that in 3-D (homogeneous) space, $b = 0$. This is in agreement with the results obtained in classical kinetics. Also, with vigorous stirring, the system is made homogeneous and again, $b = 0$. However, for diffusion-limited reactions occurring in fractal spaces, $b > 0$; this yields a time-dependent rate coefficient, k.

Figure 2.3a shows the influence of a decreasing forward reaction rate constant, k_1, for a first-order reaction, on the concentration of the antigen close to the surface, c_s. The decreasing and increasing adsorption rate coefficients are assumed to exhibit the exponential forms (Cuypers *et al.*, 1987)

$$k_1 = k_{1,0}\exp(-\beta t) \tag{2.8b}$$

and

$$k_1 = k_{1,0}\exp(\beta t). \tag{2.8c}$$

Here, β and $k_{1,0}$ are constants.

Cuypers *et al.* proposed only a decreasing adsorption rate coefficient with time. For a decreasing k_1, as time increases, the Damkohler number, Da,

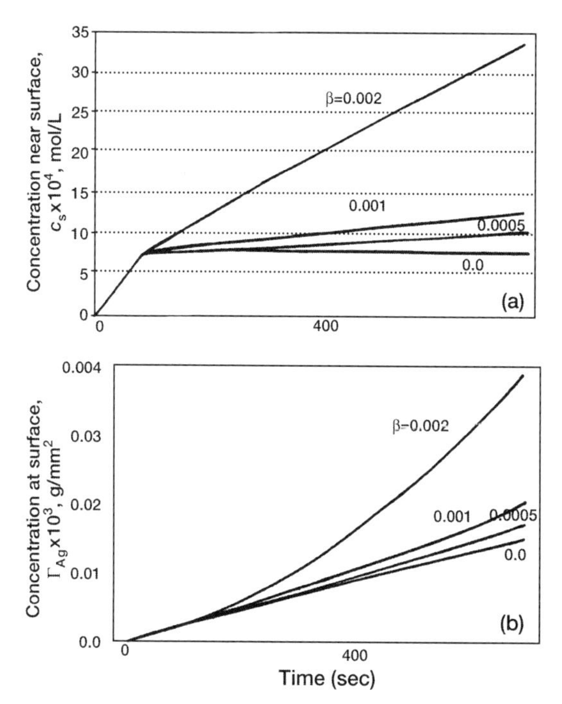

FIGURE 2.3 Influence of a decreasing forward reaction rate coefficient, k_1, for a first-order reaction on (a) the concentration of the antigen close to the surface, c_s, and (b) the amount of antigen attached to the antibody on the surface, Γ_{Ag}.

decreases. Then, according to Fig. 2.2a, the concentration of the antigen close to the surface, c_s, should increase. This is exactly what is seen in Fig. 2.3a. Another way of looking at this is that as k_1 and Da decrease, the diffusion limitation becomes smaller and therefore c_s increases.

Figure 2.3b shows the influence of a decreasing k_1 for a first-order reaction on the amount of antigen attached to the antibody on the surface. Once again, from Fig. 2.2b we note that for the same time, as the Da decreases, the amount of protein adsorbed on the surface increases. This is seen clearly in Fig. 2.3b. Note that a decreasing adsorption rate coefficient introduces nonlinearity in the attachment of the antigen to the antibody immobilized to the surface.

Nygren and Stenberg (1990) analyzed the surface-induced aggregation of ferritin onto a hydrophobic surface. Initially, the adsorption of new ferritin molecules continues mainly via a growth of clusters. This leads to an increasing rate coefficient with time. Figure 2.4a shows the influence of an increasing forward reaction (adsorption) rate coefficient, k_1, on the concentration of the antigen close to the surface, c_s. For an increasing k_1, as time increases, the Da increases and, as expected, c_s decreases. This is clearly seen in Fig. 2.4a. For this same reason, as time increases, the amount of protein adsorbed on the surface decreases. This is clearly seen in Fig. 2.4b.

The single-step binding and the first-order kinetics that result can be extended to dual-step binding.

2.2.1. Second-Order Reaction Kinetics

To see how dual-step binding can be extended from the single-step binding analysis, let us analyze a specific binding case. The elementary steps involved in the reaction scheme are given in Fig. 2.5. The rate of binding of the single antigen to the bound antibody is given by

$$\frac{d\Gamma_1}{dt} = k_1 c_s(2\Gamma_0 - \Gamma_1 - 2\Gamma_2) - k_2\Gamma_2 c_s + 2k_{-2}\Gamma_2 - k_{-1}\Gamma_1. \qquad (2.9a)$$

Here, Γ_2 is the surface concentration of the antibody that binds two antigens. Antibodies have two antigen binding sites. Thus, antibodies with no antigen bound to them have two binding sites available, antibodies with one antigen bound have one site available, and antibodies with two antigens bound have no sites available. The stoichiometric coefficient is also added to reflect the possibility that either of the two antigens may dissociate.

The rate at which the antibody binds two antigens is given by

$$\frac{d\Gamma_2}{dt} = k_2 c_s \Gamma_1 - 2k_{-2}\Gamma_2. \qquad (2.9b)$$

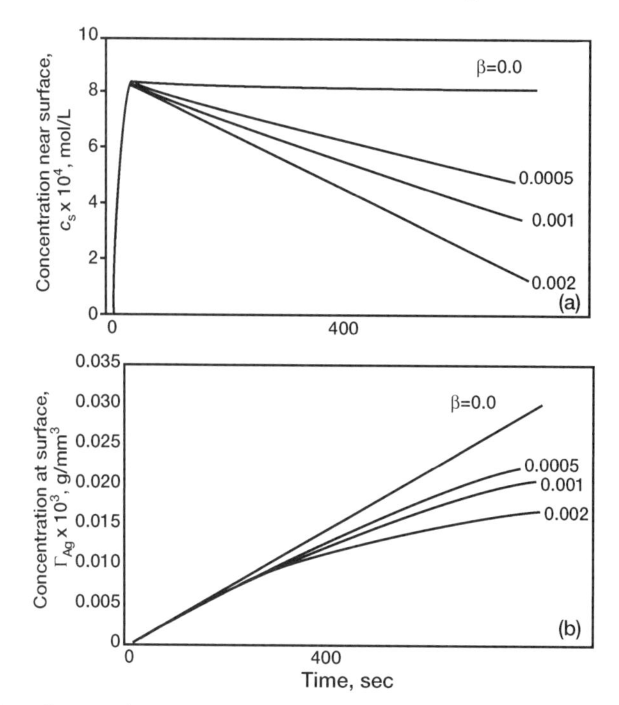

FIGURE 2.4 Influence of an increasing forward reaction rate coefficient, k_1, for a first-order reaction on (a) the concentration of the antigen close to the surface, c_s, and (b) the amount of antigen attached to the antibody on the surface, Γ_{Ag}.
Sadana, A. and Sii, D., *J. Colloid & Interface Science*, 151(1), 166 (1992). Reprinted with permission from Academic Press.

Let $\Gamma_t = 2\Gamma_2 + \Gamma_1$, the total number of bound sites. Then Eqs. (2.9a) and (2.9b) may be added to yield the general expression

$$\frac{d\Gamma_t}{dt} = k_1 c_s(2\Gamma_0 - \Gamma_1 - 2\Gamma_2) + k_2\Gamma_1 c_s - k_{-1}\Gamma_1 - 2k_{-2}\Gamma_2. \tag{2.9c}$$

Furthermore, we may write $k_2 = ck_1$, where c is a "cooperativity parameter."

We are interested in initial binding kinetics. Therefore, $\Gamma_1 \ll \Gamma_0$ and $\Gamma_2 \ll \Gamma_0$. Also, $k_1 c_s \Gamma_0 \gg k_2 c_s \Gamma_1$, or in effect, $k_1\Gamma_0 \gg k_2\Gamma_1$. Also, $k_{-2}\Gamma_2$ and $k_{-1}\Gamma_1$ are very small.

Then Eq. (2.9a) reduces to

$$\frac{d\Gamma_1}{dt} = 2k_1 c_s \Gamma_0. \tag{2.10}$$

Equation (2.10) is identical in form to Eq. (2.3).

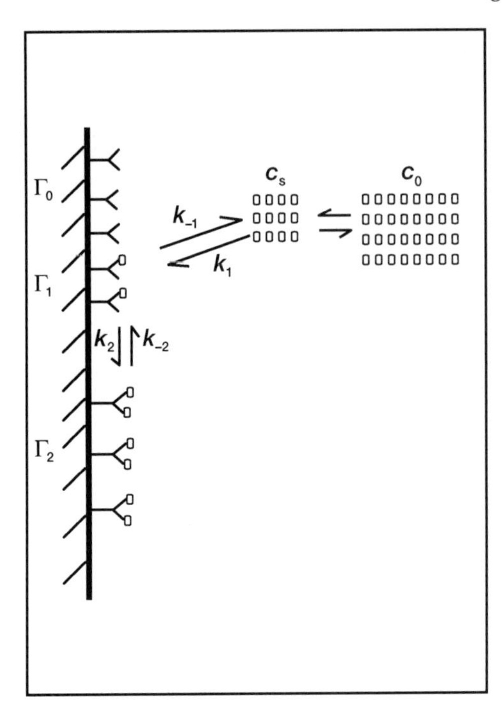

FIGURE 2.5 Elementary steps involved in the binding of the antigen to the antibody covalently attached to the surface (dual-step binding).
Sadana, A. and Sii, D., *J. Colloid & Interface Science*, 151(1), 166 (1992). Reprinted with permission from Academic Press.

The reaction scheme in Fig. 2.5 may be lumped together as

$$\mathrm{Ab} + \mathrm{Ag} \underset{k_r}{\overset{k_f;\mathrm{Ag}}{\rightleftharpoons}} \mathrm{Ag{\cdot}Ab{\cdot}Ag}, \tag{2.11a}$$

where Ab is the antibody molecule and Ag is the antigen molecule.

We now want to relate k_f and k_r to the elementary steps involved in the reaction scheme given in Fig. 2.5.

Consider

$$\mathrm{Ab} + \mathrm{Ag} \underset{k_{-1}}{\overset{k_1}{\rightleftharpoons}} \mathrm{Ab{\cdot}Ag} \underset{k_{-2}}{\overset{k_2;\mathrm{Ag}}{\rightleftharpoons}} \mathrm{Ag{\cdot}Ab{\cdot}Ag}. \tag{2.11b}$$

From Eq. (2.11b),

$$\frac{d[\mathrm{Ab{\cdot}Ag}]}{dt} = 2k_1[\mathrm{Ab}][\mathrm{Ag}] + k_{-2}[\mathrm{Ag{\cdot}Ab{\cdot}Ag}] - k_2[\mathrm{Ab{\cdot}Ag}][\mathrm{Ag}] - k_{-1}[\mathrm{Ab{\cdot}Ag}] = 0, \tag{2.12}$$

from the steady-state assumption. Here [Ag] is the concentration of the antigen close to the surface. c_s and [Ag] may be used interchangeably. Then, the concentration of the Ab•Ag complex at steady state is given by

$$[\mathrm{Ab{\cdot}Ag}] = \frac{2k_{-1}[\mathrm{Ab}][\mathrm{Ag}] + k_{-2}[\mathrm{Ag{\cdot}Ab{\cdot}Ag}]}{k_2[\mathrm{Ag}] + k_{-1}}. \tag{2.13}$$

From Eqs. (2.11a) and (2.11b), the rate of formation of the Ag•Ab•Ag complex is given by

$$\frac{d[\mathrm{Ag{\cdot}Ab{\cdot}Ag}]}{dt} = k_f[\mathrm{Ab}][\mathrm{Ag}]^2 - k_r[\mathrm{Ag{\cdot}Ab{\cdot}Ag}]. \tag{2.14a}$$

and

$$\frac{d[\mathrm{Ag{\cdot}Ab{\cdot}Ag}]}{dt} = k_2[\mathrm{Ag{\cdot}Ab}][\mathrm{Ag}] - k_{-2}[\mathrm{Ag{\cdot}Ab{\cdot}Ag}]. \tag{2.14b}$$

Substituting [Ag•Ab] from Eq. (2.13) into Eq. (2.14b) and comparing Eqs. (2.14a) and (2.14b), we get

$$k_f = \frac{2k_1k_2}{k_2[\mathrm{Ag}] + k_{-1}}$$
$$k_r = \frac{k_{-1}k_{-2}}{k_2[\mathrm{Ag}] + k_{-1}}. \tag{2.15}$$

Case I. If $k_{-1} \ll k_2[\mathrm{Ag}]$ in Eq. (2.15) (a highly probable case), then $k_r = 2k_1/[\mathrm{Ag}]$. Also, if $k_{-1} \approx k_{-2}$, then $k_r \approx 0$, which indicates negligible dissociation and matches the previous conclusion (Stenberg *et al.*, 1986).

Case II. If there is rapid dissociation (not very likely), then $k_2[\mathrm{Ag}] \ll k_{-1}$, $k_f = 2k_1k_2$ (where $K_1 = k_1/k_{-1}$), and $k_r = k_{-2}$.

Let us analyze the more probable case (case I). From Eq. (2.9a),

$$\frac{d\Gamma_1}{dt} = \frac{k_f}{2} c_s^2 \Gamma_0 = \frac{k_f}{2} [\mathrm{Ag}]^2 \Gamma_0 = \frac{d\Gamma_{\mathrm{Ag}}}{dt}, \tag{2.16}$$

where Γ_{Ag} is the concentration of the bound reactant. The second-order dependence on antigen concentration is not very surprising since two molecules of the antigen can bind to two binding sites on the same antibody molecule.

The diffusion-limited simplified reaction scheme can be represented by

$$\frac{\partial c}{\partial t} = D \frac{\partial^2 c}{\partial x^2} \tag{2.4b}$$

and

$$\frac{d\Gamma_{\mathrm{Ag}}}{dt} = \frac{k_f}{2} c_s^2 \Gamma_0 = \frac{k_f}{2} \Gamma_0 c^2(0,t). \tag{2.17}$$

The boundary condition for Eq. (2.4b) is given by Eq. (2.5a). From Eqs. (2.17) and (2.5a),

$$\left.\frac{\partial c}{\partial x}\right|_{x=0} = \frac{k_f}{2} \frac{\Gamma_0}{D} c^2(0,t). \tag{2.18}$$

Equation (2.18) may be rewritten in dimensionless form as:

$$\left.\frac{\partial y}{\partial z}\right|_{z=0} = \mathrm{Da} u^2, \tag{2.5c}$$

where the Damkohler number now is equal to $Lk_1\Gamma_0 c_0/D$. For nth-order reaction kinetics, $\mathrm{Da} = Lk_1\Gamma_0 c_0^{n-1}/D$. The appropriate initial condition is given in Eq. (2.5d).

Equation (2.4b) may be solved to yield concentration profiles for $c(L/2=R,t)$ using the boundary condition Eq. (2.17) and the initial condition Eq. (2.5d). The boundary condition is nonlinear, and the solution to Eq. (2.4b) is obtained by a numerical method of solution (Patankar, 1980). The solution is obtained by a finite difference method with specific points in the domain called grid points. The advantage of the finite difference method is that it is easy to formulate to the discretization equations. Also, it requires a large number of grid points for high accuracy. The solution of the discretization equation can be obtained by the standard Gaussian elimination method. Because of the particularly simple form of the equation, Patankar turns it into a convenient algorithm called the Thomas algorithm or the

Tridiagonal-Matrix Algorithm (TDMA). The method computes the concentration of a configuration with a given initial condition and two other boundary conditions.

The influence of the Damkohler number on c_s and Γ_{Ag} for a second-order reaction is shown in Figs. 2.6a and 2.6b, respectively. Clearly, an increase in the Damkohler number substantially decreases substantially the saturation level of c_s. An increase in the Damkohler number also decreases the rate of increase of Γ_{Ag}/c_0.

Figure 2.7a shows the influence of a decreasing forward reaction rate coefficient, k_1, for a second-order reaction on the concentration of antigen close to the surface, c_s. As time increases, the Da decreases and c_s increases. Figure 2.7b shows the influence of a decreasing k_1 for a second-order reaction on the amount of antigen attached to the antibody on the surface, Γ_{Ag}. As time increases, the Da decreases and Γ_{Ag} increases.

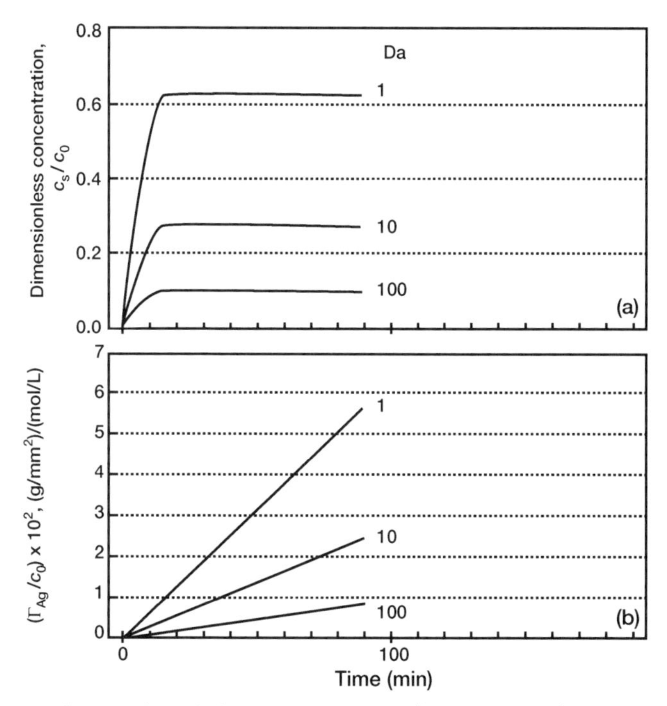

FIGURE 2.6 Influence of Damkohler number on (a) c_s/c_0 and (b) Γ_{Ag}/c_0 for a second-order reaction.

Sadana, A. and Sii, D., *J. Colloid & Interface Science*, 151(1), 166 (1992). Reprinted with permission from Academic Press.

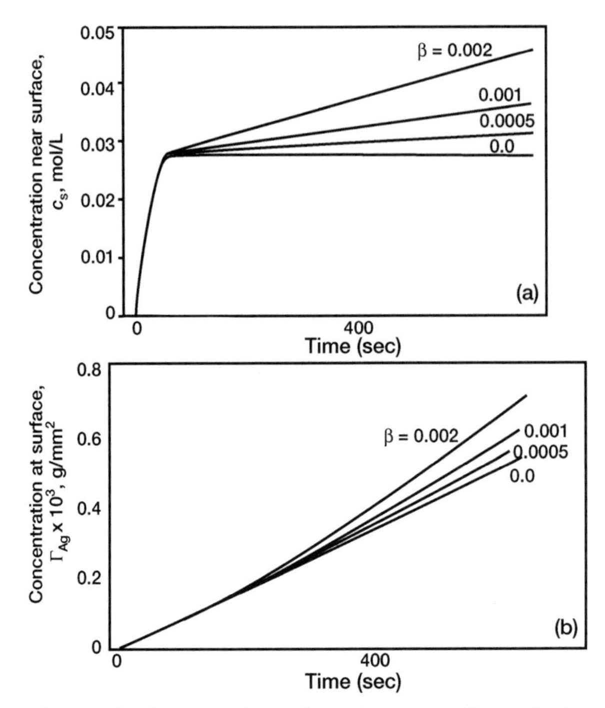

FIGURE 2.7 Influence of a decreasing forward reaction rate coefficient, k_1, for a second-order reaction on (a) the concentration of the antigen close to the surface, c_s, and (b) the amount of antigen attached to the antibody on the surface, Γ_{Ag}.
Sadana, A. and Sii, D., *J. Colloid & Interface Science*, 151(1), 166 (1992). Reprinted with permission from Academic Press.

Figures 2.8a and 2.8b show the influence of an increasing reaction rate coefficient, k_1, for a second-order reaction on (a) the concentration of the antigen close to the surface, c_s, and on (b) the amount of antigen attached to the antibody on the surface, Γ_{Ag}.

2.2.2. Other-Order Reaction Kinetics

For the present, no reaction mechanisms are proposed for one-and-one-half-order reaction kinetics. Nevertheless, it is useful to display curves of $c(0,t)$ and $\Gamma_{Ag}(t)$ for this reaction order. One possible explanation for these types of kinetics could be fractal-like kinetics (Kopelman, 1988; Avnir, 1989). The

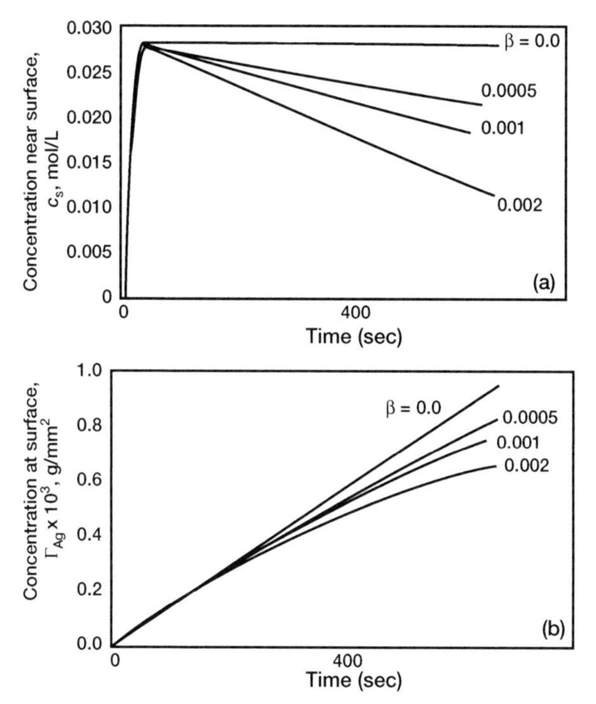

FIGURE 2.8 Influence of an increasing forward reaction rate coefficient, k_f, for a second-order reaction on (a) the concentration of the antigen close to the surface, c_s, and on (b) the amount of antigen attached to the antibody on the surface, Γ_{Ag}.
Sadana, A. and Sii, D., *J. Colloid & Interface Science*, 151(1), 166 (1992). Reprinted with permission from Academic Press.

fractal-like nature expresses itself in an anomalous reaction order, d. For our case, $d = 1.5$.

For a reaction order of one and one-half, the equations that need to be solved are Eqs. (2.4b), (2.5d), and

$$\left.\frac{\partial c}{dx}\right|_{x=0} = \frac{k_f}{2}\frac{\Gamma_0}{D}c^{3/2}(0,t). \tag{2.19}$$

The solutions for $c(0,t)$ and $\Gamma_{Ag}(t)$ are once again obtained by a numerical method of solution (Patankar, 1980). Figure 2.9a shows a plot of $c(0,t)$ versus the time for different initial solution concentrations of the antigen and different Da numbers for a one-and-one-half-order reaction. Figure 2.9b shows the concentration of the antigen attached to the antibody on the

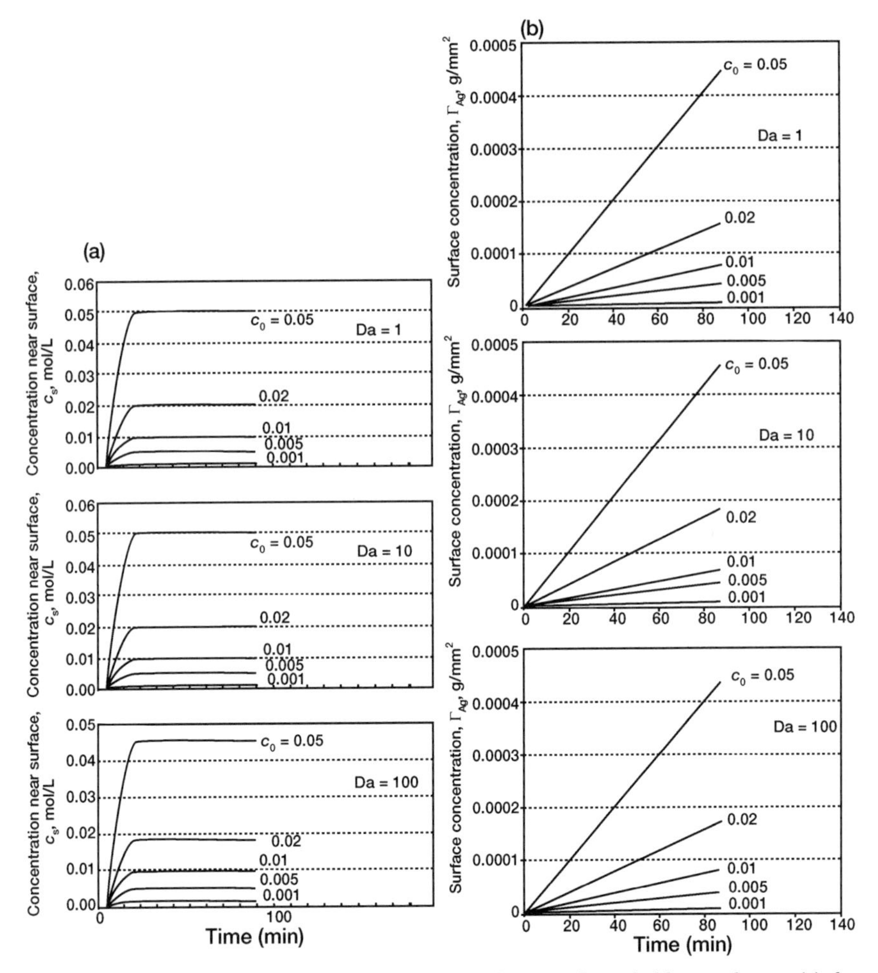

FIGURE 2.9 Influence of antigen concentration in solution and Damkohler number on (a) the concentration of antigen close to the surface, c_s, for a one-and-one-half-order reaction for Da = 1, Da = 10, and Da = 100; and (b) the amount of antigen attached to the antibody on the surface for a one-and-one-half order reaction. The c_0 values for these three Da figures are in gmol/L. Sadana, A. and Sii, D., *J. Colloid & Interface Science*, 151(1), 166 (1992). Reprinted with permission from Academic Press.

surface, Γ_{Ag}, for different initial concentrations of antigen in solution, c_0, and Da values for a one-and-one-half-order reaction.

Figures 2.10a and 2.10b show the influence of a decreasing forward reaction rate coefficient, k_1, for a one-and-one-half-order reaction on the

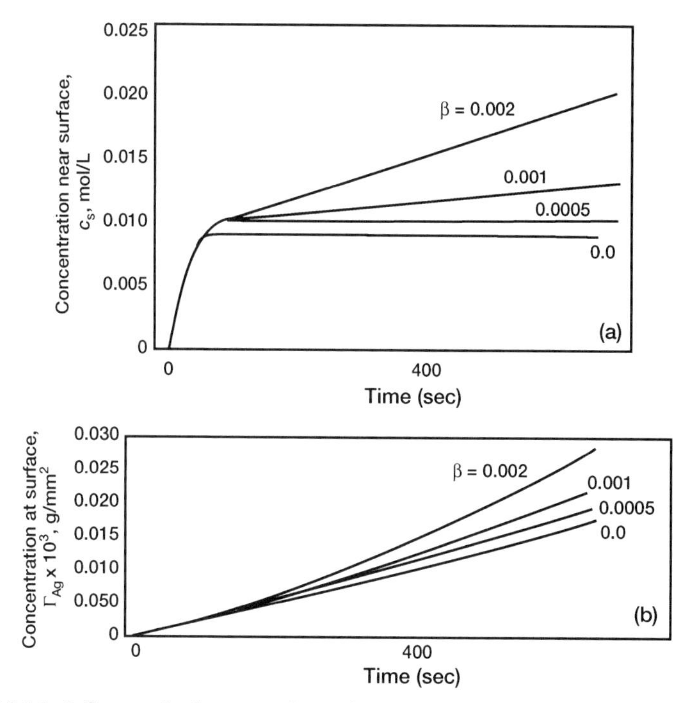

FIGURE 2.10 Influence of a decreasing forward reaction rate coefficient, k_f, for a one-and-one-half-order reaction on (a) the concentration of the antigen close to the surface, c_s, and (b) the amount of antigen attached to the antibody on the surface, Γ_{Ag}.

concentration of the antigen close to the surface, c_s, and on the amount of antigen attached to the surface, Γ_{Ag}, respectively. Figures 2.11a and 2.11b show the influence of an increasing forward reaction rate coefficient, k_1, for a one-and-one-half-order reaction.

Since it would be naive to assume that there is no nonspecific adsorption, Figure 2.12 shows that nonspecific adsorption changes (with respect to the specific adsorption) with antigen in solution and with the antibodies covalently attached to the surface. Thus, nonspecific adsorption should be considered in the development of an appropriate model. This factor should be carefully examined, and if it is negligible (say, less than 5 to 10%), it should be explicitly stated. (Nonspecific adsorption is covered in more detail in

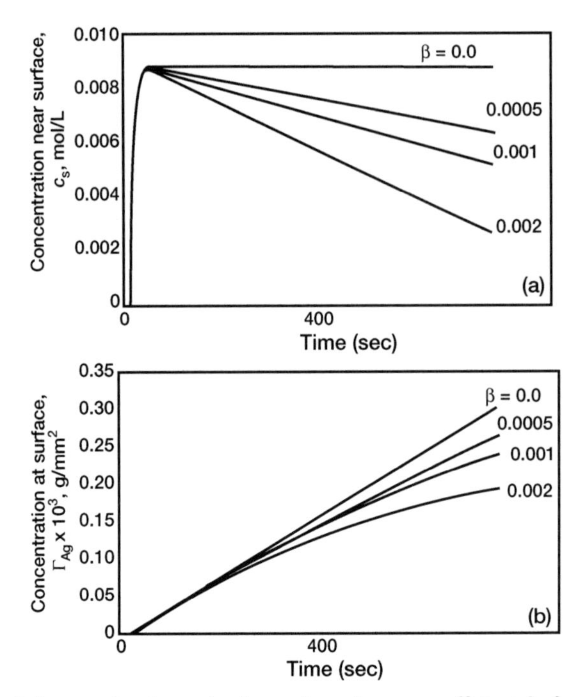

FIGURE 2.11 Influence of an increasing forward reaction rate coefficient, k_f, for a one-and-one-half-order reaction on (a) the concentration of the antigen close to the surface, c_s, and (b) the amount of antigen attached to the antibody on the surface, Γ_{Ag}.
Sadana, A. and Sii, D., *J. Colloid & Interface Science*, 151(1), 166 (1992). Reprinted with permission from Academic Press.

Chapters 8 and 9.) The nonspecific adsorption would also lead to a net time-dependent adsorption rate coefficient.

As mentioned earlier, another reason for a time-dependent adsorption rate coefficient is heterogeneity in adsorption. Thus, the importance of the analysis of a time-dependent adsorption rate coefficient and its influence on the amount of antigen in solution attached to the antibody on the surface for different reaction orders is apparent. Clearly, the factors that influence the time-dependent adsorption rate coefficient, and its eventual influence on the amount of antigen attached to the antibody on the surface, need to be further studied.

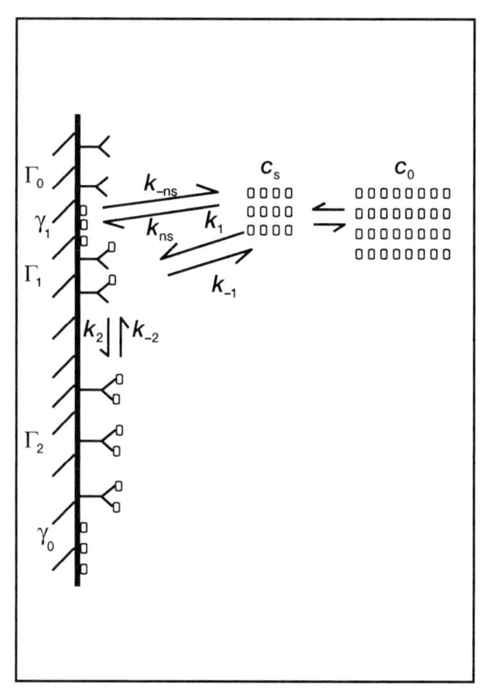

FIGURE 2.12. Elementary steps involved in the binding of the antigen to the antibody covalently attached to the surface. Nonspecific binding is also considered. γ_0 is the total of nonspecific binding sites on the surface; γ_1 is the number of filled nonspecific binding sites on the surface.
Sadana, A. and Sii, D., *J. Colloid & Interface Science*, 151(1), 166 (1992). Reprinted with permission from Academic Press.

REFERENCES

Avnir, D., *The Fractal Approach to Heterogeneous Chemistry: Surfaces, Colloids, and Polymers*, Wiley, New York, 1989.

Berg, H. C. and Purcell, E. M., *Biophys. J.* **20**, 193 (1977).

Bhatia, S. K., Shriver-Lake, L. C., Prior, K. J., George, J. H., Calvert, J. M., Brederhorst, R., and Ligler, F. S., *Anal. Biochem.* **178**, 408 (1989).

Carslaw, H. S. and Jaeger, J. C., *Conduction of Heat in Solids*, 2nd ed., Clarendon, Oxford, p. 72, 1959.

Cuypers, P. A., Willems, G. M., Hemker, H. C., and Hermens, W. T., in *Blood in Contact with Natural and Artificial Surfaces* (E. F. Leonard, V. T. Turitto, and L. Vroman, eds.), *Annals NY Acad. Sci.*, **516**, 244, 1987.

DeLisi, C. A., in *Lecture Notes in Biomathematics*, Vol. 8., 1976. Springer-Verlag, Berlin, 1976.

Giaever, I., German patent, 2,638,207; March 10, 1977.

Humphrey, A. E., *Lecture Notes in Biotechnology*, 1972. University of Pennsylvania Press, Philadelphia, 1972.

Kopelman, R., *Science* **241**, 1620 (1988).
Nygren, H. and Stenberg, M., *J. Immunol. Methods* **80**, 15 (1985a).
Nygren, H. and Stenberg, M., *J. Colloid & Interface Sci.* **107**, 560 (1985b).
Nygren, H. and Stenberg, M., *Biophys. Chem.* **38**, 67 (1990).
Patankar, S. V., *Numerical Heat Transfer and Fluid Flow,* McGraw-Hill, New York, 1980.
Place, J. F., Sutherland, R. M., Riley, A., and Mangan, C., in *Biosensors with Fiberoptics* (D. Wise, and L. B. Wingard Jr., eds.), Humana Press, New York. p. 253, 1991
Shoup, D. and Szabo, A., *Biophys. J.* **40**, 33–39 (1982).
Stenberg, M., Elwing, H., and Nygren, H., *J. Theor. Biol.* **98**, 307 (1982).
Stenberg, M. and Nygren, H., *Anal. Biochem.* **127**, 183 (1982).
Stenberg, M. and Nygren, H., *J. Immunol. Methods* **113**, 2 (1988).
Stenberg, M., Stiblert, L., and Nygren, H., *J. Theor. Biol.* **120**, 129 (1986).

CHAPTER 3

Influence of Diffusional Limitations and Lateral Interactions on Antigen–Antibody Binding Kinetics

3.1. INTRODUCTION

The success of the detection scheme using biosensors will be significantly improved if one obtains a better understanding of the different steps involved in the sensing process. Nygren and Stenberg (1989) have indicated the importance of intermolecular interactions in the reaction layer to the binding kinetics of antigen in solution to the immobilized antibody. Also available is experimental evidence that provides for attractive interactions in the surface-bound antibodies (Uzgiris and Kornberg, 1983; Nygren, 1988). Repulsive interactions in the reaction layer have been related to the passivation of surfaces (Cuypers *et al.*, 1987). Data has been presented by Nygren (1988) that indicates that cohesive forces between macromolecules help stabilize the adsorbed proteins and antigen–antibody complexes at solid surfaces. Furthermore, Nygren and Stenberg (1991) have studied to some extent the influence of lateral interactions in antigen–antibody systems. In this chapter, we will analyze the influence of lateral interactions, diffusion rate limitation, and variable adsorption rate coefficients for dual-step binding on the kinetics of antigen–antibody reactions. This should help in the development of conceptual models for these reactions occurring in the reaction layer, thereby providing physical insights to better understand and control these reactions to advantage.

3.2. THEORY

3.2.1. ANTIGEN IN SOLUTION/ANTIBODY ON THE SURFACE

The analysis of the binding of an antigen in solution to an antibody immobilized on the surface was presented in Chapter 2 and will not be repeated here, although the same equations apply. We now discuss the additional equations that arise due to the lateral interactions and give some examples of lateral interactions. The derivations provided for the simpler cases may be extended to the more complex or sophisticated cases.

Influence of Lateral Interactions

As previously mentioned, repulsive interactions in the reaction layer have been related to the passivation of surfaces by Cuypers *et al.* (1987). These authors state that an initial rapid, often diffusion-limited, rate may be followed by a continuously decreasing rate. According to Nygren (1988), this may arise due to nonidealities or heterogeneity on the surface. Nygren determined and analyzed a heterogeneous distribution of adsorbed protein molecules over the surface, indicating that cohesive forces act on the molecules. The attractive interactions would help stabilize the reaction complexes (antibody–antigen) on the surface. This would also lead to complexities on the surface that need to be characterized. One way to do this is to use fractals (or objects that exhibit self-similarity), over different lengths of scale (that is, scale invariant). Fractals are discussed in detail in later chapters. The receptors on the surface may form fractal clusters, and in these closely packed or tightly organized clusters lateral interactions may be facilitated.

More recently, Nygren and Stenberg (1990) analyzed surface-induced aggregation of ferritin onto a hydrophobic surface. Initially, the adsorption of new ferritin molecules continues mainly via a growth of clusters. This leads to an increasing adsorption rate coefficient with time. In other cases, an apparent positive cooperative effect has been noted by Werthen *et al.* (1988) in the forward binding rate coefficient. Also, Nygren and Stenberg (1985) noted a decrease in binding rate with time while studying the kinetics of antibody binding to surface-immobilized bovine serum albumin. Lateral interactions may be treated in two unrelated ways: first, by the introduction of new state and rate constants, (k_3, k_{-3}), and then by a phenomenological model (a power law) for k_1. Γ_3 represents the surface concentration of the antibody–antigen complexes that are involved in lateral interactions on the surface.

The rate of binding of the single arm of the covalently bound antibody to the antigen in solution is given by Fig. 3.1.

$$\frac{d\Gamma_1}{dt} = 2k_1 c_s(\Gamma_0 - \Gamma_1 - \Gamma_2 - 2\Gamma_3) - \Gamma_1(k_{-1} + k_2 c_s) - k_3\Gamma_1^2 + 2k_{-2}\Gamma_2 + 2k_{-3}\Gamma_3. \quad (3.1a)$$

The rate at which the antibody binds two antigens is given by

$$\frac{d\Gamma_2}{dt} = k_2\Gamma_1 c_s - 2k_{-2}\Gamma_2. \quad (3.1b)$$

The rate at which the antibody–antigen complex molecules laterally interact is

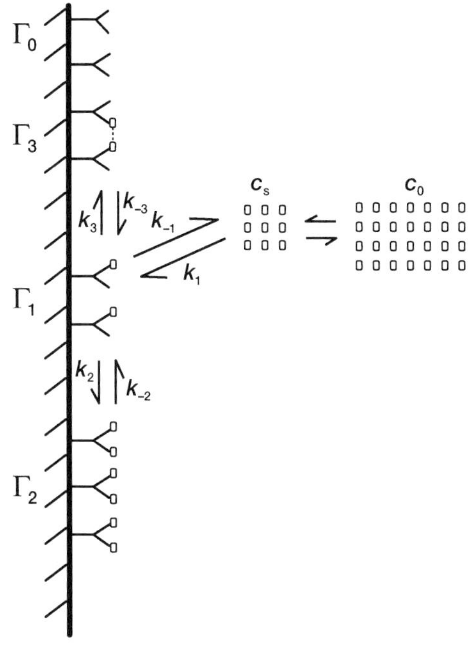

FIGURE 3.1 Elementary steps involved in the binding of antigen in solution to the antibody covalently attached to the surface (dual-step binding). Involvement of lateral interactions between antibody–antigen complexes on the surface is also shown (k_3, k_{-3}).
Sadana, A. and Madugula, A., *Biotechnology Progress*, 9, 259–266 (1993). Reprinted with permission from Academic Press.

given by

$$\frac{d\Gamma_3}{dt} = k_3\Gamma_1^2 - k_{-3}\Gamma_3. \tag{3.2}$$

The reaction schemes showing lateral interactions may be lumped together as

$$\text{Ab} + \text{Ag} \underset{k_r}{\overset{k_f;\text{Ag}}{\rightleftharpoons}} \text{Ag}\cdot\text{Ab}\cdot\text{Ag} \tag{3.3a}$$

and

$$\text{Ab} + \text{Ag} \underset{k_{-1}}{\overset{k_1}{\rightleftharpoons}} \text{Ab}\cdot\text{Ag} \underset{k_{-2}}{\overset{k_2;\text{Ag}}{\rightleftharpoons}} \text{Ag}\cdot\text{Ab}\cdot\text{Ag}. \tag{3.3b}$$

With the inclusion of the lateral step, the lumped reaction scheme is

$$\text{Ab} + \text{Ag} \underset{k_r'}{\overset{k_f';(\text{Ab}\cdot\text{Ag})}{\rightleftharpoons}} [(\text{Ag}\cdot\text{Ab})\ldots(\text{Ab}\cdot\text{Ag})] \tag{3.4a}$$

and

$$\text{Ab} + \text{Ag} \underset{k_{-1}}{\overset{k_1}{\rightleftharpoons}} \text{Ab}\cdot\text{Ag} \underset{k_{-3}}{\overset{k_3;(\text{Ab}\cdot\text{Ag})}{\rightleftharpoons}} [(\text{Ag}\cdot\text{Ab})\ldots(\text{Ab}\cdot\text{Ag})]. \tag{3.4b}$$

From Eqs. (3.3b) and (3.4b),

$$\begin{aligned}\frac{d\Gamma_1}{dt} &= 2k_1[\text{Ab}]c_s + 2k_{-2}\Gamma_2 - k_{-1}\Gamma_1 - k_2\Gamma_1 \\ &\quad + 2k_{-3}[(\text{Ag}\cdot\text{Ab})\ldots(\text{Ab}\cdot\text{Ag})] - k_3\Gamma_1^2 = 0\end{aligned} \tag{3.5a}$$

and

$$\frac{d\Gamma_2}{dt} = k_2\Gamma_1 c_s - k_{-2}\Gamma_2 = 0. \tag{3.5b}$$

From Eq. (3.3a),

$$\frac{d\Gamma_2}{dt} = k_f[\text{Ab}]c_s^2 - k_r\Gamma_2 \tag{3.5c}$$

From Eqs. (3.4a) and (3.4b),

$$\frac{d}{dt}[(\text{Ag}\cdot\text{Ab})]\ldots(\text{Ab}\cdot\text{Ag})] = k_f'[\text{Ab}]c_s\Gamma_1 - k_r'[(\text{Ag}\cdot\text{Ab})\ldots(\text{Ab}\cdot\text{Ag})] \tag{3.6a}$$

and

$$\frac{d}{dt}[(\mathrm{Ag}\cdot\mathrm{Ab})\ldots(\mathrm{Ab}\cdot\mathrm{Ag})] = k_3\Gamma_1^2 - k_{-3}[(\mathrm{Ag}\cdot\mathrm{Ab})\ldots(\mathrm{Ab}\cdot\mathrm{Ag})]. \quad (3.6b)$$

To solve for the intermediate Γ_1, we have from Eq. (3.5a),

$$\begin{aligned} &k_3\Gamma_1^2 + (k_{-1}+k_2)\Gamma_1 - 2k_1[\mathrm{Ab}]c_s - 2k_{-2}\Gamma_2 - 2k_{-3} \\ &\quad \times [(\mathrm{Ag}\cdot\mathrm{Ab})\ldots(\mathrm{Ab}\cdot\mathrm{Ag})] = 0. \end{aligned} \quad (3.7a)$$

Then, the concentration of the Γ_1 complex is given by

$$\Gamma_1 = \frac{1}{2k_3}\left[-(k_{-1}+k_2) \pm \sqrt{M+N}\right], \quad (3.7b)$$

where

$$\begin{aligned} M &= (k_{-1}+k_2)^2 + 8k_1k_3[\mathrm{Ab}]c_s \\ N &= 8k_3[k_{-2}\Gamma_2 + k_{-3}[(\mathrm{Ag}\cdot\mathrm{Ab})\ldots(\mathrm{Ab}\cdot\mathrm{Ag})]]. \end{aligned}$$

The negative root is unphysical; therefore it is of no use. Then, Γ_1 may be given by

$$\Gamma_1 = \frac{1}{2k_3}\left[(-k_{-1}+k_2) + \sqrt{1+y}\right], \quad (3.8a)$$

where

$$y = \frac{4k_3}{(k_{-1}+k_2)^2}\{2k_1[\mathrm{Ab}]c_s + 2k_{-2}\Gamma_2 + 2k_{-3}[(\mathrm{Ag}\cdot\mathrm{Ab})\ldots(\mathrm{Ab}\cdot\mathrm{Ag})]\} \quad (3.8b)$$

If $y \gg 1$, then Eq. (3.8a) reduces to

$$\Gamma_1 = [\mathrm{Ab}]^{1/2}c_s^{1/2}\sqrt{\frac{2k_1}{k_3}} \quad (3.8c)$$

on the assumptions that $2k_1[\mathrm{Ab}]c_s \gg 2k_{-2}\Gamma_2$ and $2k_1[\mathrm{Ab}]c_s \gg 2k_{-3}[(\mathrm{Ag}\cdot\mathrm{Ab})\ldots(\mathrm{Ab}\cdot\mathrm{Ag})]$.

On substituting for Γ_1 in Eq. (3.5b), we get

$$\frac{d\Gamma_2}{dt} = k_2(2k_1/k_3)^{1/2}[\mathrm{Ab}]^{1/2}c_s^{1/2}. \quad (3.9)$$

Comparing Eqs. (3.5c) and (3.14), we get

$$k_1 = \frac{k_f^2 k_3}{k_2^2}[\text{Ab}]c_s \tag{3.10}$$

Here we assume that $k_f[\text{Ab}]c_s^2 \gg k_r\Gamma_2$. Substitution for k_1 from Eq. (3.10) into the initial binding kinetics form of Eq. (3.1) (Sadana and Sii, 1992b) yields

$$\frac{d\Gamma_1}{dt} = \frac{k_f^2 k_3}{k_2^2} c_s^2[\text{Ab}]\Gamma_0. \tag{3.11}$$

It is also worthwhile to compare the kinetic forms of expression for dual-step binding [Eq. (2.16)] and for dual-step binding with lateral interactions [Eq. (3.11)]. For dual-step binding, the kinetic expression involves the square of the concentration of antigen near the surface and the initial antibody concentration on the surface. Since a single antibody has two arms, this makes sense: An antigen can attach on each arm. For dual-step binding with lateral interactions, the kinetic expression involves the square of the antigen concentration close to the surface, the initial antibody concentration on the surface, and the antibody concentration on the surface available for binding. Since the lateral interactions involve two antibody (really an antibody–antigen complex) molecules, this shows up as the initial antibody concentration on the surface, Γ_0, and the antibody concentration on the surface available for binding, [Ab]. There is also some difference in the kinetic constants.

Figure 3.2 shows the influence of lateral interactions with a different Damkohler number and constant forward (adsorption) reaction rate constant k_1 for a second-order reaction on the concentration of the antigen near the surface. The concentration of the antigen near the surface is higher when lateral interactions are present than when these interactions are absent. Also, as shown in the figure, as the value of the Damkohler number increases (mass transfer increases), the concentration of antigen near the surface decreases, as expected. The initial rate of adsorption also decreases as the Da increases. This is shown for both cases—with and without the presence of lateral interactions. Therefore, a decrease in the external diffusional limitations and an increase in lateral interactions increases the concentration of the antigen near the surface. Also, a decrease in the Da increases the initial rate of adsorption when lateral interactions are absent or present.

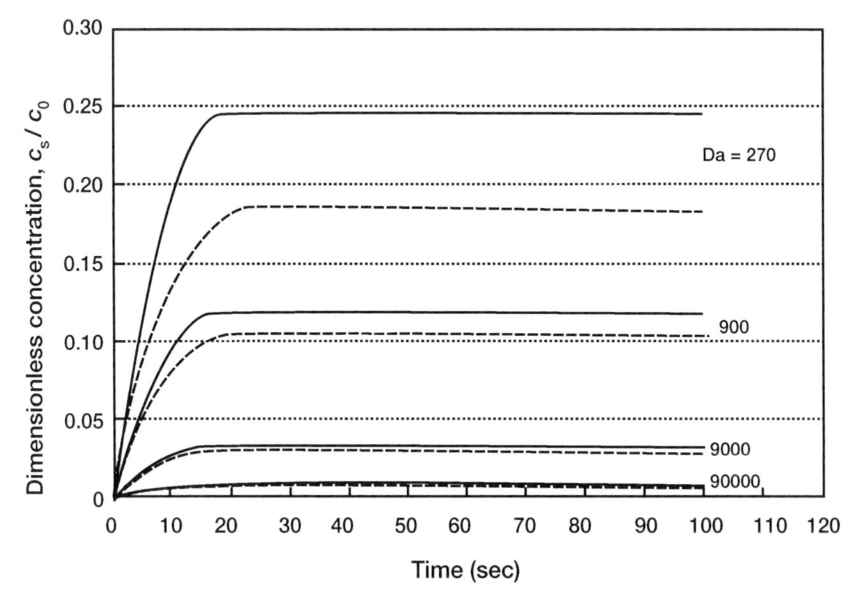

FIGURE 3.2 Concentration of the antigen near the surface versus time for different Damkohler numbers (Da) with (—) and without (----) the influence of lateral interactions.
Sadana, A. and Madugula, A., *Biotechnology Progress*, 9, 259–266 (1993). Reprinted with permission from Academic Press.

3.2.2. Variable Rate Coefficients

Variable rate binding coefficients were introduced in Chapter 2 [see for example, Eqs. (2.8a–2.8c)] and are not repeated here. Figures 3.3a and 3.3b show the influence of a variable adsorption (forward) reaction (or binding) rate coefficient, k_1, for a second-order reaction on the concentration of the antigen near the surface. The adsorption (or binding) rate coefficient, k_1, is of the form $k_1 = k' t^{-b}$. This figure shows that, as time (t) and coefficient b increase, the concentration of the antigen near the surface increases both when lateral interactions are absent (Figure 3.3a) and when they are present (Figure 3.3b). This is to be expected since with an increase in either time (t) or b, k_1 decreases. This results in an increase in the antigen near the surface. Once again, the concentration of the antigen near the surface is higher when lateral interactions are present than when they are absent, everything else being the same (Figure 3.4).

It is worthwhile to compare the influence of an increasing and a decreasing adsorption reaction rate coefficient [of the form $k_1 = k_{1,0} \exp(\pm \beta t)$] for a second-order reaction on the concentration of antigen near the surface when lateral interactions are present and when they are absent. Figure 3.5a shows

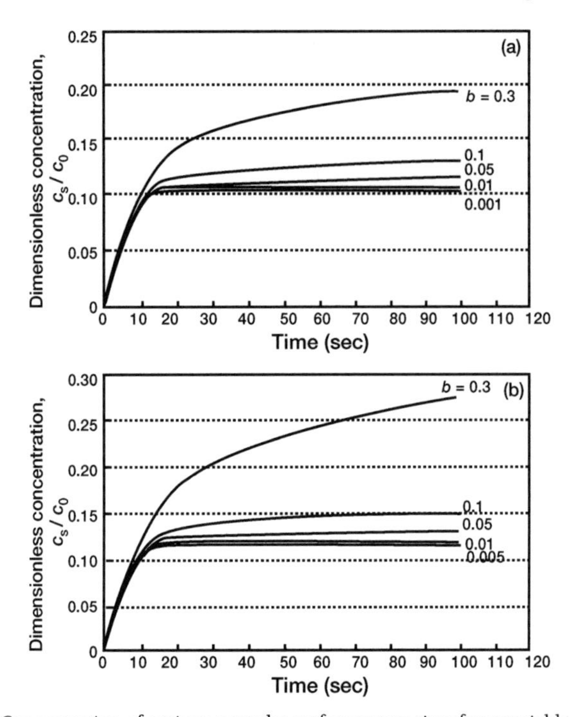

FIGURE 3.3 Concentration of antigen near the surface versus time for a variable adsorption rate coefficient ($k_1 = k't^{-b}$) (a) without and (b) with lateral interactions.
Sadana, A. and Madugula, A., *Biotechnology Progress*, 9, 259–266 (1993). Reprinted with permission from Academic Press.

that for an increasing adsorption rate coefficient, after a brief time interval as time increases the concentration of the antigen near the surface decreases, as expected for the cases when lateral interactions are absent or present. Note that this relatively low value of the variable rate coefficient ($\beta = 0.01$) introduces nonlinearity in the attachment of the antigen to the antibody immobilized to the surface. Also, with the influence of lateral interactions, the concentration of the antigen near the surface increases faster and reaches a higher value than when lateral interactions are absent. No explanation is offered at present for the maximum exhibited in Figure 3.5a for $\beta = 0.01$. Figure 3.5b shows that, for a decreasing adsorption rate coefficient as time increases, the concentration of the antigen near the surface increases continuously for the cases when lateral interactions are present or absent. Note that this relatively low value of the variable rate coefficient ($\beta = -0.01$) also introduces nonlinearity in the attachment of the antigen to the antibody immobilized to the surface (see Figure 3.5a). Also, with the influence of

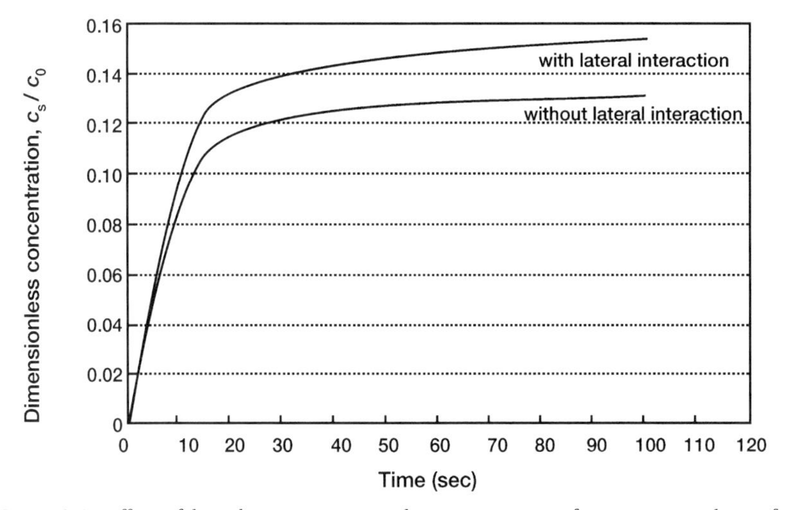

FIGURE 3.4 Effect of lateral interactions on the concentration of antigen near the surface, $b = 0.1$.
Sadana, A. and Madugula, A., *Biotechnology Progress*, 9, 259–266 (1993). Reprinted with permission from Academic Press.

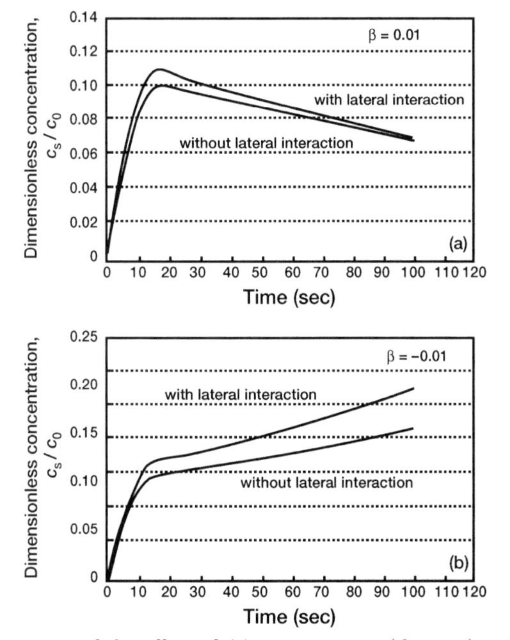

FIGURE 3.5 Comparison of the effect of (a) an increasing ($\beta = 0.01$) and (b) a decreasing variable adsorption rate coefficient ($\beta = -0.01$) on antigen concentration near the surface when lateral interactions are present and when they are absent.
Sadana, A. and Madugula, A., *Biotechnology Progress*, 9, 259–266 (1993). Reprinted with permission from Academic Press.

lateral interactions the concentration of the antigen near the surface increases faster than when lateral interactions are absent.

Figure 3.6a shows the influence of a decreasing and an increasing adsorption rate coefficient on the antigen concentration near the surface when lateral interactions are present. As expected, with a decrease in the adsorption rate coefficient, the concentration of the antigen near the surface increases. Similarly, with an increase in the adsorption rate coefficient, the antigen concentration near the surface decreases.

Figure 3.6b shows the influence of an increasing and a decreasing adsorption rate coefficient on the antigen concentration near the surface when lateral interactions are absent. The trends in antigen concentration near the surface are the same when lateral interactions are present (Figure 3.6a). Note, however, that the concentration of antigen near the surface is higher when lateral interactions are present than when they are absent.

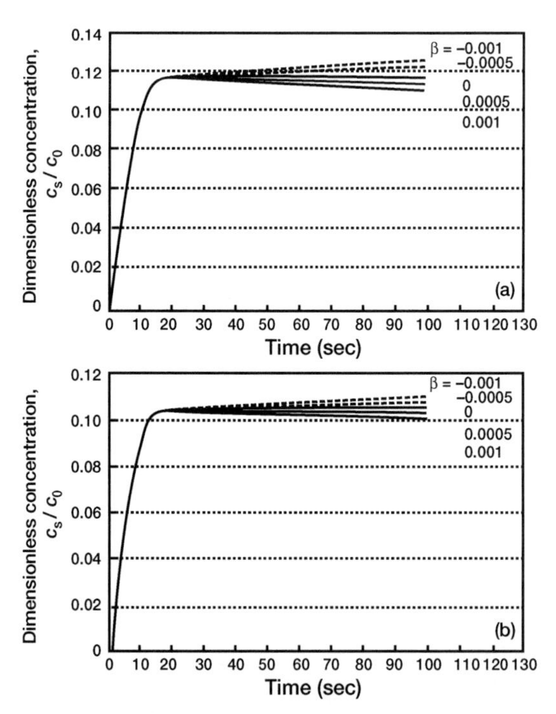

FIGURE 3.6 Concentration of antigen near the surface versus time for a decreasing $[k_1 = k_{1,0}\exp(-\beta t)]$ and an increasing $[k_1 = k_{1,0}\exp(\beta t)]$ variable rate coefficient (a) when lateral interactions are present and (b) when they are absent.

Sadana, A. and Madugula, A., *Biotechnology Progress*, 9, 259–266 (1993). Reprinted with permission from Academic Press.

We now develop the kinetic expressions for the case where the antibodies are in solution and the antigen is covalently attached to the surface. It would be of interest to determine whether the kinetic expressions for the two different cases have dissimilar or similar forms.

3.2.3. Antibody in Solution/Antigen on the Surface

We did not present this case in Chapter 2, so we analyze all the steps in detail here, starting with second-order kinetics. Then we analyze the influence of lateral interactions on second-order kinetics.

Second-Order Reaction Kinetics

Consider dual-step binding. Figure 3.7 shows the steps involved in the binding of the antibody in solution to the antigen covalently attached to the surface. The rate of binding of a single arm of the antibody to an antigen attached to the surface is given by

$$\frac{d\Gamma_1}{dt} = 2k_1 c_s(\Gamma_0 - \Gamma_1 - 2\Gamma_2) - k_2\Gamma_1(\Gamma_0 - \Gamma_1 - 2\Gamma_2) + 2k_{-2}\Gamma_2 - k_{-1}\Gamma_1. \tag{3.12}$$

Here, Γ_0 is the total concentration of the antigen sites on the surface; Γ_1 is the surface concentration of antigen sites that are bound by a single arm of the antibody; and Γ_2 is the surface concentration of the doubly bound antibody concentrations. The rate at which both arms of the antibody in solution are bound to two antigens on the surface is given by

$$\frac{d\Gamma_2}{dt} = k_2\Gamma_1(\Gamma_0 - \Gamma_1 - 2\Gamma_2) - 2k_{-2}\Gamma_2. \tag{3.13}$$

We are interested in the initial binding kinetics. Therefore, $\Gamma_1 \ll \Gamma_0$ and $\Gamma_2 \ll \Gamma_0$ (Stenberg *et al.*, 1986; Sadana and Sii, 1992a, 1992b). Also, $k_1 c_s \Gamma_0 \gg k_2\Gamma_1\Gamma_0$, or in effect, $k_1 c_s \gg k_2\Gamma_1$.

For low values of time t, $k_{-2}\Gamma_2$ and $k_{-1}\Gamma_1$ are also very small (Stenberg *et al.*, 1986; Sadana and Sii, 1992a, 1992b). Then, Eqs. (3.12) and (3.13) reduce to

$$\frac{d\Gamma_1}{dt} = 2k_1 c_s \Gamma_0 \tag{3.14a}$$

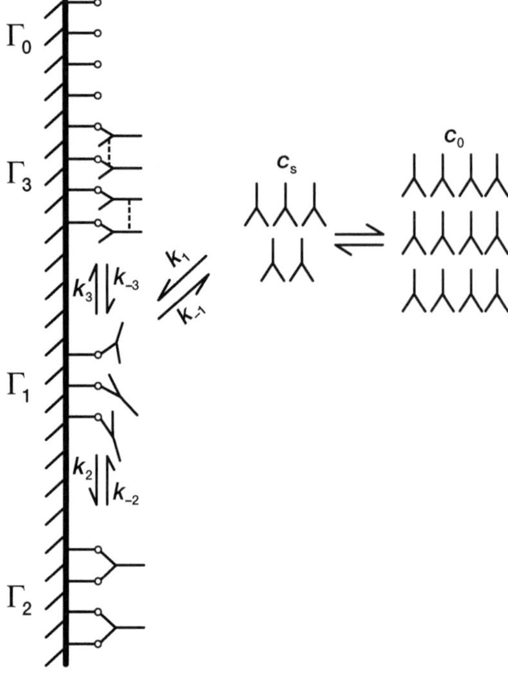

FIGURE 3.7 Elementary steps involved in the binding of the antibody in solution to the antigen covalently attached to the surface (dual-step binding), including involvement of lateral interactions between antibody–antigen complexes on the surface.
Sadana, A. and Madugula, A., *Biotechnology Progress*, 9, 259–266 (1993). Reprinted with permission from Academic Press.

and

$$\frac{d\Gamma_2}{dt} = k_2\Gamma_1\Gamma_0, \tag{3.14b}$$

respectively.

The equations for lumping the reaction scheme are identical in form for the two cases analyzed—antibody in solution/antigen on the surface (Sadana and Sii, 1992b) and antigen in solution/antibody on the surface—and are not repeated here. Bear in mind, however, that now Γ_1 is the surface concentration of the antibodies that are bound by a single arm to antigens at any time t, c_s is the concentration of the antibodies close to the surface, k_1 and k_2 are the forward reaction rate coefficients, and k_{-1} and k_{-2} are the reverse reaction rate coefficients. In this case, [Ab] also is the concentration of antibody close to the surface; c_s and [Ab] may be used interchangeably.

Substitution of $k_1 = k_f[\mathrm{Ag}]/2$ into Eq. (3.14a) yields

$$\frac{d\Gamma_1}{dt} = k_f[\mathrm{Ag}]c_s\Gamma_0. \tag{3.15}$$

Note that in this case the concentration of the bound antibody on the surface exhibits a first-order dependence on both the antibody close to the surface, c_s, and on the antigen on the fiber surface available for binding.

Influence of Lateral Interactions

Figure 3.7 shows the lateral interactions between the antibody–antigen complex molecules on the surface. Here, Γ_3 represents the surface concentration of antibody–antigen complexes that are involved in lateral interactions on the surface.

The rate of binding of a single arm of the antibody in solution to the antigen molecule covalently bound to the surface is given by

$$\frac{d\Gamma_1}{dt} = k_1c_s(\Gamma_0 - \Gamma_1 - 2\Gamma_2 - 2\Gamma_3) - k_{-1}\Gamma_1 - k_2\Gamma_1(\Gamma_0 - \Gamma_1 - 2\Gamma_2 - \Gamma_3) + 2k_{-2}\Gamma_2. \tag{3.16}$$

The rate at which the antibody molecule in solution binds two antigens on the surface is given by

$$\frac{d\Gamma_2}{dt} = k_2\Gamma_1(\Gamma_0 - \Gamma_1 - 2\Gamma_2 - 2\Gamma_3) - 2k_{-2}\Gamma_2. \tag{3.17}$$

The rate at which the antibody–antigen complex molecules laterally interact is given by

$$\frac{d\Gamma_3}{dt} = k_3\Gamma_1^2 - k_{-3}\Gamma_3. \tag{3.18}$$

The nomenclature is intentionally the same as for the case where the antigen is in solution and the antibody is on the surface.

Once again, we are interested in initial binding kinetics. Therefore, $\Gamma_1 \ll \Gamma_0$, $\Gamma_2 \ll \Gamma_0$, and $\Gamma_3 \ll \Gamma_0$ (Stenberg *et al.*, 1986; Sadana and Sii, 1992a, 1992b). Also, $k_{-1}c_s\Gamma_0 \gg k_2\Gamma_1\Gamma_0$, or in effect, $k_1c_s \gg k_2\Gamma_1$. Finally, $k_{-1}\Gamma_1$ and $k_{-2}\Gamma_2$ are very small. Then Eq. (3.16) reduces to

$$\frac{d\Gamma_1}{dt} = 2k_1c_s\Gamma_0,$$

which is Eq. (3.14a).

The reaction schemes shown in Figure 3.7 may be lumped together as

$$\mathrm{Ab} + \mathrm{Ag} \underset{k_r}{\overset{k_f;(\mathrm{Ab}\cdot\mathrm{Ag})}{\rightleftharpoons}} [(\mathrm{Ab}\cdot\mathrm{Ag})\ldots(\mathrm{Ag}\cdot\mathrm{Ab})] \tag{3.19}$$

and

$$\mathrm{Ab} + \mathrm{Ag} \underset{k_{-1}}{\overset{k_1}{\rightleftharpoons}} \mathrm{Ab}\cdot\mathrm{Ag} \underset{k_{-2}}{\overset{k_2;(\mathrm{Ab}\cdot\mathrm{Ag})}{\rightleftharpoons}} [(\mathrm{Ab}\cdot\mathrm{Ag})\ldots(\mathrm{Ag}\cdot\mathrm{Ab})]. \tag{3.20}$$

From Eq. (3.19), at steady state

$$\frac{d[(\mathrm{Ab}\cdot\mathrm{Ag})\ldots(\mathrm{Ag}\cdot\mathrm{Ab})]}{dt} = k_f c_s[\mathrm{Ag}]\Gamma_1 - k_r[(\mathrm{Ab}\cdot\mathrm{Ag})\ldots(\mathrm{Ag}\cdot\mathrm{Ab})], \tag{3.21}$$

and from Eq. (3.20),

$$\frac{d[(\mathrm{Ab}\cdot\mathrm{Ag})\ldots(\mathrm{Ag}\cdot\mathrm{Ab})]}{dt} = k_2\Gamma_1^2 - k_{-2}[(\mathrm{Ab}\cdot\mathrm{Ag})\ldots(\mathrm{Ag}\cdot\mathrm{Ab})]. \tag{3.22}$$

From Eq. (3.20), the steady-state concentration of Γ_1 is given by

$$\frac{d\Gamma_1}{dt} = k_1 c_s[\mathrm{Ag}] - k_{-1}\Gamma_1 - k_2\Gamma_1^2 + 2k_{-2}[(\mathrm{Ab}\cdot\mathrm{Ag})\ldots(\mathrm{Ag}\cdot\mathrm{Ab})] = 0. \tag{3.23}$$

Equation (3.23) is a quadratic equation in Γ_1, which may be rearranged to give

$$k_2\Gamma_1^2 + k_{-1}\Gamma_1 - k_1[\mathrm{Ag}]c_s - 2k_{-2}[(\mathrm{Ab}\cdot\mathrm{Ag})\ldots(\mathrm{Ag}\cdot\mathrm{Ab})] = 0. \tag{3.24}$$

Then, the concentration of Γ_1 from Eq. (3.24) is given by

$$\Gamma_1 = \frac{-k_{-1} \pm \sqrt{k_{-1}^2 + 4k_2\{k_1 c_s[\mathrm{Ag}] + 2k_{-2}[(\mathrm{Ab}\cdot\mathrm{Ag})\ldots(\mathrm{Ag}\cdot\mathrm{Ab})]\}}}{2k_2}. \tag{3.25}$$

The negative root is unphysical; therefore, it is of no use. The positive root may be written as

$$\Gamma_1 = \frac{-k_{-1} + k_{-1}\sqrt{1+y}}{2k_2}, \tag{3.26a}$$

where

$$y = \frac{4k_2}{(k_{-1})^2}\{k_1 c_s[\mathrm{Ag}] + 2k_{-2}[(\mathrm{Ab}\cdot\mathrm{Ag})\ldots(\mathrm{Ag}\cdot\mathrm{Ab})]\}. \tag{3.26b}$$

If $y \gg 1$, then Eq. (3.26b) reduces to

$$\Gamma_1 = c_s^{1/2}[\mathrm{Ag}]^{1/2}\sqrt{\frac{k_1}{k_2}} \tag{3.26c}$$

on the assumption that $k_1 c_s[\mathrm{Ag}] \gg 2_{k_{-2}}[(\mathrm{Ab}\cdot\mathrm{Ag})\ldots(\mathrm{Ag}\cdot\mathrm{Ab})]$.

Substitution for Γ_1 in Eqs. (3.21) and (3.22) yields

$$\frac{d[(\mathrm{Ab}\cdot\mathrm{Ag})\ldots(\mathrm{Ag}\cdot\mathrm{Ab})]}{dt} = k_f c_s^{3/2}[\mathrm{Ag}]^{3/2}\sqrt{\frac{k_1}{k_2}} - k_r \times [(\mathrm{Ab}\cdot\mathrm{Ag})\ldots(\mathrm{Ag}\cdot\mathrm{Ab})] \tag{3.27a}$$

and

$$\frac{d[(\mathrm{Ab}\cdot\mathrm{Ag})\ldots(\mathrm{Ag}\cdot\mathrm{Ab})]}{dt} = k_1 c_s[\mathrm{Ag}] - 2k_{-2}[(\mathrm{Ab}\cdot\mathrm{Ag})\ldots(\mathrm{Ag}\cdot\mathrm{Ab})]. \tag{3.27b}$$

Comparison of Eqs. (3.27a) and (3.27b) yields

$$k_1 = \frac{k_f^2}{k_2}[\mathrm{Ag}]c_s \tag{3.28a}$$

and

$$2k_{-2} = k_r. \tag{3.28b}$$

Substitution for k_1 from Eq. (3.28a) into (3.14a) yields

$$\frac{d\Gamma_1}{dt} = \frac{k_f^2}{k_2}[\mathrm{Ag}]c_s^2\Gamma_0. \tag{3.29}$$

Compare Eqs. (3.15) and (3.29). Equation (3.15) represents the kinetic expression for dual-step binding without lateral interactions, while Eq. (3.29) represents the kinetic expression for dual-step binding with lateral interactions. The forms of the two kinetic expressions are very similar. In both cases there is a first-order dependence on the antigen concentration available for binding on the surface, [Ag], and on the initial concentration of the antigen on the surface. Also, when there are no lateral interactions involved there is a first-order dependence on antibody concentration in solution near the surface. However, when lateral interactions are involved, the kinetic expression shows a second-order dependence on antibody concentration in solution near the surface. This is to be expected since two antigen–antibody complex molecules are involved in lateral interactions on the surface.

Table 3.1 compares the kinetic expressions for dual-step binding and dual-step binding with lateral interactions for both cases—when the antigen is in solution and the antibody is covalently attached to the surface and when the antibody is in solution and the antigen is covalently attached to the surface.

3.2.4. MULTIVALENCY ANTIBODIES FOR LARGE ANTIGEN SYSTEMS

Sadana and Vo-Dinh (1997) have developed a model for multivalency antibodies for large antigen systems. These authors presented a theoretical analysis of the influence of multivalency of antigen on external mass transfer-limited binding kinetics to divalent antibodies for biosensor applications to polycyclic-aromatic systems. The design of antibody-targeted agents for a large class of chemical species such as polycyclic-aromatic compounds (PACs) could be an important development for biosensors. Such biosensors could be used to screen samples for their overall content of PACs rather than for specific PACs.

Let us investigate the reaction mechanisms that would be involved in a situation in which an antibody is targeted to a group of antigens having multiple-antigenic sites. This model is relevant to the situation in which the antibody is designed to have a paratope targeted to only a monocyclic aromatic-or part of a monocyclic ring. Such an antibody would be capable of recognizing not only one PAC, but a family of PACs. Figure 3.8 schematically depicts such an antibody targeted to a family of PACs. The concept of multivalency for antibodies requires certain conditions. In general, antibodies are larger than antigens. Therefore, certain size and steric conditions must be fulfilled to allow more than one antibody to be attached to an antigen. This could occur for antigens with sufficiently large size or with antibodies specifically designed to have a small size or sterically favorably paratope

Table 3.1 Kinetic Expressions for Attachment of Antigen in Solution to Antibody on the Surface or of Antibody in Solution to Antigen on the Surface

Type of binding	Antigen in solution/antibody on surface	Antibody in solution/antigen on surface
Dual-step	$\frac{d\Gamma_1}{dt} = k_1[\mathrm{Ag}]^2\Gamma_0$	$\frac{d\Gamma_1}{dt} = k_f[\mathrm{Ag}][\mathrm{Ab}]\Gamma_0$
Dual-step with lateral interactions	$\frac{d\Gamma_1}{dt} = \frac{k_1^2 k_3}{k_2^2}[\mathrm{Ag}]^2[\mathrm{Ab}]\Gamma_0$	$\frac{d\Gamma_1}{dt} = \frac{k_f^2}{k_2}[\mathrm{Ag}][\mathrm{Ab}]^2\Gamma_0$

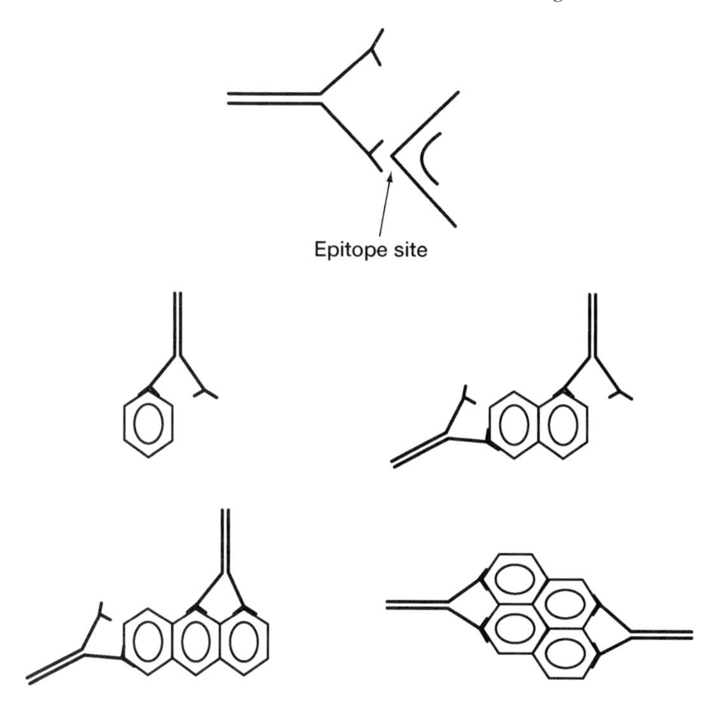

FIGURE 3.8 Schematic diagram of antibodies having paratopes targeted to the antigen series of polycyclic aromatic compounds.
Sadana, A. and Madugula, A., *Biotechnology Progress*, 9, 259–266 (1993). Reprinted with permission from Academic Press.

geometry. Of course, the combining site on the antibody should not be so large that it completely encloses the PAC (antigen). Note that steric hindrance may be particularly significant if the binding pockets are generally deep. Also, it may be challenging to design an antibody to have a PAC-combining site smaller than a PAC that would have a useful binding affinity. Another approach is to design systems consisting of parts of the antibody by cleaving and combining the appropriate paratopes.

The analysis of multivalent antigen–antibody binding is still in the initial stage. We now briefly present some possible mechanisms of multivalent antigen–antibody binding involving lateral interactions. Figure 3.9a shows the elementary steps involved in the binding of divalent antigen in solution to divalent antibody noncovalently or covalently attached to the surface. The dotted lines indicate the lateral interactions involved. Sadana and Vo-Dinh (1997) have presented the reaction scheme involved without the lateral

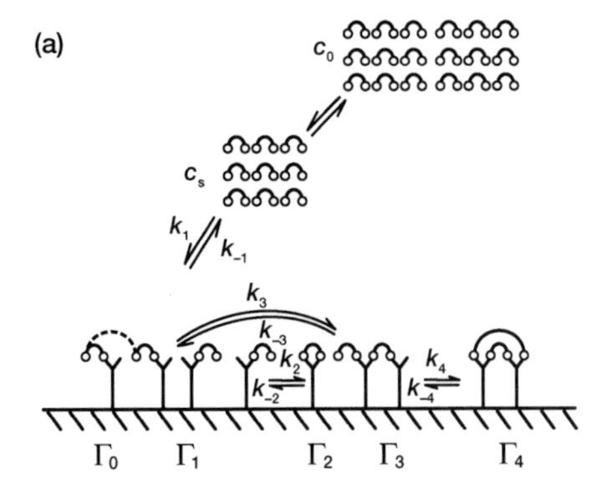

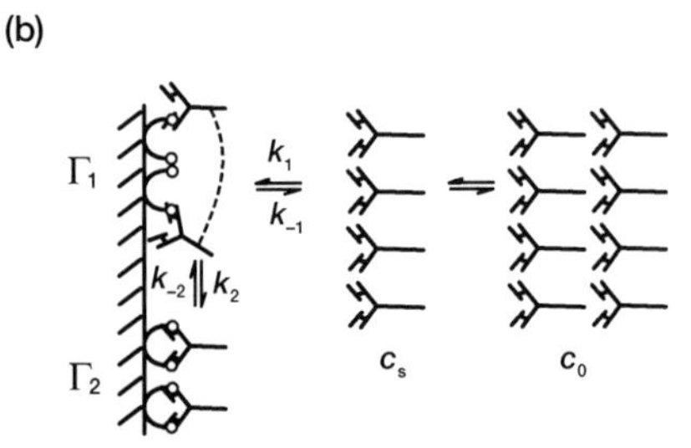

FIGURE 3.9 (a) Elementary steps involved in the binding of divalent antigen in solution to divalent antibody noncovalently or covalently attached to the surface. Lateral interactions are shown with dotted lines (---). (b) Elementary steps involved in the dual-step binding of divalent antibody in solution to divalent antigen noncovalently or covalently attached to the surface. Lateral interactions are shown with dotted lines (---).
Sadana, A. and Madugula, A., *Biotechnology Progress*, 9, 259–266 (1993). Reprinted with permission from Academic Press.

interactions step as

$$\mathrm{Ab} + \mathrm{Ag} \underset{k_{-1}}{\overset{k_1}{\rightleftharpoons}} \mathrm{Ab} \cdot \mathrm{Ag} \underset{k_{-2}}{\overset{k_2}{\rightleftharpoons}} \mathrm{Ab} \cdot \mathrm{Ag}'$$

$$\mathrm{Ab} \cdot \mathrm{Ag}, \underset{k_{-3}}{\overset{k_3}{\rightleftharpoons}} \tag{3.30a}$$

$$\mathrm{Ab} \cdot \mathrm{Ag} \cdot \mathrm{Ab} \cdot \mathrm{Ag} \underset{k_{-4}}{\overset{k_4}{\rightleftharpoons}} \mathrm{Ab} \cdot \mathrm{Ag} \cdot \mathrm{Ab} \cdot \mathrm{Ag}'.$$

The reaction scheme shown in Figure 3.9a may be combined as

$$\mathrm{Ab} + \mathrm{Ag} \underset{k_r}{\overset{\mathrm{Ab;Ag};k_f}{\leftrightarrow}} \mathrm{Ab} \cdot \mathrm{Ag} \cdot \mathrm{Ab} \cdot \mathrm{Ag}'. \tag{3.30b}$$

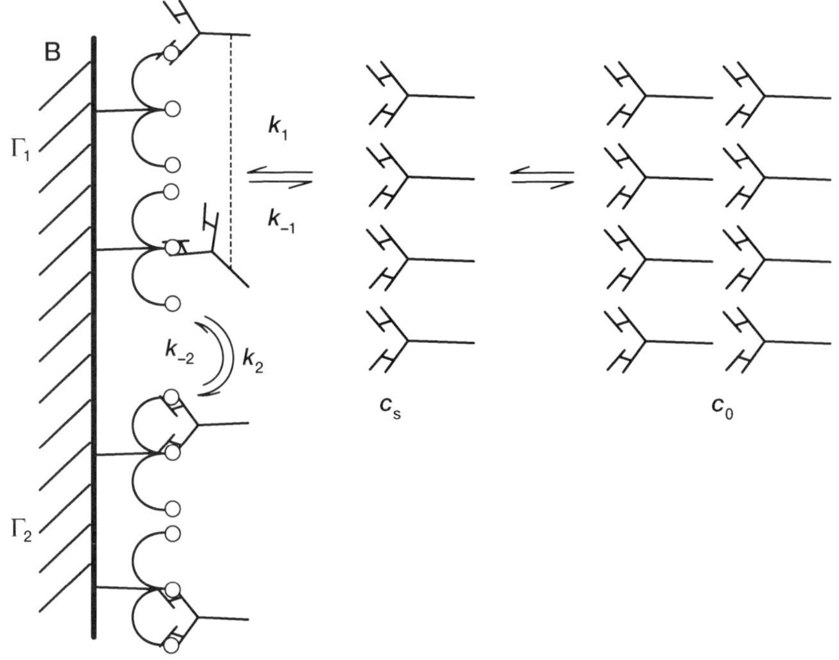

FIGURE 3.10 Elementary steps involved in the dual-step binding of divalent antibody in solution to trivalent antigen noncovalently or covalently attached to the surface.
Sadana, A. and Madugula, A., *Biotechnology Progress*, 9, 259–266 (1993). Reprinted with permission from Academic Press.

The inclusion of the lateral steps to obtain expressions similar to Eqs. (3.30a) and (3.30b) is left as an exercise for the reader.

Figure 3.10 shows the binding of divalent antibody in solution to trivalent antigen immobilized on a sensor surface. Once again, the dotted lines show the lateral interactions involved. The derivation of reaction rate expressions for this case when lateral interactions are absent and when they are present is left as an exercise for the reader.

3.3. CONCLUSIONS

In this chapter, we have developed kinetic expressions for the binding of an antigen in solution by an immobilized antibody and for the binding of an antibody in solution by an immobilized antigen. We analyzed the similarities and dissimilarities in the different rate forms for the two different types of systems. In general, the results are applicable to analyte-receptor reaction systems on biosensor as well as other surfaces, such as cellular surfaces.

The dual-step binding expression for antigen in solution and antibody immobilized on the surface exhibits a second-order dependence on antigen concentration close to the surface. This is to be expected since two antigen molecules are involved in dual-step binding. The dual-step binding rate expression is easily extended to the case where lateral interactions are involved between two antibody–antigen complexes. As expected, there is now an additional first-order dependence on the antibody concentration available on the surface for binding.

The dual-step binding expression for antibody in solution to antigen immobilized on the surface exhibits a first-order dependence both on the antibody concentration close to the surface and on the antigen on the surface available for binding, [Ag]. This is to be expected since one antibody molecule with two arms is involved in dual-step binding. Once again, the dual-step binding rate expression is easily extended to the case where lateral interactions are involved between two antigen–antibody complexes. As expected, now there is a second-order dependence on the antibody concentration close to the surface.

As seen in the reactions with lateral interactions, there is an increase in the rate of binding and an increase in the concentration of antigen near the surface when compared to the reaction with no lateral interactions. Nygren (1988) suggests that the lateral interactions between macromolecules help stabilize the adsorbed protein and the antigen–antibody complexes at the solid surface, which might contribute to the increase in concentration of antigen near the surface. Also, Werthen *et al.* (1988) indicate that the binding isotherm of antibody to the immobilized antigen at the solid surface is caused not only by intrinsic antibody affinity but also by other macromolecular properties such as lateral intermolecular interactions.

An increase in the adsorption rate coefficient with time decreases the concentration of antigen near the surface as expected. Similarly, a decrease in the adsorption rate coefficient with time increases the concentration of the antigen near the surface. In general, the conclusions that apply for a time-variant adsorption rate coefficient as far as lateral interactions are concerned also apply for a time-dependent adsorption rate coefficient. The similarities and dissimilarities in the kinetic rate forms for two different systems (see Table 3.1) provide physical insights into the reactions occurring and the stability at the surface. In addition, the influence of lateral interaction between molecules for antigen–antibody reactions near solid–liquid interfaces has been analyzed to some extent here. This should facilitate a better understanding and control of these reactions, which may result in manipulating the reactions in desired directions. For example, higher concentrations of antigen near the surface would enhance the sensitivity of biosensors; thus every attempt may be made to enhance lateral interactions. Also, a decreasing adsorption rate coefficient with time would enhance sensitivity. Note that sensitivity would primarily depend on the concentration of the analyte on the surface, not just near the surface. Finally, the complexities at the reaction surface, and the tightly organized molecular reaction complexes at the surface (with lateral interactions contributing to this tight organization) presumably leads to the irreversible nature of these reaction sensor systems.

REFERENCES

Cuypers, P. A., Willems, G. M., Kop, J. M., Corsel, J. W., Jansen, M. P., and Hermes, W. T., in *Proteins at Interfaces. Physicochemical and Biochemical Studies*, J. L. Brash, and T. A. Horbett, eds. American Chemical Society, Washington, DC, pp. 208–211, 1987.
Di Cera, E., *J. Chem. Phys.* **95**(7), 5082–5086 (1991).
Eddowes, M. J., *Biosensors* **3**, 1–15 (1988).
Giaever, I., *J. Immunol.* **116**, 766–771 (1976).
Kopelman, R., *Science* **241**, 1620–1626 (1988).
Lewis, A. N., *Nucleation and Growth of Thin Films*, Academic Press, New York, 1978.
Nygren, H. A., *J. Immunol Methods*, **114**, 107–114 (1988).
Nygren, H. A. and Stenberg, M., *J. Colloid & Interface Sci.* **107**, 560–566 (1985).
Nygren, H. A. and Stenbereg, M., *Immunology* **66**(3), 321–327 (1989).
Nygren, H. A. and Stenberg, M., *Biophys. Chem.*, **38**, 67–75 (1990).
Nygren, H. A. and Stenberg, M., in *Immunochemistry of Solid Phase Immunoassay* (J. Butler, ed.), CRC Press, Boca Raton, FL, Chapter 19, 1991.
Patankar, S. V., *Numerical Heat Transfer and Fluid Flow*, McGraw-Hill, New York, 1980.
Sadana, A. and Sii, D., *J. Colloid & Interface Sci.* **151**(1), 166–177 (1992a).
Sadana, A. and Sii, D., *Biosen. & Bioelectr.* **7**, 559–568 (1992b).
Sadana, A. and Vo-Dinh., T., *Appl. Biochem. & Biotech.* **67**(1), 1 (1997).
Shoup, D. and Szabo, A., *Biophys. J.* **40**, 33–39 (1982).
Stenberg, H. A. and Nygren, H. A., *Anal. Biochem.* **127**, 183–192 (1982).
Stenberg, M., Elwing, H. B., and Nygren, H. A., *J. Theor. Biol.* **98**, 307–316 (1982).
Stenberg, M., Stiblert, L., and Nygren, H. A., *J. Theor. Biol.* **120**, 129–142 (1986).
Uzgiris, E. E. and Kornberg, R. D., *Nature (London)* **301**, 125–128 (1983).
Werthen, M. and Nygren, H. A., *J. Immunol. Methods* **115**, 71–78 (1988).

CHAPTER 4

Fractal Reaction Kinetics

4.1. INTRODUCTION

Biosensors, as the name indicates, use biologically derived molecules as sensing elements. Biosensors should be sensitive, specific, and stable (Scheller *et al.*, 1991). Their sensitivity and stability can be improved by a better understanding of their mode of operation. Eddowes (1987) emphasizes the balance inherent in the practical utility of biosensor systems. He estimates that although acceptable response times on the order of minutes or less should be obtainable at μM concentration levels, inconveniently lengthy response times will be found at nM or lower concentrations. The success of the detection scheme will be significantly enhanced from physical insights into the different steps involved in the sensing process. One such detection scheme is the solid-phase immunoassay technique that has already gained importance.

The solid-phase immunoassay technique provides a convenient means for the separation of reactants (for example, antigen) in a solution. Such a separation is possible because of a high specificity of the analyte for the immobilized antibody. External diffusional limitations play a role in the analysis of such assays (Giaever, 1977; Eddowes, 1987; Bluestein *et al.*, 1991; Place *et al.*, 1991). The influence of diffusion in such systems has been analyzed to some extent (Stenberg *et al.*, 1986; Nygren and Stenberg, 1985; Stenberg and Nygren, 1982; Sadana and Sii, 1992a, 1992b).

In protein adsorption systems, which exhibit behavior similar to that of antibody–antigen systems at the solid–liquid interface (Stenberg and Nygren, 1988), the influence of the surface-dependent intrinsic adsorption and desorption rate constants on the amount of protein adsorption has been analyzed (Cuypers *et al.*, 1987; Nygren and Stenberg, 1990). Nygren and Stenberg (1990), while studying the adsorption of ferritin from a water solution to a hydrophobic surface, noted that initially the adsorption rate coefficient of new ferritin molecules increased with time. Nygren and Stenberg (1985) also noted a decrease in binding rate with time while studying the kinetics of antibody binding to surface-immobilized bovine serum albumin (antigen) by ellipsometry. They indicated that the decrease in binding rate with time is probably due to a saturation rate through steric hindrance at the surface.

In addition to these typical examples, we will discuss examples where such variability in the adsorption rate coefficients is exhibited and the possible reasons for such variability are given. Nygren and Stenberg (1989) have indicated the importance of intermolecular interactions in the reaction layer to the binding kinetics of an antigen to an immobilized antibody. Experimental evidence is also available that provides for attractive interactions in the surface-bound antibodies (Uzgiris and Kornberg, 1983; Nygren, 1988). Cuypers *et al.* (1987) related the repulsive interactions in the reaction layer to the passivation of surfaces and analyzed the influence of a variable adsorption rate coefficient on protein adsorption. Nygren (1988) presented data indicating that cohesive forces between macromolecules help stabilize the adsorbed proteins and antigen–antibody complexes at solid surfaces. Nygren and Stenberg (1991) and Madagula and Sadana (1993) have also studied to some extent the influence of lateral interactions in antibody–antigen systems.

For protein adsorption, Guzmann *et al.* (1986) initially proposed that the activation energies for adsorption and desorption are dependent on surface coverage. Similarly, Hunter *et al.* (1990) suggest that as the surface coverage increases, the activation energies for adsorption and desorption of proteins increase and decrease, respectively. Tilton *et al.* (1990) postulate that surface diffusion of randomly adsorbed proteins at an interface allows the proteins to assemble in organized structures. This produces nonrandom orientations and patchwise aggregation, which should lead to fractal formation. The authors noted that the surface diffusion of bovine serum albumin adsorbed from aqueous solution onto poly(methylmethacrylate) surfaces is hindered by lateral–lateral interactions.

A decrease in the self-diffusion coefficient through protein–protein lateral interaction has been noted by other researchers also (Scalettar *et al.*, 1988; Abney *et al.*, 1989). Kondo and Hagashitani (1992) examined the adsorption isotherms of ribonuclease A (Rnase A), cytochrome c, lysozyme, α-lactalbumin, and bovine serum albumin on colloidal particles of polystyrene, styrene/2-hydroxyethyl methacrylate, and silica as a function of pH and ionic strength. The authors propose that lateral interactions between the larger protein molecules are stronger because of the thicker adsorption layers at the solid–liquid interfaces. Thus, one needs methods to describe both the complex reactions and interactions occurring on the surface and, particularly, the nature of the surface.

4.2. FRACTAL KINETICS

Kopelman (1988) indicates that surface diffusion-controlled reactions that occur on clusters or islands are expected to exhibit anomalous and fractal-like kinetics. These fractal kinetics exhibit anomalous reaction orders and time-dependent rate (for example, binding) coefficients. Mandelbrot (1975, 1983) initially introduced fractals, or self-similar objects that exhibit dilatational symmetry. The word *fractal* was taken from the Latin word *fractus*, meaning broken. Fractals have details on all scales; therefore euclidean geometry and classical calculus are insufficient for their description; fractal geometry is required. Markel *et al.* (1991) indicate that fractals are widespread in nature. They indicate that the products of a wide class of diffusion-controlled aggregation reactions in solutions and in gases may be labeled as fractals. Thus, rough surfaces, disordered layers on surfaces, and porous objects (such as heterogeneous catalysts) possess fractal structure. Furthermore, gels, soot and smoke, and most macromolecules are fractals.

Fractals are disordered systems; the disorder is described by nonintegral dimensions (Pfeifer and Obert, 1989). As long as surface irregularities show scale invariance—that is, dilatational symmetry—they can be characterized by a single number, the fractal dimension. This means that the surface exhibits self-similarity over certain-length scales. In other words, the structure exhibited at the scale of the basic building blocks is reproduced at the level of larger and larger conglomerates. Fractals possess nontrivial geometrical properties; in other words, they are geometrical structures with noninteger dimensions. A consequence of the fractal nature is a power-law dependence of a correlation function (in our case, the analyte-receptor complex on the biosensor or cell surface) on a coordinate (for example, time).

The repeating shape or form does not have to be identical. An increase in the disorder on the surface leads to higher values of the fractal dimensions. For example, a very ordered "assembly" of objects along a straight line should yield a fractal dimension of 1 (ideally). If there is some disorder or degree of heterogeneity along this straight line, a slightly higher value of the fractal dimension will be found. If there are holes along this straight line, the fractal dimension will be less than 1. Similarly, if the assembly of objects under consideration are very organized on a surface, the fractal dimension is close to 2 or exactly equal to 2 (ideally). A fractal dimension value different from 2 provides a quantitative measure of how far the surface is from an ideal or homogeneous surface exhibiting a fractal dimension of 2. Thinking along the same lines, we may have two-dimensional surfaces exhibiting fractal dimensions greater than or less than 2. We may consider the fractal dimension (loosely) as a "space-filling" ability of a system. Thus the highest value of the fractal dimension exhibited is 3, since we are restricted to three-dimensional space.

In a review of the heterogeneity of materials and multifractality, Lee and Lee (1996) note that the fractal approach provides a convenient means to quantitatively represent the different structures and morphologies at the reaction interface. The authors analyzed simulations of Eley–Rideal diffusion-limited reactions on different objects. The primary advantage is that this permits the development of a predictive approach in the field of catalysis. Lee and Lee emphasize using the fractal approach to develop optimal structures, noting that today's sensors tend to be costly, cumbersome, and specialized (Service, 1997). Service indicates that it would be helpful to develop new sensors that are based on dirt-cheap starting materials. Such sensors could then be effectively used as low-cost detectors for medical diagnostics, industrial monitoring, and environmental testing.

Avnir *et al.* (1998) emphasize that the power law utilized in describing the fractal nature of systems very appropriately condenses the complex nature of the system being analyzed. Furthermore, it provides a simple picture of the correlation between the system structure and the dynamics of its formation. This type of information is particularly relevant in the study of analyte-receptor binding reactions occurring on surfaces. In analyzing the optical amplification of ligand-receptor binding using liquid crystals, Gupta *et al.* (1998) schematically show the change in the surface heterogeneity (or the fractal dimension) as avidin or IgG molecules in solution bind to ligands attached to self-assembled monolayers of molecules supported on a gold film. Their schematic indicates that the surface roughness increases on the binding of the analyte (Av or IgG) in solution to the ligands on the surface.

Fractal kinetics exhibit anomalous reaction orders and time-dependent rate (for example, binding) coefficients. These are unlike "regular" reaction kinetics, which exhibit integer orders of reaction, such as zero, first, second, etc. The time-dependent adsorption rate coefficients observed experimentally, as indicated above, may also be due to nonidealities or heterogeneity on the surface. Antibodies are heterogeneous and their immobilization on a fiber-optic surface, for example, will definitely exhibit a degree of heterogeneity. This is a good example of a disordered system, and a fractal analysis is appropriate for such systems. In addition, the antibody–antigen reaction on the surface is a good example of a low-dimension reaction system in which the distribution tends to be "less random" (Kopelman, 1988), and a fractal analysis would provide novel physical insights into the diffusion-controlled reactions occurring at the surface.

Matuishita (1989) indicates that the irreversible aggregation of small particles occurs in many natural processes, such as polymer science, material science, and immunology. These aggregation processes frequently result in the formation of complex materials that can be described by fractals (Mandelbrot, 1983). Daccord (1989) emphasizes that when too many parameters are involved in a reaction, the fractal dimension for reactivity may be a useful global parameter. Since biosensor performance is constrained by chemical binding kinetics, equilibrium, and mass transport of the analyte to the biosensor surface, it behooves one to pay particular care to the design of such systems and to explore new avenues by which further insight or knowledge may be obtained in these systems. Fractal analysis is one such avenue by which one may obtain physical clarification of the diffusion-controlled reactions at the surface.

Havlin (1989), in a brief discussion of the diffusion of reactants on and toward fractal surfaces, indicates that although diffusion toward fractal surfaces has been studied experimentally more extensively than diffusion on fractal surfaces (owing to the number of applications, such as catalytic reactions), diffusion toward fractal surfaces has been analyzed theoretically much less. Some studies are available, however. For example, Giona (1992), reporting on first-order reaction-diffusion kinetics in complex fractal media, emphasizes that the exploration of the temporal nature of the diffusion-limited reaction on the surface could play an important role in understanding the reaction kinetics as well as the reaction itself. We now examine some typical (adsorption) studies where fractal dimension values have been obtained.

Adsorption of molecules of different diameters on a solid surface exhibit fractal characteristics (Avnir *et al.*, 1983, 1984; Van Damme and Fripiat, 1985). The number of molecules of A, n_A, adsorbed on a surface may be given by

$$n_A \approx [d_{\mathrm{A,eff}}]^{-D_{f,\mathrm{ads}}}, \tag{4.1}$$

where $d_{A,\mathrm{eff}}$ is the effective molecular diameter, and $D_{f,\mathrm{ads}}$ is the fractal dimension for adsorption and lies between a value of 2 and 3. Another method of determining the fractal dimension for adsorption studies (Demertzis and Pomonis, 1997) uses particles of adsorbent of varying size (diameter) d onto which a single molecule is adsorbed. Then, the number of species adsorbed N per unit mass of the particles is given by

$$N \propto d^{D-3} \tag{4.2}$$

Nitrogen is typically the material adsorbed, and the adsorbent, for example, may be natural rocks (quartz, feldspar) and various coals. Here D is the fractal dimension for adsorption. The number 3 corresponds to the three-dimensional space in which the system is embedded.

Fractal kinetics also have been reported in such biochemical reactions as the gating of ion channels (Liebovitch *et al.*, 1987; Liebovitch and Sullivan, 1987), enzyme reactions (Li *et al.*, 1990), and protein dynamics (Dewey and Bann, 1992). Li *et al.* establish that the nonintegral dimensions of the Hill coefficient, used to describe the allosteric effects of proteins and enzymes, are a direct consequence of the fractal properties of proteins as biological macromolecules composed of amino acid residues whose branches form fractals. The substrate molecules "randomly walk" on the enzyme surface until they "hit," or react on, an active site. For a better physical understanding of reaction at interfaces, fractal analysis may be used to model the behavior of diffusion-limited antigen–antibody or, in general, analyte-receptor binding kinetics on biosensor surfaces.

Let us now look at some other examples available in the biology, biotechnology, and biomedical literature that exhibit fractal characteristics. Proteins have a hierarchical structure; during protein folding subdomains are initially formed. These subdomains then combine to yield domains, which eventually combine with other domains to produce the final active structure of the protein. This process involves many similar (though not identical) repeating biochemical units. Even in the complex protein structure there is a repeating pattern. This repeating pattern and the characteristic heterogeneity of the protein structure could be aptly described by fractals (Sadana and Vo-Dinh, 2001). It would seem appropriate to represent the different folding stages using a fractal analysis. The fractal nature is also associated with DNA, the gene frequency of which determines the protein structure.

Repeating patterns are also present in signals emanating from biological systems such as those traced by ECGs (electrocardiograms) and EEGs (electoencephalograms), as well as in the basic structures of some human organs such as the lungs and in the way that arteries divide and subdivide (Zamir, 1999). Furthermore, allometric scaling laws, including the metabolic reactions, have been analyzed by West *et al.* (1997), who indicate that these laws are characteristic of all organisms. For example, the authors were able to describe the 3/4 power law for metabolic reactions using a model of transport of essential materials through space-filling fractal networks of branching tubes. Note once again that a characteristic feature of fractals is the self-similarity at different levels of scale. Self-similarity implies that the features of a structure or process look alike at different levels of length or time.

Goldberger *et al.* (1990) have indicated that when the heart rate (beats per minute) of a healthy individual is recorded for three, thirty, and three hundred minutes, the quick erratic fluctuations seem to vary in a manner similar to the slower fluctuations. This indicates a self-similarity. Note that self-similarity implies that the features of a structure look alike at different scales of length or time. This self-similarity of a process at different scales of time can be characterized with a fractal dimension: A higher value of the fractal dimension indicates a higher level of heterogeneity or state of disorder.

Pfeifer and Avnir (1983) refer to the fractal dimension as the hidden symmetry of irregular (self-similar) surfaces. In trying to determine whether there were systematic trends in the fractal dimension as the size of the protein molecule was changed, Goetze and Brickmann (1992) analyzed the self-similarity of protein surfaces and found that the fractal dimension of a protein surface increases when the size of the protein molecule is increased. Apparently, larger molecules are "rougher" (to molecular partners) than smaller molecules. Feder (1988) has defined the fractal dimension (surface dimension) as a local surface property and has attempted to associate high receptor selectivity with high values of the fractal dimension. Pfeifer *et al.* (1984, 1985) indicate there is a balance involved for surface fractal dimension values greater than 2. Although such values would promote the transport of the analyte in solution *to* the surface, they would hinder the transport *along* the surface.

Consider the binding of an analyte of fractal nature (such as a protein or a macromolecule) in solution to the receptor immobilized on a biosensor surface. It is not unreasonable to assume that the receptor, like the analyte, would exhibit fractal characteristics. Proteins are also known to adsorb on "receptorless" surfaces. However, these surfaces themselves may or may not exhibit fractal characteristics. Low-dimension fractals have been observed for analyte-receptorless systems, for example, during the computer-simulated aggregation of ferritin (Stenberg and Nygren, 1991), the adsorption of ferritin on a quartz surface (Nygren, 1993), and polymer adsorption (Douglas *et al.*, 1993). Note that the antibody is not fractal with binding sites on randomly distributed branches, but has only one or two binding sites on well-defined and unique parts of the molecule.

For ligand-receptor systems, it is recognized that the population of receptors for a given ligand may be represented by several subpopulations with different affinities (Lord *et al.*, 1977; Eriksson *et al.*, 1978; Agarwal and Phillipe, 1977; Barnett *et al.*, 1978). Jose (1985) has developed a model for ligand-binding systems at equilibrium and has analyzed the influence of heterogeneity, cross-reactivity, and site–site interactions on this system. Site–site interactions are themselves a source of affinity heterogeneity (Jose and Larralde, 1982), and their binding to different types of ligands may effectively be described by fractal systems. Swalen *et al.* (1987) have indicated that the control of the structural organization of molecules at an interface are a key to understanding the reactions at interfaces and for the design of advanced materials. The characterization of a solid surface for antibody–antigen, ligand-receptor, and analyte-receptorless binding is of importance (Ebersole *et al.*, 1990). Losche *et al.* (1993) have analyzed the influence of surface chemistry on the adsorption of protein layers on aqueous interfaces and structural organization. Axelrod and Wang (1994) have indicated the importance of reduction of dimensionality kinetics, wherein reaction between ligands and cell-surface receptors can be enhanced by nonspecific adsorption followed by two-dimensional diffusion to a cell-surface receptor.

In their analysis of the quenching of fluorescein-conjugated lipids, Ahlers *et al.* (1992) indicate (1) the binding of lipid-bound haptens in biomembrane models and (2) the formation of two-dimensional protein domains. These authors emphasize that the basis of drug delivery strategies and immunoassays is the specific recognition of cell membrane epitopes by antibodies or specific sections of antibodies. They further indicate that proteins self-organize into two-dimensional crystals at the interface (lipid monolayer), for example during the high-affinity binding of antibodies to lipid-bound haptens. This self-organization of proteins into two-dimensional crystals at the surface is characteristic of fractal aggregation and formation.

Hsieh and Thompson (1994) indicate that in addition to other factors, ligand-receptor (binding and dissociation) kinetics depends on (1) the receptor density, (2) the diffusion coefficient if the ligand is bivalent or multivalent for the receptor, (3) whether the ligand induces receptor clustering, (4) and the influence of receptor clustering (Berg and Purcell, 1977; Dembo and Goldstein, 1978; Kaufmann and Jain, 1991; Goldstein *et al.*, 1989). Factors (3) and (4) lead to heterogeneities on the surface and would contribute toward a fractal surface at the reaction interface leading to fractal kinetics.

Baish and Jain (2000) have recently advocated utilizing fractal principles in cancer study and its treatment. They indicate, for example, that the tumor vessels yield fractal dimensions of 1.89 ± 0.04, while normal arteries and veins yield fractal dimensions of 1.70 ± 0.03. They emphasize the potential of fractal analysis in both treatment delivery and the diagnosis of cancer. Furthermore, these same authors (along with Losa, 1995; Cross, 1997; Coffey, 1998) indicate the widespread applications of fractals in pathology.

We now describe a typical example where fractal properties of both the analyte and the receptor are exhibited. Peng *et al.* (1992) analyzed the nucleotide sequences in DNA using an *n*-step Markov chain, noting the presence of long-range correlations in nucleotide sequences. This indicated to them the presence of scale-free (fractal) phenomena. In hybridization reactions on biosensor surfaces, the analyte is typically a DNA in solution, and the receptor is a complementary DNA that is immobilized on the biosensor surface. In this case both the DNA in solution and the complementary DNA immobilized on the surface would seem to exhibit fractal characteristics. If the DNA immobilized on the biosensor surface is not complementary to the DNA in solution, effective binding does not take place.

One reason for analyzing antigen–antibody (or, in general, analyte-receptor) binding data is to provide a better physical understanding of the underlying mechanisms. We will illustrate by analyzing fractal dimensions for marine particles. This is not a directly related example, but the basic principles for using the fractal analysis should be the same. In his analysis of the correlation of fractal dimension of marine particles with ocean depth, Risovic (1998) indicates that the average fractal dimension of marine particles/aggregates changes from 2.9 ± 0.1 just beneath the surface to 2.0 ± 0.1 at 800 m down. This correlates with a decrease in the turbulent energy dissipation rate with depth. His results indicate that there is a domination of shear coagulation for depths less than (or equal to) 400 m (fractal dimension $= 2.7 \pm 0.3$) and coagulation due to a differential sedimentation rate at greater depths (fractal dimension $= 2.1 \pm 0.3$).

Rice (1994), in his review of Kaandrop's (1994) text on fractal modeling, emphasizes that one should be able to relate the rules by which the fractal structures are generated to the underlying processes by which these structures develop. This then provides fundamental insights into the basic mechanisms involved in our case for the analyte-receptor binding process. It would be worthwhile to develop a relationship between surface roughness (characterized by a fractal dimension) and the rate of binding. This is in view of the different (statistical) fractal growth laws that are prevalent in the literature. These laws include invasion percolation, kinetic gelation, and diffusion-limited aggregation (DLA) (Viscek, 1989). These laws (or models) permit the computer simulation of the shape and the growth of natural processes. For example, in the DLA model introduced by Witten and Sander (1981), a randomly diffusing particle (seed) collides with a surface and stops. Another particle (from far away) diffuses to the surface and arrives at a site close (adjacent) to the first particle and stops. Another particle follows, and so on. In this way clusters are generated and exhibit the randon branching and open structures that are self-similar in nature.

The analyte or, in general, the receptor has to be immobilized or adsorbed to the surface. Heterogeneity of adsorption is a more realistic picture of the actual situation and should be carefully examined to determine its influence on external mass transfer limitations and on the ultimate analytical procedure. Heterogeneity in the covalent attachment of the antibody (or receptor) to the surface probably can be accounted for and needs to be considered in the analysis.

Heterogeneity may arise due to several different factors. For instance, antibodies, especially polyclonal antibodies, possess an inherent heterogeneity in that the antibodies in a particular sample are not identical. Furthermore, different sites on the antibody may become covalently bound to the surface. As a result, especially in large antibodies, steric factors will play a significant role in determining the Ag/Ab (or, in general, the receptor/analyte) ratio. It would be helpful to make the influence of heterogeneity on the kinetics of antibody–antigen interactions more quantitative. We are now implicitly indicating and associating heterogeneity on the surface with a fractal dimension [the Kopelman approach (1988)], with changes in the heterogeneity on the surface leading to changes in the fractal dimension.

Note that antigen–antibody binding is unlike reactions in which the reactant reacts with the active site on the surface and the product is released. In this sense the catalytic surface exhibits an unchanging fractal surface to the reactant in the absence of fouling and other complications. In the case of the antigen–antibody binding, the biosensor surface exhibits a changing fractal surface to the antigen or antibody (analyte) in solution. This occurs since as each binding reaction takes place, smaller and smaller amounts of binding sites are available on the biosensor surface to which the analyte may bound. This is in accord with Le Brecque's (1992) comment that the active sites on a surface may themselves form a fractal surface. Furthermore, the inclusion of nonspecific binding sites on the surface would increase the fractal dimension of the surface.

In general, a log–log plot of the distribution of molecules $M(r)$ as a function of the radial distance (r) from a given molecule is required to demonstrate fractal-like behavior (Nygren, 1993). This plot should be close to a straight line. The slope of the log $M(r)$ versus $\log(r)$ plot determines the fractal dimension. This is the classical definition and means of demonstrating fractal behavior.

One way of introducing heterogeneity into the analysis is to consider a time-dependent rate coefficient. Classical reaction kinetics is sometimes unsatisfactory when the reactants are spatially constrained on the microscopic level by walls, phase boundaries, or force fields (Kopelman, 1988). The types of heterogeneous reactions—for example, bioenzymatic reactions—that occur at interfaces of different phases exhibit fractal orders for elementary reactions and rate coefficients with temporal memories. In these types of reactions the rate coefficient exhibits a form given by

$$k = k' t^{-b}, \quad 0 \leq b \leq 1 \quad (t \geq 1). \tag{4.3}$$

Note that Eq. (4.3) fails in short time frames. In general, k depends on time, whereas $k' = k(t=1)$ does not. Kopelman indicates that in three dimensions (homogeneous space), $b = 0$. This is in agreement with the results obtained in classical kinetics. Also, with vigorous stirring the system is made homogeneous, and again, $b = 0$. However, for diffusion-limited reactions occurring in fractal spaces, $b > 0$; this yields a time-dependent rate coefficient.

The time dependence of the adsorption rate coefficient, k_1, may be due to a mathematical poisoning that is created through self-ordering (Kopelman, 1988). Kopelman emphasizes that since Eq. (4.3) fails in short time frames, the equation may be rewritten as

$$k_1 = k_1'(t+1)^b, \quad t \geq 0. \tag{4.4}$$

The range of b chosen is 0 to 1, as indicated by Kopelman. It is possible that for the reactions occurring at the interface, the values of b may be greater than 1 for antibody–antigen reactions.

The random fluctuations on a two-state process in ligand-binding kinetics can be analyzed (Di Cera, 1991). The stochastic approach can be used as a means to explain the variable adsorption rate coefficient. The simplest way to model these fluctuations is to assume that the adsorption rate coefficient, $k_1(t)$, is the sum of its deterministic value (invariant) and the fluctuation, $z(t)$. This $z(t)$ is a random function with a zero mean. The decreasing and increasing adsorption rate coefficients can be assumed to exhibit an exponential form (Cuypers *et al.*, 1987) as follows.

$$k_f = k_{f,0} \exp(-\beta t) \tag{4.5a}$$

$$k_f = k_{f,0} \exp(\beta t). \tag{4.5b}$$

For A + A type of reactions, Kopelman (1988) indicates that $b = 1 - (d_s/2)$ (Kopelman, 1986; Klymko and Kopelman, 1982, 1983), where d_s is the spectral (or random-walk occurrence) dimension defined by

$$p \approx t_s^{-d/2}. \tag{4.6}$$

Here, p is the probability of the random walker returning to its origin after time, t. Kopelman (1988) emphasizes that for the whole class of random fractals, all in embedded euclidean dimensions (two, three, or higher), d_s is always $\approx$ 1.33 (Kopelman, 1986; Alexander and Orbach, 1982). Then b equals 0.33 for A + A reactions. The self-ordering effect is much more prominent for the two-reactant case (A + B), which is closer to our case.

For the diffusion-limited case, Kopelman (1986) indicates that the reaction order, n, is given by

$$n = 1 + (2/d_s). \tag{4.7}$$

Then, a d_s value of 4/3 yields a value of 5/2 for n. Kopelman (1988) emphasizes that, semantically, any binary reaction kinetics with $b > 0$ or $n > 2$ may be referred to as fractal-like kinetics. As b increases from 0 to 1, n

increases slowly at first but more rapidly as $b \to 1$. For b equal to 0.25, 0.5, and 0.75, n equals 2.33, 3, and 5, respectively.

Reactions such as antibody–antigen interactions on a fiber-optic surface will be diffusion controlled and may be expected to occur on clusters or islands (indicating some measure of heterogeneity at the reaction surface). This leads to anomalous reaction orders and time-dependent adsorption (or binding) rate coefficients. It appears that the nonrandomness of the reactant distributions in low dimensions leads to an apparent "disguise" in the reaction kinetics. This disguise in the diffusion-controlled reaction kinetics is manifested through changes in both the reaction coefficient and the order of the reaction. Examples of reaction-disguised and deactivation-disguised kinetics due to diffusion are available in the literature (Malhotra and Sadana, 1989; Sadana, 1988; Sadana and Henley, 1987).

It would be of interest to obtain a characteristic value for the fractal parameter b (or perhaps a range for the fractal parameter b) for fiber-optic systems involving antibody–antigen interactions. This would be of tremendous help in analyzing these systems, in addition to providing novel physical insights into the reactions occurring at the interface. Techniques for obtaining values of fractal parameters from reaction systems are available, though they may have to be modified for fiber-optic biosensor systems. The discovery of ways to relate the fractal parameter as a measure of heterogeneity at the reaction interface would facilitate the manipulation of the interface reaction in desired directions.

Kopelman (1988) emphasizes that in a classical reaction system the distribution stays uniformly random, and in a fractal-like reaction system the distribution tends to be less random; that is, it is actually more ordered. Also, initial conditions that are usually of little importance in "re-randomizing" classical kinetics may become more important in fractal kinetics. One may wish to examine the effect of fractal-like systems of gaussian and other distributions. Finally, fractal kinetics are not the only way to obtain time-dependent adsorption rate coefficients in antibody–antigen (or, more generally, protein) interactions. As indicated in Eqs. (4.5a) and (4.5b), the influence of decreasing and increasing adsorption rate coefficients on external diffusion-limited kinetics may be analyzed.

REFERENCES

Abney, J. R., Scalettar, B. A., and Owicki, J. C., *Biophys. J.* 55, 817–833 (1989).

Agarwal, M. K. and Phillipe, M., *Biochim. Biophys.* 500, 47 (1977).

Ahlers, M., Grainger, D. W., Herron, J. N., Ringsdorf, H., and Salesse, C. *Biophys. J.* 63, 823–838 (1992).

Alexander, S. and Orbach, R., *J. Phys. (Revs.) Lett.* 43, 6625–6628 (1982).

Avnir, D., Biham, O., Lidar, D., and Malcai, O., *Science* **279**, 39–40 (2 January 1998).
Avnir, D., Farrin, D., and Pfeifer, P., *J. Chem. Phys.* **79**(7), 3566–3571 (1983).
Avnir, D., Farrin, D., and Pfeifer, P., *Nature* **308**(5956), 261–263 (1984).
Axelrod, D. and Wang, M. D., *Biophys. J.* **64**, 588 (1994).
Baish, J. W. and Jain, R. K., *Cancer Res.* **60**, 3683–3688 (2000).
Barnett, D. B., Rugg, E., and Nahorski, S. R., *Nature (London)* **273**, 166 (1978).
Berg, H. C. and Purcell, E. M., *Biophys. J.* **20**, 193–219 (1977).
Bluestein, B. I., Craig, M., Slovacek, G., Stundtner, L., Uricouli, C., Walczak, I., and Luderer, A. in *Biosensors with Fiberoptics* (D. Wise and L. B. Wingard, Jr., eds.), Humana Press, New York, p. 181, 1991.
Coffey, D. S., *Nat. Med.* **4**, 882 (1998).
Cross, S. S., *J. Pathol.* **182**, 1–8 (1997).
Cuypers, P. A., Willems, G. M., Hemker, H. C., and Hermans, W. T., in *Blood in Contact with Natural and Artificial Surfaces* (E. F. Leonard, V. T. Turitto, and C. Vroman, eds.), *Annals N.Y. Acad. Sci.* **516**, 244–252 (1987).
Daccord, G. in *The Fractal Approach to Heterogeneous Chemistry: Surfaces, Colloids, and Polymers* (D. Avnir, ed.), Wiley, New York, pp. 181–197, 1989.
Dembo, M. and Goldstein, B., *J. Immunol.* **121**, 345–353 (1978).
Demertzis, P. G. and Pomonis, P. G., *J. Colloid & Interface Sci.* **186**, 410–413 (1997).
Dewey, T. G. and Bann, J. G., *Biophys. J.* **63**, 594–598 (1992).
Di Cera, E., *J. Chem. Phys.* **95**, 5082–5086 (1991).
Douglas, J. F., Johnson, H. E., and Granick, S., *Science* **262**, 2010 (1993).
Ebersole, R. C., Miller, J. A., and Ward, M. D., *Biophys. J.* **59**, 387 (1990).
Eddowes, M. J., *Biosensors* **3**, 1–15 (1987).
Eriksson, H., Upchurch, S., Hardin, S. W., Peck, Jr., E. J., and Clark, J. H. *Biochem. Biophys. Res. Commun.* **81**, 1 (1978).
Feder, J., *Fractals*, Plenum Press, New York, pp. 236–243 1988.
Giaever, I., German patent, 2,638,207; March 10, 1977.
Giona, M., *Chem. Eng. Sci.* **47**, 1503–1515 (1992).
Goetze, T. and Brickmann, J., *Biophysic. J.* **64**, 109–118 (1992).
Goldberger, A. L., Rigney, D. R., and West, B. R., *Sci. Amer.* February, 43–49 (1990).
Goldstein, B., Posner, R. G., Torney, D. C., Erickson, J., Holowka, D., and Baird, B., *Biophys. J.* **56**, 955–966 (1989).
Gupta, V. K., Skaife, J. J., Dubrovsky, T. B. and Abbott, N. L., *Science* **279**, 2077 (1998).
Guzmann, R. Z., Carbonell, R. G., and Kilpatrick, P. K., *J. Colloid & Interface Sci.* **114**, 536–547 (1986).
Havlin, S., in *The Fractal Approach to Heterogeneous Chemistry: Surfaces, Colloids, and Polymers* (D. Avnir, ed.), Wiley, New York, pp. 181–197, 1989.
Hsieh, H. V. and Thompson, N., *Biophys. J.* **66**, 898–911 (1994).
Hunter, J. R., Kilpatrick, P. K., and Carbonell, R. G., *J. Colloid & Interface Sci.* **137**, 462–472 (1990).
Jose, M. V., *Anal. Biochem.* **144**, 494 (1985).
Jose, M. V. and Larralde, C., *Math. Biosci.* **58**, 159 (1982).
Kaandorp, J. A., *Fractal Modeling: Growth and Form in Biology*, Springer-Verlag, New York, 1994.
Kaufman, E. N. and Jain, R. K., *Biophys. J.* **60**, 596–610 (1991).
Klymko, P. and Kopelman, R., *J. Phys. Chem.* **86**, 3686–3688 (1982).
Klymko, P. and Kopelman, R., *J. Phys. Chem.* **87**, 4565–4567 (1983).
Kondo, A. and Hagashitani, K., *J. Colloid & Interface Sci.* **150**, 344–351 (1992).
Kopelman, R., *J. Stat. Phys.* **42**, 185–192 (1986).
Kopelman, R., *Science* **241**, 1620–1626 (1988).

Le Brecque, M., *Mosaic* **23**, 12–15 (1992).
Lee, C. K. and Lee, S. L., *Heterog. Chem. Rev.* **3**, 269–302 (1996).
Li, H., Chen, S., and Zhao, H., *Biophys. J.* **58**, 1313–1320 (1990).
Liebovitch, L. S., Fischbarg, J., and Koniarek, J. P., *Math. Biosci.* **84**, 37–68 (1987).
Liebovitch, L. S. and Sullivan, J. M., *Biophys. J.* **52**, 979–988 (1987).
Lord, J. A. H ., Waterfield, A. A., and Hughes, J., *Nature (London)* **267**, 495 (1977).
Losa, G. A., *Pathologica* **87**, 310–317 (1995).
Losche, M., Pieponstock, M., Diederich, A., Grunewald, T., Kjaer, K., and Vaknin, D., *Biophys. J.* **65**, 2160 (1993).
Madagula, A. and Sadana, A., *Biotech. Progr.* **9**, 259–268 (1993).
Malhotra, A. and Sadana, A., *Biotech. Bioeng.* **34**, 725–728 (1989).
Mandelbrot, B. B., *Les Objects Fractals. Forme, Hasard, et Dimension*, Paris: Flammarion, 1975.
Mandelbrot, B. B., *The Fractal Geometry of Nature*, New York: Freeman (1983).
Markel, V. A., Muratov, L. S., Stockman, M. I., and George, T. F., *Phys. Rev. B.* **43** (10), 8183–8195 (1991).
Matuishita, M., in *The Fractal Approach to Heterogeneous Chemistry: Surfaces, Colloids, and Polymers.* (D. Avnir, ed.) J. Wiley, New York, pp. 161–179, 1989.
Nygren, H. A., *J. Immunol. Methods* **114**, 107–114 (1988).
Nygren, H. A., *Biophys. J.* **65**, 1508–1512 (1993).
Nygren, H. A., and Stenberg, M., *J. Immunol. Methods* **80**, 15–24 (1985).
Nygren, H. and Stenberg, M., *Immunology* **66**(3), 321–327 (1989).
Nygren, H. A., and Stenberg, M., *Biophys. Chem.* **38**, 67–75 (1990).
Nygren, H. and Stenberg, M., in *Immunochemistry of Solid Phase Immunoassay*, Butler J., ed., CRC Press, Boca, Raton, FL, Chapter 19, 1991.
Pajkossy, T. and Nyikos, L., *Electrochim. Acta.* **34**, 171–179 (1989).
Peng, C. K., Buldyrev, S. V., Goldberger, A. L., Havlin, S., Sciortino, F., Simons, M., and Stanley, H. E., *Nature* **356**, 168–170 (1992).
Pfeifer, P., *Appl. Surf. Sci.* **18**, 146–164 (1984).
Pfeifer, P. and Avnir, D., *J. Chem. Phys.* **79**, 3558–3565 (1983).
Pfeifer, P. and Obert, M., in *The Fractal Approach to Heterogeneous Chemistry: Surfaces, Colloids, and Polymers* (D. Avnir, ed.), Wiley, New York, pp. 11–44, 1989.
Pfeifer, P., Welz, U., and Wippermann, A., *Chem. Phys. Lett.* **113**, 535–540 (1985).
Place, J. F., Sutherland, R. M., Riley, A., and Mangan, C., in *Biosensors with Fiberoptics* (D. Wise and L. B. Wingard Jr., eds.), Humana Press, New York, p. 253, 1991.
Rice, S., *Science* **266**, 664–664 (1994).
Risovic, D., *J. Colloid & Interface Sci.* **197**, 391–394 (1998).
Sadana, A., *Trends in Biotech.* **6**(5), 84–89 (1988).
Sadana, A. and Henley, J. P., *J. Biotech.* **5**, 67–73 (1987).
Sadana, A. and Sii, D., *J. Colloid & Interface Sci.* **151**(1), 166–177 (1992a).
Sadana, A. and Sii, D., *Biosen. & Bioelectr.* **7**, 559–568 (1992b).
Sadana, A. and Vo-Dinh, T., *Biotech. Appl. Biochem.* **33**, 17–28 (2001).
Scalettar, B. A., Abney, J. R., and Owicki, J. C., *Proc. Natl. Acad. Sci.* **85**, 6726–6730 (1988).
Scheller, F. W., Hintsche, R., Pfeiffer, P., Schubert, F., Rebel, K., and Kindervater, R., *Sensors & Actuators* **4**, 197–206 (1991).
Service, R. F., *Science* **278**, 806–806 (1997).
Stenberg, M. and Nygren, H. A., *J. Immunol. Meth.* **113**, 3–15 (1988).
Stenberg, M. and Nygren, H. A., *Anal. Biochem.* **127**, 183–192 (1982).
Stenberg, M. and Nygren, H. A., *Biophys. Chem.* **41**, 131 (1991).
Stenberg, M., Stiblert, L., and Nygren, H. A., *J. Theor. Biol.* **120**, 129–140 (1986).

Swalen, J. D., Allara, J., Andrade, J. D., Chandross, E. A., Garoff, S., Israelachvili, J., McCarthy, T. J., Murray, R., Pease, R. F., Rabott, J. F., Wynne, K. J., and Yu, H., *Langmuir* **3**, 932 (1987).
Tilton, R. D., Gast, A. P., and Robertson, C. R., *Biophys. J.* **58**, 1321–1326 (1990).
Uzguris, E. E. and Kornberg, R. D., *Nature (London)* **301**, 125–128 (1983).
Van Damme, H. and Fripiat, J. J., *J. Chem. Phys.* **82**(6), 2785–2789 (1985).
Viscek, T., *Fractal Growth Phenomena*, World Scientific Publishing, Singapore, 1989.
West, G. B., Brown, J. H, Enquist, B. J., *Science* **276** pp. 122–126. (4th April, 1997).
Witten, T. A. and Sander, L. M., *Phys. Rev. Lett.* **47**, 1400 (1981).
Zamir, M., *J. Theor. Biol.* **197**, 517–526 (1999).

CHAPTER 5

Influence of Different Parameters on Fractal Dimension Values During the Binding Phase

5.1. INTRODUCTION

A promising area in the investigation of biomolecular interactions is the development of biosensors, which are finding application in the areas of biotechnology, physics, chemistry, medicine, aviation, oceanography, and environmental control. One advantage of these biosensors is that they can be used to monitor the analyte-receptor reactions in real time (Myszka *et al.*, 1997). In addition, some techniques—like the surface plasmon resonance (SPR) biosensor—do not require radio labeling or biochemical tagging (Jonsson *et al.*, 1991), are reusable, have a flexible experimental design, provide a rapid and automated analysis, and have a completely integrated system. Moreover, the SPR combined with mass spectrometry (MS) exhibits the potential to provide a protemic analysis (Williams and Addona, 2000). In addition to evaluating affinities and interactions, the SPR can be used to determine unknown concentrations, to determine specificity, for kinetic analysis, to check for allosteric effects, and to compare binding patterns of different species. Of course, the SPR is not the only biosensor available, but it has gained increasing popularity as it has demonstrated the potential to be applied to the detection of different analytes in a wide variety of areas.

There is a need to characterize the reactions occurring at the biosensor surface in the presence of diffusional limitations that are inevitably present in these types of systems. It is essential to characterize not only the binding, or associative, reaction (by a binding rate coefficient) but also the desorption, or dissociation, reaction (by a desorption rate coefficient). This significantly assists in enhancing the biosensor performance parameters—such as reliability, multiple usage for the same analyte, and stability—as well as providing further insights into the sensitivity, reproducibility, and specificity of the biosensor. However, in this chapter we will analyze only the binding rate coefficient. In later chapters, we will examine the desorption rate coefficient.

The details of the association of the analyte (antibody or substrate) to a receptor (antigen or enzyme) immobilized on a surface is of tremendous significance for the development of immunodiagnostic devices as well as for biosensors (Pisarchick *et al.*, 1992). The analysis we will present is, in general, applicable to ligand-receptor and analyte-receptorless systems for biosensor and other applications (e.g., membrane-surface reactions). External diffusional limitations play a role in the analysis of immunodiagnostic assays (Bluestein *et al.*, 1987; Eddowes, 1987/1988; Place *et al.*, 1991; Giaver, 1976; Glaser, 1993; Fischer *et al.*, 1994). The influence of diffusion in such systems has been analyzed to some extent (Place *et al.*, 1991; Stenberg *et al.*, 1986; Nygren and Stenberg, 1985; Stenberg and Nygren, 1982; Morton *et al.*, 1995; Sjolander and Urbaniczky, 1991; Sadana and Sii, 1992a, 1992b; Sadana and Madagula, 1994; Sadana and Beelaram, 1995). Chapters 2 and 3 in this book also discuss and analyze the importance of diffusional limitations in biosensor analyte-receptor binding kinetics. The influence of partial (Christensen, 1997) and total (Matsuda, 1976; Elbicki *et al.*, 1984; Edwards *et al.*, 1995) mass transport on analyte-receptor binding kinetics is also available. The analysis presented for partial mass transport limitation (Christensen, 1997) is applicable to simple one-to-one association as well as to cases in which there is heterogeneity of the analyte in the liquid. This applies to the different types of biosensors utilized for the detection of different analytes.

Kopelman (1988) indicates that surface diffusion-limited reactions that occur on clusters (or islands) are expected to exhibit anomalous and fractal-like kinetics. These fractal kinetics exhibit anomalous reaction orders and time-dependent (e.g., binding) rate coefficients. (Since this topic was discussed in detail in Chapter 4, the discussion will not be repeated here.) Kopelman further indicates that as long as surface irregularities show dilatational symmetry scale invariance, such irregularities can be characterized by a single number, the fractal dimension. A consequence of the fractal nature is a power-law dependence of a correlation function (in our case, the analyte-receptor complex on the surface) on a coordinate (e.g., time). This fractal nature or power-law dependence is exhibited during the association (or binding) phase. This fractal power-law dependence has been shown for the binding of antigen–antibody (Sadana and Madagula, 1994; Sadana and Beelaram, 1995; Sadana, 1999) as well as for analyte-receptor (Ramakrishnan and Sadana, 2000) and analyte-receptorless (protein) systems (Sadana and Sutaria, 1997).

Fractal aggregate scaling relationships have been determined for both diffusion-limited processes and diffusion-limited scaling aggregation processes in spatial dimensions 2, 3, 4, and 5 by Sorenson and Roberts (1997). These authors noted that the prefactor (in our case, the binding rate coefficient) displays uniform trends with the fractal dimension, D_f. Fractal dimension values for the kinetics of antigen–antibody binding (Sadana, 1997; Milum and Sadana, 1997) and analyte-receptor binding (Sadana and Sutaria, 1997) are available.

In this chapter, we delineate (1) the role of analyte concentration, (2) the effect of different surfaces, and (3) the influence of regeneration on binding rate coefficients and fractal dimensions during analyte-receptor binding in different biosensor systems. We also discuss the role of surface roughness on the speed of response, specificity, sensitivity, and the regenerability or reusability of fiber-optic and other biosensors. As we present the fractal dimension and rate coefficient values for the binding phase, the noninteger orders of dependence obtained for the binding rate coefficient(s) on their respective fractal dimension(s) should further reinforce the fractal nature of these analyte-receptor binding systems.

5.2. THEORY

In the analysis to be presented we will assume that we have a heterogeneous surface that exists at the reaction interface. The heterogeneity on the surface may be due various factors, such as the inherent surface roughness, the heterogeneity of the receptors on the surface, the manner in which the receptors are immobilized on the surface, steric hindrances, nonspecific binding, the inappropriate or incorrect binding of the analyte in solution to the receptor immobilized on the surface, and the binding of impurities along with the steric hindrances caused by this (which minimizes the correct binding of the "regular" analyte to the receptor on the surface). Utilizing a fractal analysis, we will model this heterogeneity on the surface that is present under diffusional limitations. We will see that as the reaction progresses on the surface an increase or a decrease in the surface roughness may result, leading in the extreme case to a temporal fractal dimension. In this case, the biosensor surface roughness is continuously increasing with time.

5.2.1. Single-Fractal Analysis

Havlin (1989) indicates that the diffusion of a particle (analyte [Ag]) from a homogeneous solution to a solid surface (e.g., receptor [Ab]-coated surface) on which it reacts to form a product [analyte-receptor complex, (Ab·Ag)] is given by

$$(\text{Analyte} \cdot \text{Receptor}) \sim \begin{cases} t^{(3-D_{\text{f, bind}})/2} = t^p & (t < t_c) \\ t^{1/2} & (t > t_c) \end{cases}. \tag{5.1}$$

Here $D_{\text{f, bind}}$ is the fractal dimension of the surface during the binding step. Equation (5.1) indicates that the concentration of the product Ab·Ag(t) in a reaction $\text{Ab} + \text{Ag} \rightarrow \text{Ab}\cdot\text{Ag}$ on a solid fractal surface scales at short and intermediate time frames as $[\text{Ab}\cdot\text{Ag}] \sim t^p$, with the coefficient $p = (3 - D_{\text{f, bind}})/2$ at short time frames and $p = \frac{1}{2}$ at intermediate time frames. This equation is associated with the short-term diffusional properties of a random walk on a fractal surface. Note that in a perfectly stirred kinetics on a regular (nonfractal) structure (or surface), k_1 is a constant; that is, it is independent of time. In all other situations, one would expect a scaling behavior given by $k_{\text{bind}} \sim k' t^{-b}$, with $-b = p < 0$. Also, the appearance of the coefficient p different from $p = 0$ is the consequence of two different phenomena—the heterogeneity (fractality) of the surface and the imperfect mixing condition.

Havlin indicates that the crossover value may be determined by $r_c^2 \sim t_c$. Above the characteristic length r_c, the self-similarity is lost. Above t_c, the surface may be considered homogeneous since the self-similarity property disappears and regular diffusion is present. For the present analysis, we chose t_c arbitrarily and assume that the value of t_c is not reached. One may consider our analysis as an intermediate "heurisitic" approach in that in the future one may also be able to develop an autonomous (and not time-dependent) model of diffusion-controlled kinetics.

5.2.2. Dual-Fractal Analysis

We can extend the preceding single-fractal analysis to include two fractal dimensions. In doing so, it is appropriate to establish a general criterion for adopting a dual- (versus a single-) fractal model in the modeling; we do not simply try the dual if the single approach does not fit. Instead, we look at the r^2 factor (goodness of fit) for a single-fractal analysis; only if it is less than 0.97 do we try a dual-fractal analysis. In general, since the dual-fractal analysis has four parameters (two for the binding rate coefficient and two for the fractal dimension), higher-multiple models will not be required unless the binding curves exhibit a very high level of complexity in their shape. Thus, the dual-fractal analysis should serve as a quantitative and physical cutoff for multiple-fractal models.

At present, the time ($t = t_1$) at which the first fractal dimension "changes" to the second fractal dimension is arbitrary and empirical. For the most part, it is dictated by the data analyzed and experience gained by handling a single-fractal analysis. A smoother curve is obtained in the transition region if care is taken to select the correct number of points for the two regions. In this case, the analyte-receptor complex is given by

$$(\text{Analyte} \cdot \text{Receptor}) \sim \begin{cases} t^{(3-D_{f_1,\text{bind}})/2} = t^{p_1} & (t < t_1) \\ t^{(3-D_{f_2,\text{bind}})/2} = t^{p_2} & (t_1 < t < t_2 = t_c). \\ t^{1/2} & (t > t_c) \end{cases} \tag{5.2}$$

Note that antigen–antibody (or, in general, analyte-receptor) binding is unlike reactions in which the reactant reacts with the active site on the surface and the product is released. In this sense the catalytic surface exhibits an unchanging fractal surface to the reactant in the absence of fouling and other complications. In the case of antigen–antibody binding, the biosensor surface exhibits a changing fractal surface to the antigen or antibody (analyte) in solution. This occurs because as each binding reaction takes place, fewer and fewer sites are available on the biosensor surface to which the analyte may bind. This is in accord with Le Brecque's comment (1992) that the active sites on a surface may themselves form a fractal surface. Furthermore, the inclusion of nonspecific binding sites on the surface would increase the fractal dimension of the reaction surface. In general, to demonstrate fractal-like behavior, log–log plots of the distribution of molecules, $M(r)$, as a function of the radial distance, (r), from a given molecule are required (Nygren, 1993). This plot should be close to a straight line. The slope of the log $M(r)$ versus log(r) plot determines the fractal dimension.

It is worthwhile to develop a relationship between surface roughness (measured by a fractal exponent, p) and the rate of binding in view of the different (statistical) fractal growth laws prevalent in nature. These laws include invasion percolation, kinetic gelation, and diffusion-limited aggregation (DLA) (Viscek, 1989). These laws can be modeled to permit computer simulation of the shape and growth of natural processes. For example, in the DLA model introduced by Witten and Sander (1981), a randomly diffusing particle (seed) collides with a surface and stops. Another particle (from far away) diffuses to the surface and arrives at a site close (adjacent) to the first particle and stops. Another particle follows, and so on. In this way clusters are generated and exhibit the random branching and open structure that are self-similar in nature.

To obtain the rate of binding, we take the time derivative of both sides of Eq. (5.1) to yield

$$d[\text{analyte} \cdot \text{receptor}]/dt = kpt^{p-1}. \tag{5.3}$$

Here, k is the proportionality constant in Eq. (5.1). This indicates that the rate of binding is directly dependent on the binding rate coefficient, k, and the fractal exponent, p. We can determine the maximum rate of binding by setting d^2 [analyte·receptor]$/dt^2 = 0$. This yields

$$kp(p-1)t^{p-2} = 0. \tag{5.4}$$

This is the location of stationary point and yields $p=1$ or $p=0$. $p=0$ is the trivial case, so it is neglected. Substituting this in Eq. (5.3) yields d[analyte·receptor]$/dt = k$. This occurs at time $t=0$, which is intuitively correct. It is apparently difficult to confirm the nature of the stationary point by taking higher-order derivatives. Perhaps, another way is possible.

Let's try again, starting with Eq. (5.3). The fractal parameter, p, equals $(3-D_f)/2$ and characterizes the degree of heterogeneity on the surface. We define the rate of binding as $r = d$[analyte·receptor]$/dt$. Let's take the derivative of r with respect to p and set it equal to zero to obtain the maximum condition. Thus,

$$(dr/dp) = kt^{p-1} + kpt^{p-2} = 0. \tag{5.5}$$

This yields $p=1$, as noted above. Once again, $p=0$ is the trivial case and is neglected. Also, solving the quadratic from Eq. (5.5), gives

$$p = [1 + (1-4t)^{1/2}]/2. \tag{5.6}$$

To prevent imaginary numbers, an appropriate range of p values are possible (for time, $t \leq 0.25$). Since $p = (3-D_f)/2$, the corresponding optimum range of fractal dimension values are $D_f = 2 \pm (1-4t)^{1/2}$.

5.3. RESULTS

In this discussion, a fractal analysis will be applied to the data obtained for analyte-receptor binding data for different biosensor systems. This is one possible explanation for analyzing the diffusion-limited kinetics assumed to be present in all of the systems to be analyzed. The parameters thus obtained would provide a useful comparison of different situations. Alternate expressions involving saturation, first-order reaction, and no diffusion limitations are possible but seem to be deficient in describing the heterogeneity that exists on the surface. The analyte-receptor binding is a complex reaction, and fractal analysis via the fractal dimension and the rate coefficient provide a useful lumped-parameter analysis of the diffusion-limited situation. Basically, we are following the Kopelman approach (1988), wherein the diffusion-limited reaction occurring on the heterogeneous surface is modeled using a fractal approach.

In all fairness, we must emphasize that we present no independent proof or physical evidence (like a classical log–log plot to help determine the fractal dimension from the slope) for the existence of fractals in the analysis of these analyte-receptor binding systems except by indicating that fractal analysis has been applied in other areas and is a convenient means to make more quantitative the degree of heterogeneity that exists on the surface. Thus, this is only one possible way by which to analyze this analyte-receptor binding data. One might justifiably argue that appropriate modeling may be achieved by using a Langmuir or other approach. However, a major drawback of the Langmuir approach is that it does not allow for the heterogeneity that exists on the surface.

Researchers in the past have successfully modeled the adsorption behavior of analytes in solution to solid surfaces using the Langmuir model even though it does not conform to theory. The Langmuir approach may be utilized to model the data presented if one assumes the presence of discrete classes of sites. Rudzinski *et al.* (1983) indicate that other appropriate "liquid" counterparts of the empirical isotherm equations have been developed. These include the Freundlich (Dabrowski and Jaroniec, 1979), Dubinin–Radushkevich (Oscik *et al.*, 1976), and Toth (Jaroniec and Derylo, 1981) equations. These studies, with their known constraints, have provided some restricted physical insights into the adsorption of adsorbates on different surfaces.

5.3.1. Effect of Analyte Concentration in Solution

Nieba *et al.* (1997) analyzed histidine-tagged proteins using a chelating nitrilotriacetic acid (NTA) sensor chip. Using a BIACORE biosensor, the authors analyzed the binding kinetics of the chaperone system of *E. coli* GroEL and GroES. Chaperones assist the proteins to fold correctly to their native and active form. Initially, 19 nM GroES was injected over a Ni^{++}-NTA surface. Then, premixed samples containing 5 mM ATP and different concentrations of GroEL (13 to 139 nM) were injected onto the GroES surface. Figure 5.1 shows the curves obtained using Eq. (5.1) for the binding of the nucleotide + GroEL premixed solution to GroES immobilized on the Ni^{++}-NTA surface. In each case, a single-fractal analysis was adequate to describe the binding kinetics. Table 5.1a shows the values of the binding rate coefficient, k, and the fractal dimension, D_f. The values of k were obtained from a regression analysis using Sigmaplot (1993) to model the experimental data using Eq. (5.1), wherein $(Ab{\cdot}Ag) = kt^p$. Both the k and the D_f values are

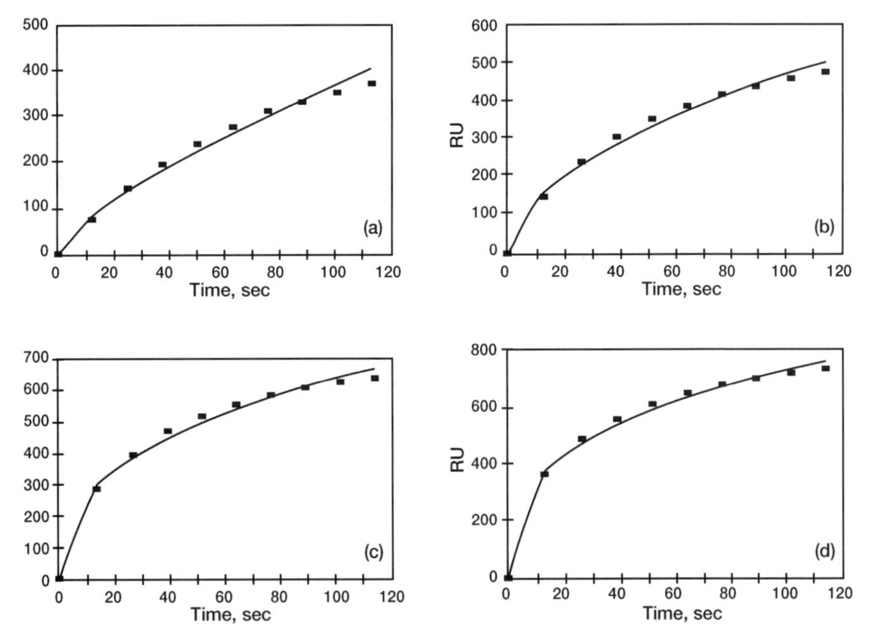

FIGURE 5.1 Binding rate curves for four different GroEL concentrations (in nM) premixed with 5 nM nucleotide in solution bound to GroES immobilized on a Ni^{++}-nitrilotriacetic acid (NTA) sensor chip (Nieba *et al.*, 1997).

within 95% confidence limit. For example, the value of k reported for the binding of 5 nM nucleotide premixed with 13 nM GroEL in solution to GroES immobilized on the Ni^{++}-NTA surface is 13.713 ± 1.010. The 95% confidence limit indicates that 95% of the k values will lie between 12.703 and 14.723. This indicates that the Table 5.1 values are precise and significant. The curves presented in the figures are theoretical curves. In all of the cases presented in Table 5.1, the regression coefficient (r-squared, r^2), was always greater than or equal to 0.97.

Note that as one increases the concentration of GroEL from 13 to 130 nM the binding rate coefficient, k, increases by a factor of 13.1—from a value of 13.71 to 179.66—and the fractal dimension, D_f, increases by 65.7%—from a value of 1.43 to 2.37. Thus an increase in the fractal dimension leads to an increase in the binding rate coefficient. Figure 5.2a shows that the binding rate coefficient, k, increases as the GroEL concentration in solution increases. In the GroEL concentration range of 13 to 130 nM, the binding rate coefficient is given by

$$k = (0.7634 \pm 0.1975)[\mathrm{GroEL}]^{1.1595 \pm 0.1347}. \tag{5.7}$$

TABLE 5.1 Influence of Different Parameters on Fractal Dimensions and Binding Rate Coefficients for Analyte-Receptor Binding Kinetics: Single-Fractal Analysis (Nieba *et al.*, 1997)

Analyte concentration in solution/receptor on surface	Binding rate coefficient, k	Fractal dimension, D_f
(a) 5 nM nucleotide premixed with 13 nM GroEL/GroES	13.71 ± 1.01	1.43 ± 0.07
5 nM nucleotide premixed with 32 nM GroEL/GroES	43.13 ± 2.44	1.95 ± 0.05
5 nM nucleotide premixed with 65 nM GroEL/GroES	124.51 ± 4.81	2.29 ± 0.04
5 nM nucleotide premixed with 130 nM GroEL/GroES	179.6 ± 6.48	2.37 ± 0.034
(b) *E. coli* maltose binding protein (MBP1)/Ni^{++}-nitriloacetic (NTA) surface	13.02 ± 0.548	1.256 ± 0.035
MBP2/Ni^{++}-(NTA) surface	33.52 ± 1.97	1.495 ± 0.048
MBP3/Ni^{++}-(NTA) surface	39.11 ± 2.23	1.593 ± 0.049
(c) GroES/Ni^{++}-(NTA) surface	14.26 ± 0.37	1.226 ± 0.035
CS-4His/Ni^{++}-(NTA) surface	8.799 ± 0.07	1.137 ± 0.011
CS-2His/Ni^{++}-(NTA) surface	14.0 ± 0.271	1.293 ± 0.026
GrpE/Ni^{++}-(NTA) surface	12.26 ± 0.177	1.457 ± 0.019
MBP/Ni^{++0}-(NTA) surface	69.46 ± 3.54	1.793 ± 0.07

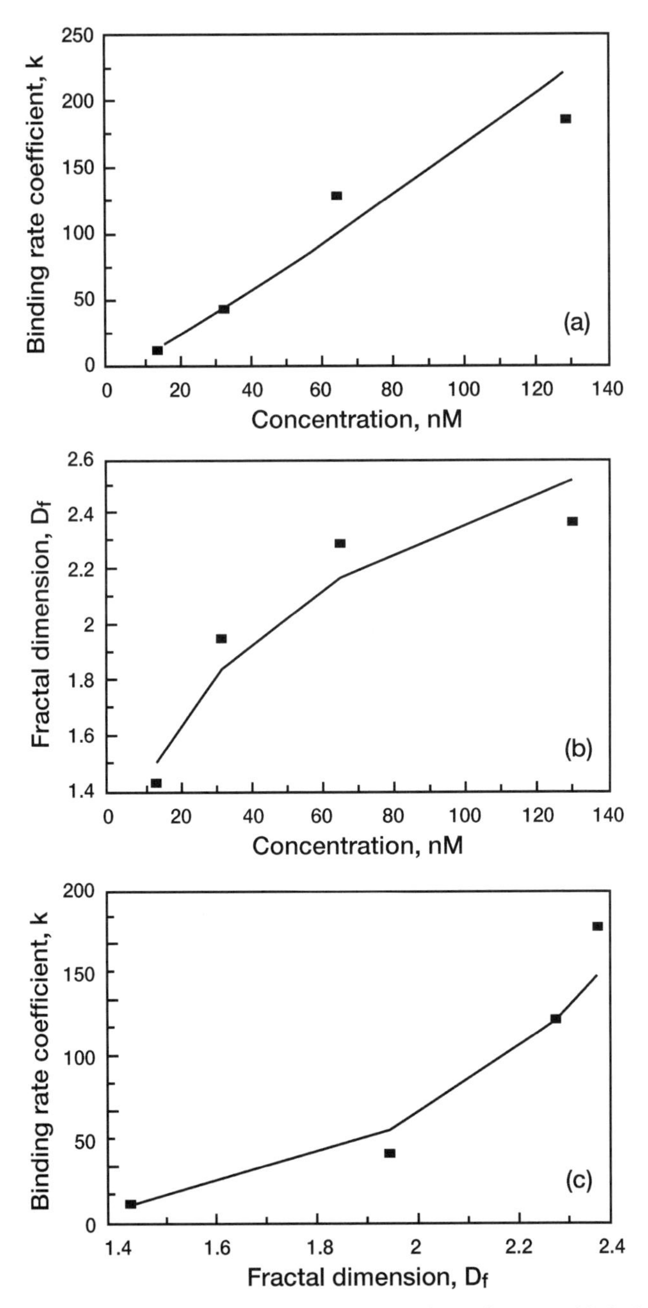

FIGURE 5.2 Influence of the GroEL concentration (in nM) in solution on (a) the binding rate coefficient, k, and (b) the fractal dimension, D_f. (c) Influence of D_f on k.

This predictive equation fits the values of k presented in Table 5.1a reasonably well. The fractional exponent dependence of the binding rate coefficient on the GroEL concentration in solution lends support to the fractal nature of the system.

Figure 5.2b shows that the fractal dimension, D_f, increases as the GroEL concentration increases. In the GroEL concentration range of 13 to 130 nM, D_f is given by

$$D_f = (0.8466 \pm 0.0709)[\text{GroEL}]^{0.2244 \pm 0.0470}. \tag{5.8}$$

This predictive equation fits the Table 5.1 data reasonably well. The fractal dimension is not very sensitive to the GroEL concentration in solution, as seen by the very low exponent dependence of D_f on GroEL concentration.

Figure 5.2c shows that k increases as D_f increases. This is in accord with the prefactor analysis of fractal aggregates (Sorenson and Roberts, 1997). Note that the fractal dimension is not an actual variable, such as temperature or concentration, that may be directly controlled; it is estimated from Eq. (5.1) or (5.2). One may consider it to be a "derived variable." In any case, it provides a quantitative estimate of the degree of heterogeneity or surface roughness. For the data presented in Table 5.1a, the binding rate coefficient is given by

$$k = (2.0993 \pm 0.5856) D_f^{4.9434 \pm 0.6158}. \tag{5.9}$$

This predictive equation fits the values of k presented in Table 5.1 reasonably well. Note the high exponent dependence of k on D_f. This underscores that k is sensitive to the surface roughness, or the degree of heterogeneity, D_f that exists on the surface.

Nieba *et al.* (1997) also analyzed the binding of three variants of maltose-binding protein (MBP) at a concentration of 120 nM in solution to the Ni^{++}-NTA surface. Figure 5.3 shows the curves obtained using Eq. (5.1) for the binding of different MBP variants to the Ni^{++}-NTA surface. MBP1 represents the variant with the N- and C-terminal histidine (His) tag; MBP2 represents the variant with the N-terminal His tag; and MBP3 is the terminally tagged His-tagged MBP. In each case, a single-fractal analysis is adequate to describe the binding kinetics.

Table 5.1b indicates that as one goes from MBP1 to MBP2 to MBP3, the fractal dimension, D_f, increases from 1.2564 to 1.4948 to 1.5928 and the binding rate coefficient, k, increases from 13.022 to 33.21 to 39.107, respectively. Note that the changes in D_f and k are in the same direction. For the data presented in Table 5.1b and in Figure 5.4, the binding rate coefficient

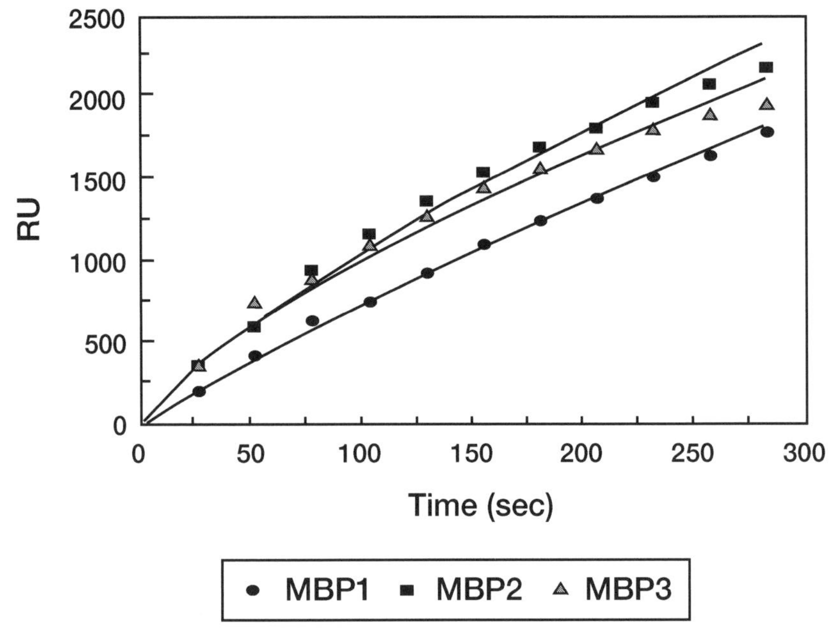

FIGURE 5.3 Binding rate curves for three variants of maltose-binding protein (MBP) in solution to a Ni^{++}-NTA surface (Nieba *et al.*, 1997).

is given by

$$k = (4.405 \pm 0.5203) D_f^{4.8059 \pm 0.6366}. \tag{5.10}$$

This predictive equation fits the values of k presented in Table 5.1b reasonably well. The high exponent dependence indicates that the binding rate coefficient is rather sensitive to the degree of heterogeneity that exists on the surface. Note that the location of the His tag on the MBP leads to different degrees of heterogeneity on the surface and subsequently to different binding rate coefficient values. Furthermore, as indicated by Eq. (5.10), the binding rate coefficient is very sensitive to the location of the His tag, as seen from the high value of the exponent. It would be valuable to know whether this (change in heterogeneity on the surface) is seen or applies to other molecules of interest.

Finally, Nieba *et al.* (1997) also analyzed the binding of five different proteins of different numbers of His tags. Figures 5.5a and 5.5b show the curves obtained using Eq. (5.1) for the binding of these different proteins to the Ni^{++}-NTA surface. In each case, a single-fractal analysis is sufficient to adequately describe the binding kinetics. Table 5.1c indicates that the fractal

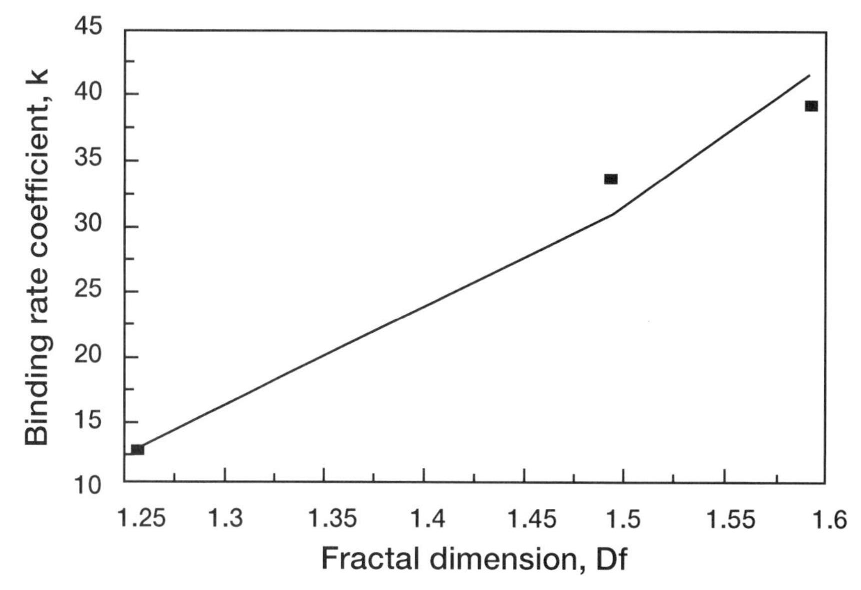

FIGURE 5.4 Influence of the fractal dimension, D_f, on the binding rate coefficient, k, for the binding of the three variants of MBP to a Ni^{++}-NTA surface.

dimension is different for the different proteins on the Ni^{++}-NTA surface, and this leads to different binding rate coefficient values. Figure 5.5c shows that an increase in the fractal dimension, D_f, leads to an increase in the binding rate coefficient, k. For the data presented in Table 5.1c, the binding rate coefficient is given by

$$k = (4.896 \pm 2.512) D_f^{4.040 \pm 1.164}. \tag{5.11}$$

This predictive equation fits the values of k presented in Table 5.1c reasonably well. There is some scatter in the data at the higher fractal dimension values. The high exponent dependence once again indicates that the binding rate coefficient is rather sensitive to the degree of heterogeneity that exists on the surface.

In this final case, Nieba *et al.* had changed the number and the positions of the histidine tags. This led to different fractal dimensions on the Ni^{++}-NTA surface and subsequently to different binding rate coefficient values. The highest fractal dimension is obtained for the MBP-Ni^{++}-NTA surface. In this case, 120 nM MBP was used along with one His tag at the C-terminal. The lowest fractal dimension was obtained for the CS-4His/Ni^{++}-NTA surface, where 15 nM CS-4His was used along with a total of four His tags at the C- and N-terminals.

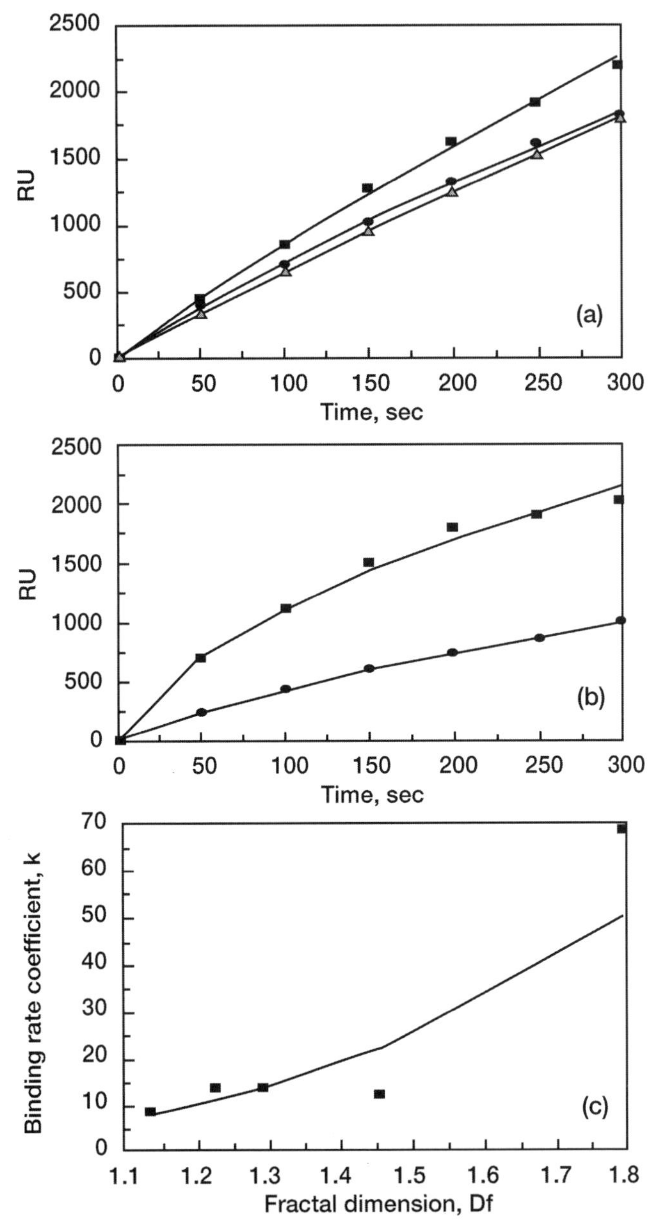

FIGURE 5.5 Binding rate curves for five different proteins with different numbers of histidine tags in solution to a Ni^{++}-NTA surface (Nieba *et al.*, 1997). (a) ■ GroES, ▲ CS-4-His, ● CS-2His; (b) ● GrpE, ■ MBP. (c) Influence of the fractal dimension, D_f, on the binding rate coefficient, k.

Direct optical techniques, such as the surface plasmon resonance (SPR) technique (Sutherland *et al.*, 1984), have been used to analyze biospecific interactions at solid–liquid interfaces. In this technique, there is a resonant coupling of the incident light to plasmons (conducting electrons) at the metal film surface. The oscillations of the plasmons give rise to an evanescent field, which extends into a sample solution. For SPR sensing, the antibody or the antigen (in general, protein) is adsorbed to the metal surface and exposed to the analyte in solution.

Fagerstam *et al.* (1997) used the SPR technique to analyze a dextran-modified sensor chip to which one of the components is attached covalently. These authors analyzed the binding of a fusion protein between the lac repressor and β-galactosidase, and between the lac operator DNA bound to the matrix of an SPR biosensor. It was found that the lac operator DNA was captured by streptavidin immobilized on the biosensor chip. This DNA is synthetic in nature, has 35 base pairs, and is biotinylated at the 5′ end. Figures 5.6a–5.6e show the curves obtained for the binding of fusion protein in the concentration range of 0.4 to 5.0 μg/ml. Table 5.2 shows the values of the binding rate coefficients and the fractal dimensions obtained from single- and dual-fractal analysis. Once again, the dual-fractal analysis provides a better fit than that obtained from a single-fractal analysis for the binding of the fusion protein in the concentration range of 0.4 to 5.0 μg/ml.

Note that for the protein fusion concentration range analyzed, an increase in the value of the fractal dimension from D_{f_1} to D_{f_2} leads to an increase in the value of the binding rate coefficient from k_1 to k_2. The magnitude of the changes in the fractal dimension that lead to changes in the binding rate coefficients for a particular fusion protein concentration are significant since they provide one means of controlling or varying the binding rate coefficient on the biosensor surface. Furthermore, these results are consistent with Fagerstam *et al.* (1997), who inferred from the shape of the binding curves that the binding interaction appears to be heterogeneous on the surface.

Figure 5.7a shows the linear increase in k_1 and k_2 with an increase in the fusion protein concentration in solution. However, the linearity shown is not convincing due to the small number of data points and the scatter in the estimated values for k_2 at different fusion protein concentrations. Nevertheless, the trend presented is useful.

Figure 5.7b shows that the fractal dimension, D_{f_1}, increases linearly as the fusion protein concentration increases in the concentration range analyzed. Once again, there is scatter in the data, and more data points would more firmly establish the trend presented. Figure 5.6b also shows that D_{f_2} exhibits a slight linearly decreasing trend with an increase in the fusion protein concentration. And again, more data points would more firmly establish the trend presented.

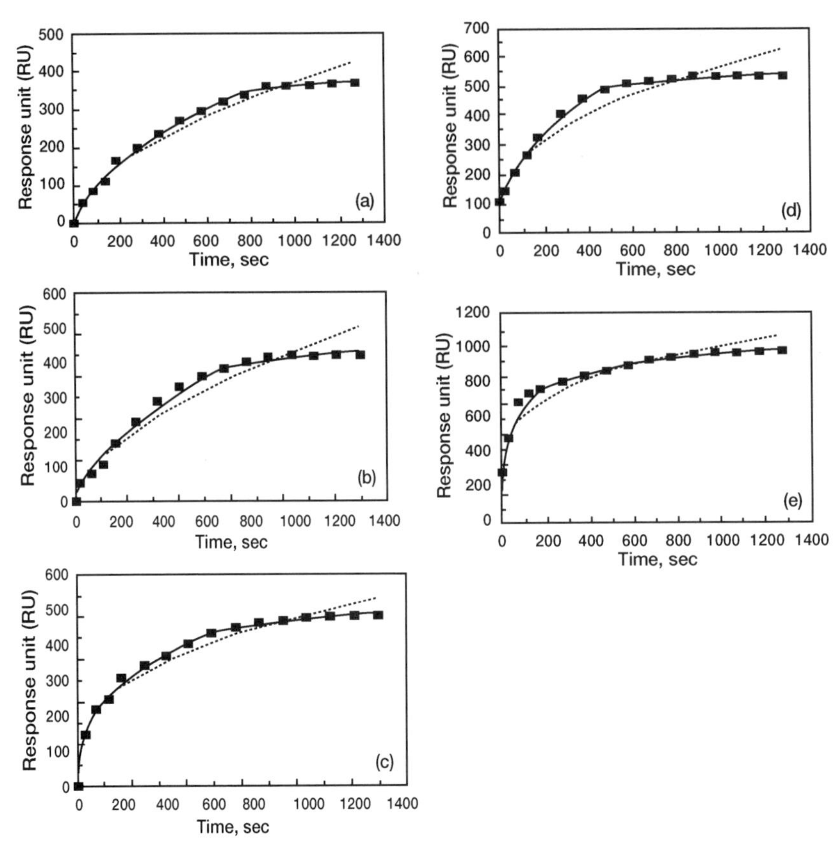

FIGURE 5.6 Binding of different concentrations (in μg/ml) of a fusion protein between the lac repressor and β-galactosidase and the lac operator DNA bound to the matrix of an SPR biosensor [(- - -, Eq. (5.1) single-fractal analysis; (—), Eq. (5.2) dual-fractal analysis] (Fagerstam *et al.*, 1997).

In spite of the scatter in the data and the few data points presented, the trends presented for the binding rate coefficient and the fractal dimension are helpful. These trends provide one possible means of controlling or changing the binding rate coefficients and the degree of heterogeneity on the reaction surface. More such analyses are required to shed further light on these binding reactions that occur on biosensor or other surfaces of immunological importance.

Nath *et al.* (1997) utilized a fiber-optic evanescent sensor to detect *L. donovani* antibodies in sera of kala azar patients. Cell surface protein of *L. donovani* was immobilized on a fiber-optic sensor. In the first step, the antigen on the fiber reacts with the *L. donovani*–infected sample. In the second step,

TABLE 5.2 Influence of Fusion Protein Concentration on the Fractal Dimensions and Binding Rate Coefficients for the Binding between the lac Repressor and β-Galactosidase and the lac Operator DNA Bound to the Matrix of an SPR Biosensor (Fagerstam *et al.*, 1997)

Fusion protein concentration, μg/ml	k	D_f	k_1	k_2	D_{f_1}	D_{f_2}
0.4	9.08 ± 0.87	1.93 ± 0.04	7.38 ± 0.51	239 ± 0.89	1.84 ± 0.04	2.88 ± 0.03
0.6	8.36 ± 1.28	1.85 ± 0.06	6.83 ± 1.18	125 ± 3.0	1.76 ± 0.09	2.65 ± 0.18
0.8	42.3 ± 2.58	2.29 ± 0.022	36.5 ± 1.18	171 ± 2.60	2.22 ± 0.02	2.69 ± 0.04
1.2	39.4 ± 4.1	2.22 ± 0.04	29.1 ± 0.42	254 ± 3.9	2.08 ± 0.03	2.78 ± 0.03
5.0	210 ± 26.4	2.54 ± 0.05	149 ± 23.1	384 ± 4.2	2.37 ± 0.10	2.73 ± 0.02

this reacts with fluorescein isothiocyanate (FITC)–labeled antihuman IgG to generate the signal.

Figures 5.8a–5.8e show the binding of parasite *L. donovani*–diluted pooled sera (1:1600 to 1:25,600) to FITC-labeled antihuman IgG immobilized on an

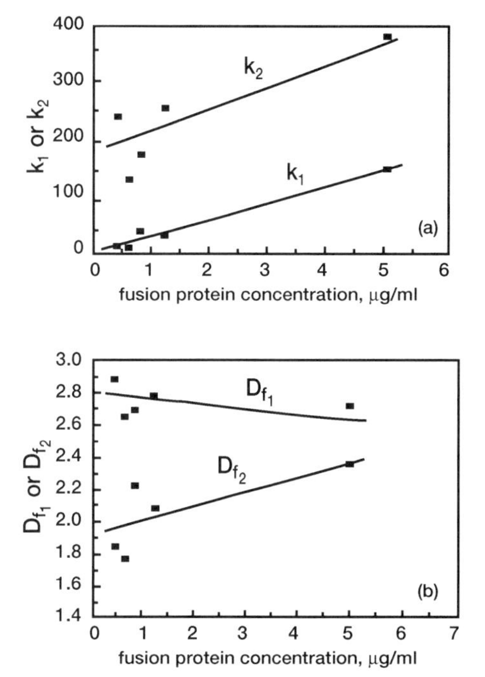

FIGURE 5.7 (a) Linear increase in the binding rate coefficients, k_1 and k_2, with an increase in the fusion protein concentration in solution (Fagerstam *et al.*, 1997); (b) linear increase and decrease in the fractal dimensions, D_{f_1} and D_{f_2}, respectively, with an increase in fusion protein concentration in solution (Fagerstam *et al.*, 1997).

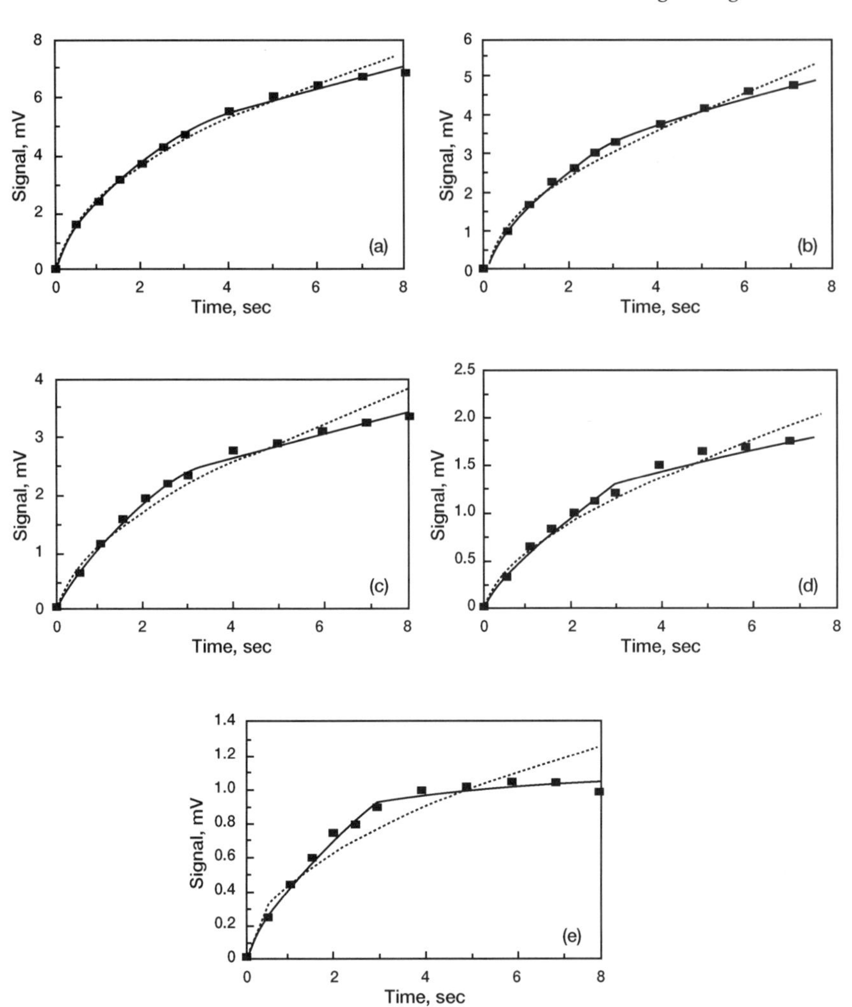

FIGURE 5.8 Binding of parasite *L. donovani*–diluted pooled sera in solution to fluorescein isothiocyanate (FITC)–labeled antihuman IgG immobilized on an optical fiber (Nath *et al.*, 1997) (a) 1:1600; (b) 1:3200; (c) 1:6400; (d) 1:12,800; (e) 1:25,600 (- - -, single-fractal analysis; —, dual-fractal analysis).

optical fiber. The diluted pooled sera was diluted by a factor of 2, starting from 1:1600. In each case, a single-fractal analysis did not provide an adequate fit, and thus a dual-fractal analysis was used. Table 5.3 shows the values of the binding rate coefficient, k, and the fractal dimension, D_f, for a single-fractal analysis, as well as the binding rate coefficients, k_1 and k_2 and the fractal dimensions, D_{f_1} and D_{f_2} for a dual-fractal analysis. In each case,

TABLE 5.3 Influence of Dilution of Parasite *L. donovani* Pooled Sera in Solution on Fractal Dimensions and Binding Rate Coefficients on Its Binding to Fluorescein Isothiocyanate (FITC)–Labeled Antihuman IgG Immobilized on an Optical Fiber (Nath *et al.*, 1997)

Parasite *L. donovani* diluted pooled sera	k	D_f	k_1	k_2	D_{f_1}	D_{f_2}
1:1600	2.445 ± 0.14	1.903 ± 0.04	2.402 ± 0.03	3.127 ± 0.09	1.756 ± 0.02	2.205 ± 0.07
1:3200	1.704 ± 0.15	1.873 ± 0.06	1.663 ± 0.06	2.271 ± 0.06	1.628 ± 0.05	2.237 ± 0.06
1:6400	1.105 ± 0.13	1.787 ± 0.08	1.071 ± 0.07	1.541 ± 0.04	1.469 ± 0.08	2.214 ± 0.06
1:12,800	0.619 ± 0.08	1.816 ± 0.09	0.598 ± 0.05	0.922 ± 0.06	1.528 ± 0.11	2.317 ± 0.16
1:25,600	0.452 ± 0.08	2.015 ± 0.01	0.432 ± 0.03	0.825 ± 0.04	1.570 ± 0.08	2.764 ± 0.10

note that (as previously indicated) an increase in the fractal dimension from D_{f_1} to D_{f_2} leads to an increase in the binding rate coefficient from k_1 to k_2. Also note that the changes in the fractal dimension and in the binding rate coefficient are in the same direction, as previously indicated.

It is interesting that k_1 and k_2 both increase as the dilution factor (defined as the reciprocal of the dilution of the pooled positive serum) increases (see Fig. 5.9), For example, the dilution factor for the 1:1600 case is 0.000625. In the dilution factor range analyzed, k_1 is given by

$$k_1 = (284.98 \pm 20.363)[\text{dilution factor}]^{0.6426 \pm 0.0314}. \tag{5.12}$$

This predictive equation fits the values of k_1 presented in Table 5.3 and in Fig. 5.9a reasonably well. The low exponent dependence of k_1 on the dilution factor indicates that the binding rate coefficient exhibits a rather low dependence on the dilution factor in this range. The fractional exponent dependence exhibited by k_1 on the dilution factor further reinforces the fractal nature of the system.

Similarly, in the dilution factor range analyzed, k_2 is given by

$$k_2 = (138.734 \pm 15.34)[\text{dilution factor}]^{0.5143 \pm 0.0478}. \tag{5.13}$$

This predictive equation fits the values of k_2 presented in Table 5.3 and in Figure 5.9b reasonably well. The low exponent dependence of k_2 on the dilution factor indicates, once again, that the binding rate coefficient exhibits a rather low dependence on the dilution factor in this range. Once again, the fractional exponent dependence exhibited by k_2 on the dilution factor further reinforces the fractal nature of the system.

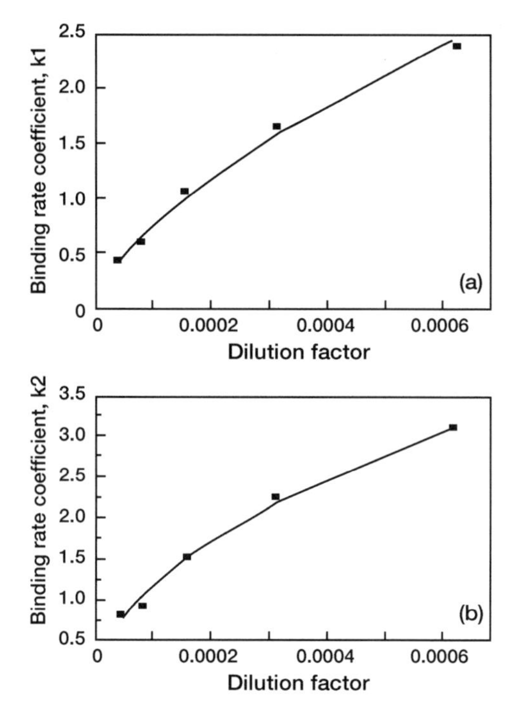

FIGURE 5.9 Influence of the dilution factor on (a) the binding rate coefficient, k_1; (b) the binding rate coefficient, k_2.

5.3.2. Effect of Different Surfaces

Zhao and Reichert (1992) analyzed the time-dependent fluorescence intensity of the binding of FITC-avidin in solution to sensor tips doped with B-DPPE and B-X-DPPE. Figures 5.10a and 5.10b show the curves obtained using Eqs. (5.1) and (5.2) for the binding of FITC-avidin. Table 5.4 shows the values of the binding rate coefficients and the fractal dimensions obtained for single- and dual-fractal analysis. Clearly, once again, the dual-fractal analysis provides a better fit for the binding of FITC-avidin to both B-DPPE and B-X-DPPE. Once again, for the binding of FITC-avidin to either B-DPPE or B-X-DPPE, an increase in the fractal dimension from D_{f_1} to D_{f_2} leads to an increase in the binding rate coefficient from k_1 to k_2. For example, for the binding of FITC-avidin to B-DPPE, an increase in the fractal dimension value by a factor of 2.36—from $D_{f_1} = 1.27$ to $D_{f_2} = 3$—leads to an increase in the binding rate coefficient value by a factor of 4.4—from $k_1 = 0.085$ to $k_2 = 0.376$. Similar trends have been observed for the previously presented cases.

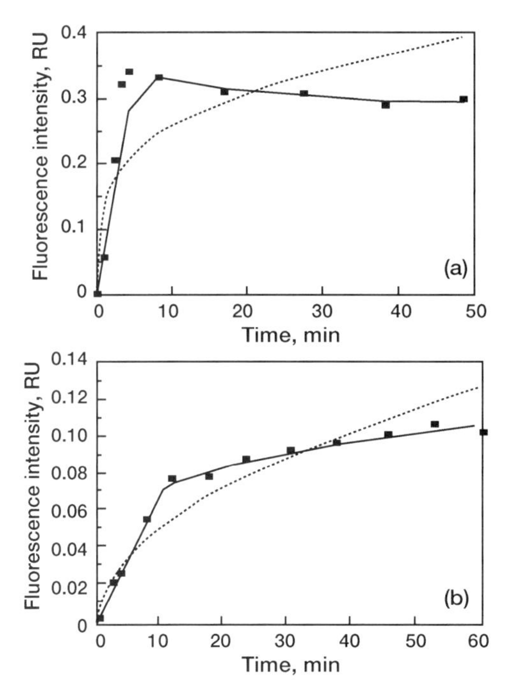

FIGURE 5.10 Binding of FITC-avidin in solution to sensor tips doped with (a) B-X-DPPE (- - -), Eq. (5.1) (single-fractal); (—), Eq. 5.2 (double-fractal); (b) B-DPPE (- - -), Eq. (5.1) (single-fractal); (—), Eq. 5.2 (double-fractal).

Nellen and Lukosz (1991) analyzed the affinity reaction between protein A adsorbed on the surface and h-IgG in solution. These authors indicate that immunoglobulins G are known to bind with their F_c part to protein A. Furthermore, this property has been used to obtain oriented immobilization of IgGs. Figure 5.11a shows the binding of 5 mg/ml IgG in solution to a single adlayer (F′) of protein A adsorbed on the surface of an optical waveguide. In this case, a single-fractal analysis is sufficient to adequately describe the binding kinetics. However, for the binding of 5 mg/ml IgG in solution to a double layer ($F' + F''$) of protein A adsorbed on the surface, a dual-fractal analysis provides a better fit. There is a corresponding change in the binding mechanism involved when the IgG binds to either a single adlayer (F′) or to a double adlayer ($F' + F''$), as the fractal dimension increases by 54.3%—from $D_{f_1} = 1.71$ to $D_{f_2} = 2.64$—and the binding rate coefficient increases by 53.5%—from $k_1 = 0.074$ to $k_2 = 0.134$. The changes in the fractal dimension and in the binding rate coefficient are in the same direction. The almost identical relative change is purely coincidental.

TABLE 5.4 Influence of Different Parameters on Fractal Dimensions and Binding Rate Coefficients for Different Analyte-Receptor Binding Kinetics: Single- and Dual-Fractal Analysis

Analyte in solution/receptor on surface	k	D_f	k_1	k_2	D_{f_1}	D_{f_2}	Reference
FITC-avidin/B-X-DPPE	0.144 ± 0.09	2.47 ± 0.25	0.085 ± 0.05	0.376 ± 0.006	1.27 ± 0.59	3 ± 0	Zhao and Reichert, 1992
FITC-avidin/B-DPPE	0.017 ± 0.003	2.018 ± 0.01	$0.011 + 0.034$	0.046 ± 0.001	1.35 ± 0.01	2.57 ± 0.04	Zhao and Reichert, 1992
5 mg/ml h-IgG/single adlayer (F′) of protein A	0.166 ± 0.01	2.56 ± 0.03	na	na	na	na	Nellen and Lukosz, 1991
5 mg/ml h-IgG/double adlayer (F′ + F″) of protein A	0.095 ± 0.02	2.41 ± 0.07	0.088 ± 0.02	0.134 ± 0.003	1.71 ± 0.22	2.64 ± 0.02	Nellen and Lukosz, 1991

na: not applicable

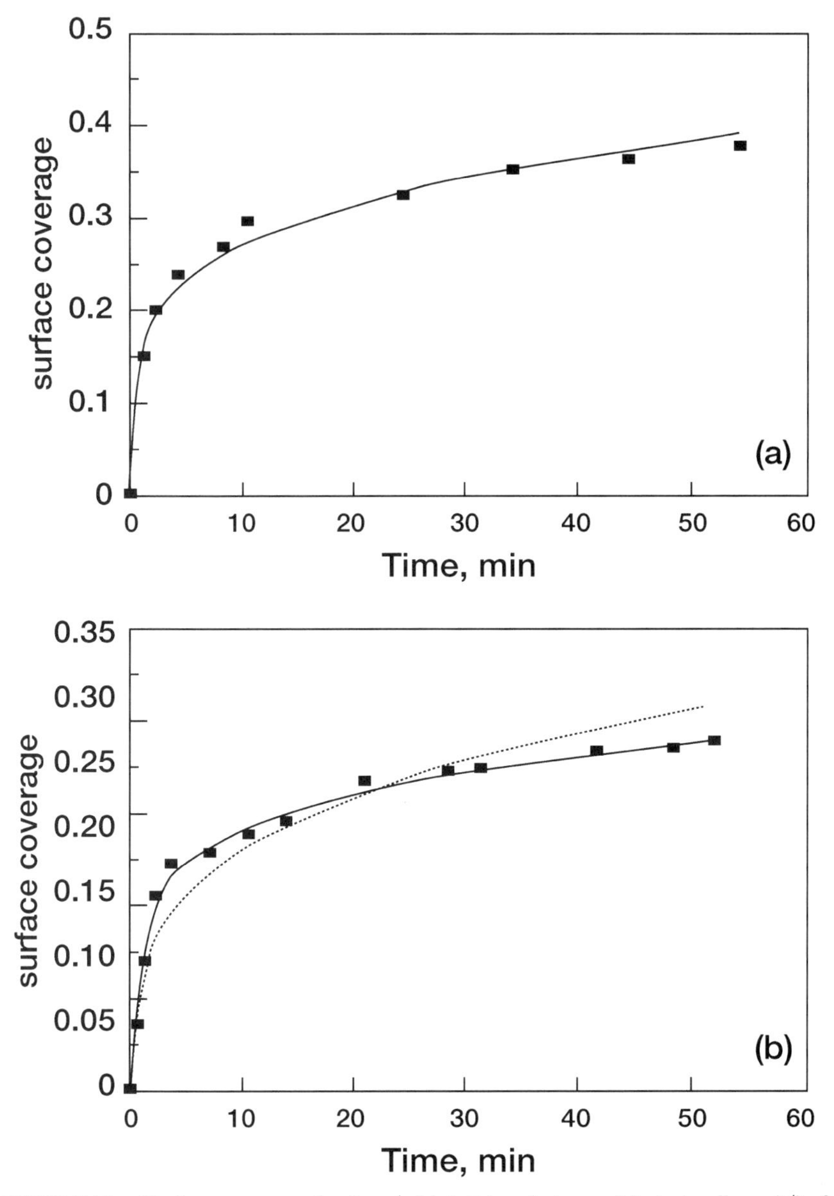

FIGURE 5.11. Binding rate curves for 5 mg/ml h-IgG in solution to (a) single adlayer (F′) of protein A (single-fractal analysis); (b) double adlayer (F′ + F″) of protein A (--- single-fractal analysis; — double-fractal analysis) adsorbed on a waveguide surface (Nellen and Lukosz, 1991).

Zhao and Reichert (1992) analyzed the time-dependent fluorescence intensity of the binding of FITC-avidin in solution to sensor tips doped with biotin lipid at different surface densities. Figure 5.12 shows the curves obtained using Eqs. (5.1) and (5.2) for the binding of FITC-avidin in solution to a biotin lipid surface with densities ranging from 0.28 to 2.70 mol%. Clearly, the dual-fractal analysis provides a better fit than that obtained from a single-fractal analysis at all four biotin lipid concentrations utilized. Note also that an increase in the fractal dimension from D_{f_1} to D_{f_2} leads to an increase in the binding rate coefficient from k_1 to k_2. For example, for the binding of FITC-avidin to a sensor tip doped with 0.28 mol% biotin (lowest surface density), an increase in the fractal dimension value by 83.8%—from D_{f_1} = 1.38 to D_{f_2} = 2.53 leads to an increase in the binding rate coefficient value by a factor of 4.8—from $k_1 = 0.010$ to $k_2 = 0.048$. Also, for the binding of FITC-avidin to a sensor doped with 2.70 mol% biotin (highest surface density), an increase in the fractal dimension value by a factor of 2—from D_{f_1} = 1.36 to D_{f_2} = 2.72—leads to an increase in the binding rate coefficient by a factor of 4.4—from $k_1 = 0.0515$ to $k_2 = 0.234$.

Also note that as the biotin lipid surface density increases, the binding rate coefficients, k_1 and k_2, exhibit increases (see Table 5.5). Figure 5.13 shows

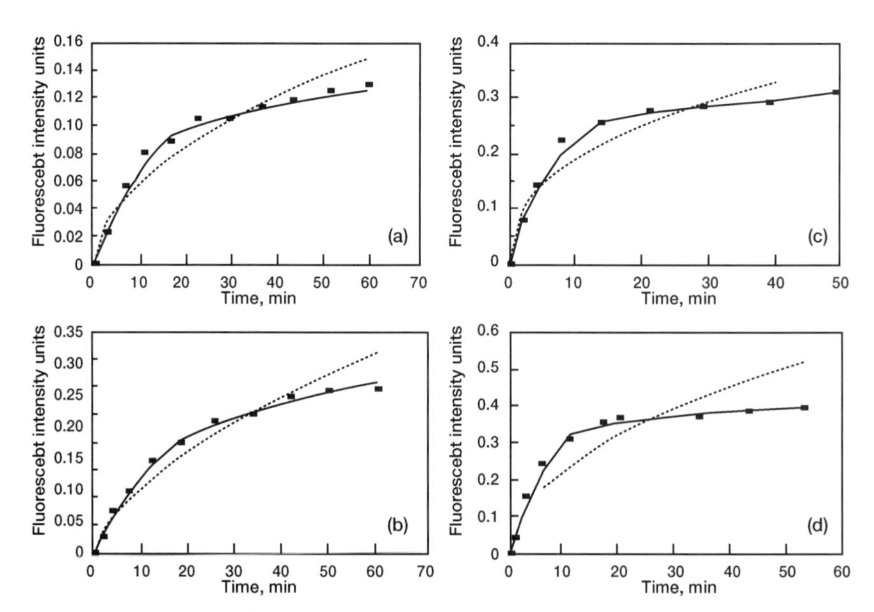

FIGURE 5.12 Theoretical curves using Eqs. (5.1) (- - -, single-fractal analysis) and (5.2 —, dual-fractal analysis) for the binding of avidin in solution to different biotin doping densities (in mol%) on a sensor tip (Zhao and Reichert, 1992): (a) 0.28, (b) 0.50, (c) 0.99, (d) 2.70.

TABLE 5.5 Influence of Different Parameters on Fractal Dimensions and Binding Rate Coefficients for Different Analyte-Receptor Reactions: Single- and Dual-Fractal Analysis

Analyte in solution/receptor on surface	k	D_f	k_1	k_2	D_{f_1}	D_{f_2}	Reference
Avidin/0.28 mol% biotin lipid surface density	0.018 ± 0.005	1.98 ± 0.128	0.010 ± 0.002	0.048 ± 0.001	1.38 ± 0.27	2.53 ± 0.06	Zhao and Reichert, 1992
Avidin/0.50 mol% biotin lipid surface density	0.025 ± 0.005	1.747 ± 0.109	0.0181 ± 0.003	0.0648 ± 0.003	1.371 ± 0.189	2.29 ± 0.085	Zhao and Reichert, 1992
Avidin/0.99 mol% biotin surface density	0.078 ± 0.016	2.205 ± 0.121	0.051 ± 0.008	0.181 ± 0.008	1.752 ± 0.187	2.72 ± 0.023	Zhao and Reichert, 1992
Avidin/2.70 mol% biotin surface density	0.074 ± 0.025	1.998 ± 0.16	0.051 ± 0.01	0.234 ± 0.008	1.358 ± 0.247	2.721 ± 0.051	Zhao and Reichert, 1992
m-Xylene-saturated STE buffer solution/EDTA-treated cell (transformed *E. coli*) suspension immobilized on fiber-optic tip with dialysis membrane	119.82 ± 17.02	0.9664 ± 0.132	na	na	na	na	Ikariyama *et al.*, 1997
m-Xylene/cell suspension immobilized on fiber-optic tip with polycarbonate membrane	959.58 ± 201.2	1.853 ± 0.189	na	na	na	na	Ikariyama *et al.*, 1997
m-Xylene/cell suspension immobilized on fiber-optic tip with polycarbonate membrane; lucerfin added after 2 h of luciferase induction; absence of methyl-benzyl alcohol	5567 ± 326	1.5378 ± 0.081	na	na	na	na	Ikariyama *et al.*, 1997
m-Xylene/cell suspension; lucerfin added; presence of methyl-benzyl alcohol	$0.0154 + 0.0432$	0	165.54 ± 83.4	1282 ± 7.25	1.210 ± 0.478	0.721 ± 0.06	Ikariyama *et al.*, 1997

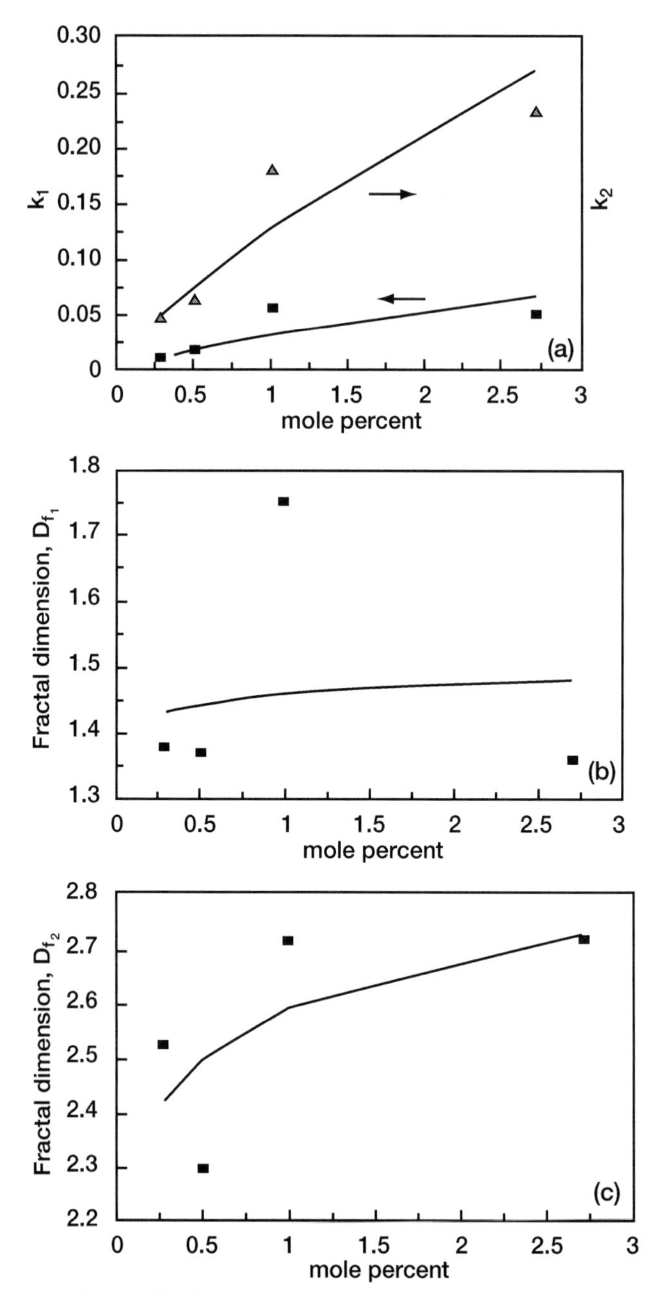

FIGURE 5.13 Influence of different doping densities (in mol%) on a sensor tip on (a) the binding rate coefficients, k_1 and k_2; (b) the fractal dimension, D_{f_1}; (c) the fractal dimension, D_{f_2}.

the best fit curve, which indicates that both k_1 and k_2 increase as the mole percentage of biotin lipid surface density increases. More data points are required to more firmly establish the trend presented. Nevertheless, the trend is of interest since it shows one possible way of changing the binding rate coefficients, k_1 and k_2. For example, as the biotin lipid surface density increases by a factor of 9.64—from 0.28 to 2.7 mol%, the binding rate coefficient, k_1, increases by a factor of 5.15—from a value of 0.010 to 0.0515—and the binding rate coefficient, k_2, increases by a factor of 4.92—from a value of 0.0475 to 0.234.

In the concentration range analyzed, the binding rate coefficient, k_1, can be described by

$$k_1 = (0.0177 \pm 0.0026)[\text{biotin surface density, mol\%}]^{0.566 \pm 0.539}, \tag{5.14}$$

and the binding rate coefficient, k_2, can be described by

$$k_2 = (0.129 \pm 0.0441)[\text{biotin surface density}]^{0.757 \pm 0.175}. \tag{5.15}$$

These predictive equations fit the Table 5.5 values of k_1 and k_2 reasonably well (Fig. 5.13a).

Similarly, Figs. 5.13b and 5.13c show that both fractal dimensions, D_{f_1} and D_{f_2}, exhibit increases as the biotin lipid surface density increases in the range from 0.28 to 2.70 mol%. There is, however, considerable scatter in the data for both D_{f_1} and D_{f_2}. An increase in biotin surface lipid density from 0.28 to 2.70 mol% leads to a small increase in the fractal dimension, D_{f_2}, by about 7.5%—from $D_{f_2} = 2.53$ (at the lowest surface biotin surface density of 0.28 mol%) to $D_{f_2} = 2.72$ (at the highest biotin surface density of 2.70 mol%).

In the concentration range analyzed, the fractal dimension, D_{f_1}, is given by

$$D_{f_1} = (1.461 \pm 0.237)[\text{biotin surface density, mol\%}]^{0.0143 \pm 0.0890}, \tag{5.16}$$

and the fractal dimension, D_{f_2}, is given by

$$D_{f_2} = (2.594 \pm 0.201)[\text{biotin surface density, mol\%}]^{0.0529 \pm 0.0441}. \tag{5.17}$$

The fractional exponent dependence on biotin surface density exhibited by the binding rate coefficients and by the fractal dimensions D_{f_1} and D_{f_2} provide further support for the fractal nature of the system.

As mentioned, there is scatter in the data. A better fit may be obtained if an expression such as

$$D_{f_1} \text{ or } D_{f2} = a[\text{biotin surface density, mol\%}]^{\mathrm{b}} + \mathrm{c}\,[\text{biotin surface density}]^{\mathrm{d}} \tag{5.18}$$

is used. Here *a*, *b*, *c*, and *d* are coefficients to be determined by regression. But, at present, this just introduces more variables, and this leads to a better fit. The fractal dimension, D_{f_1}, exhibits a complex dependence on biotin surface density. Apparently, the D_{f_1} versus biotin surface density curve exhibits a maximum. More data points are required to more firmly describe the trend exhibited.

Ikariyama *et al.* (1997) developed and analyzed a biosensor to detect environmental pollutants. These authors indicate that some microorganisms can assimilate benzene-related and other compounds since they possess a series of enzymes that can digest these chemicals (Koga *et al.*, 1985; Yen and Gunsalus, 1982). Ikariyama *et al.* indicate that the genetic information is encoded in a series of degradation plasmids and that the TOL plasmid in *Pseudomonas putida* mt-2 contains a series of genes that can degrade xylene and toluene. The authors utilized a fiber-optic biosensor to monitor benzene derivatives by recombinant *E. coli* that contained the luciferase gene. They constructed a fusion gene between TOL plasmid and the luciferase gene. Recombinant *E. coli* bearing this fusion gene was then immobilized on the fiber-optic end.

Figure 5.14a shows the curve obtained using Eq. (5.1) for the binding of m-xylene-saturated STE buffer solution to the microorganism immobilized to the fiber-optic tip and covered with a polycarbonate membrane. In this case, a single-fractal analysis is sufficient to adequately describe the binding kinetics. Table 5.5 shows the values of the binding rate coefficient and the fractal dimension. Ikariyama *et al.* indicate that there is a fluctuating relationship between m-xylene and the luminescence, which is reflected in the "error" observed for estimating the binding rate coefficient, *k*. Note that this fluctuating relationship does not significantly affect the error in the estimated value of the fractal dimension or the degree of heterogeneity that exists on the biosensor surface.

Figure 5.14b shows the curve obtained using Eq. (5.1) for the binding of m-xylene-saturated STE buffer solution to the immobilized microorganism immobilized to the fiber-optic tip and covered with a dialysis membrane. In this case too, a single-fractal analysis is sufficient to adequately describe the binding kinetics. Table 5.5 shows the values of the binding rate coefficient and the fractal dimension. In this case, the fluctuating relationship observed when the polycarbonate membrane was used was not present. Ikariyama *et al.* indicate that the less-hydrophilic property of the polycarbonate membrane is the reason for the fluctuations. Note that there is an increase in the fractal dimension, D_f, and a corresponding increase in the binding rate coefficient, *k*, as one goes from the dialysis membrane to the polycarbonate membrane. An 89.6% increase in D_f, from 0.9664 (dialysis membrane) to 1.8532 (polycarbonate membrane), leads to an increase in *k* by a factor of about

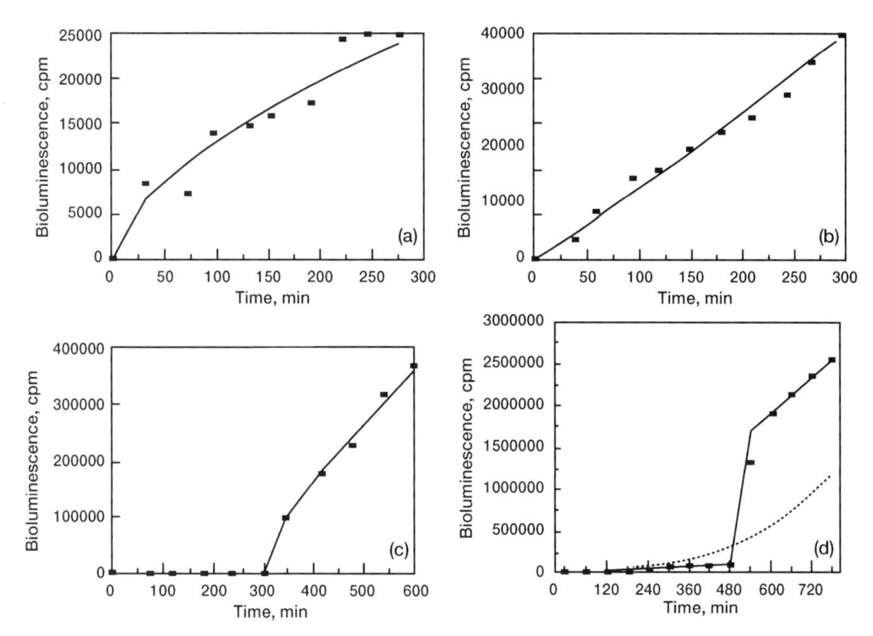

FIGURE 5.14 Binding of m-xylene-saturated STE buffer solution to the immobilized microorganism immobilized on the fiber-optic tip (Ikariyama *et al.*, 1997), and covered with (a) polycarbonate membrane and (b) dialysis membrane. Influence of the absence (c) and presence (d) of 1 mM methylbenzyl alcohol on the binding of m-xylene in solution to cell suspension immobilized on the fiber-optic tip with a polycarbonate membrane. Luciferin added after 2 h of luciferase induction. For (d): --- single-fractal analysis; —, dual-fractal analysis.

8—from a value of 119.82 to 959.58. This indicates that the binding rate coefficient is rather sensitive to the fractal dimension or the degree of heterogeneity that exists on the biosensor surface.

Figures 5.14c and 5.14d show the influence of the absence and presence of 1 mM methylbenzyl alcohol on the binding kinetics using a polycarbonate membrane. Ikariyama *et al.* wanted to analyze the influence of induction time on luminescence intensity. Two hours after luciferase induction, luciferin was added. The authors noted that few ppm of methylbenzyl alcohol could be detected in an hour. Table 5.5 shows the values of the binding rate coefficients and the fractal dimensions in the absence and in the presence of 1 mM methylbenzyl alcohol. In the absence of methylbenzyl alcohol, there was no detectable luminescence for about 300 min. After that time period, the binding kinetics could be described by a single-fractal analysis. The values of the binding rate coefficient, k, and the fractal dimension, D_f, are presented in Table 5.5. In the presence of 1 mM methylbenzylalcohol (Fig. 5.14d) a dual-fractal analysis clearly provides a better fit. The parameter values for both analyses are presented in Table 5.5.

It would be of interest to determine the influence of the fractal dimension (or the degree of heterogeneity that exists on the surface) on the binding rate coefficient. However, not enough data is available for a particular set of conditions. In lieu of that, Figure 5.15 plots values of the binding rate coefficient as a function of the fractal dimension for two different sets of conditions. Two points are taken when a single-fractal analysis was applicable. One point is taken when a dual-fractal analysis was applicable. For this case, the first set of parameter values (k_1 and D_{f_1}) are plotted as k and D_f, respectively. Because of this, the result that follows should be viewed with caution. Nevertheless, the binding rate coefficient is given by

$$k = (114.04 \pm 43.58) D_f^{3.314 \pm 0.692}. \tag{5.19}$$

Figure 5.15 indicates that this predictive equation is very reasonable, considering that data were plotted from two different sets of conditions and that the final analysis also includes both a single- and a dual-fractal analysis. However, the predictive equation does indicate that the binding rate coefficient is very sensitive to the surface roughness or the degree of heterogeneity that exists on the biosensor surface. This is because of the high exponent dependence of the binding rate coefficient on the fractal dimension.

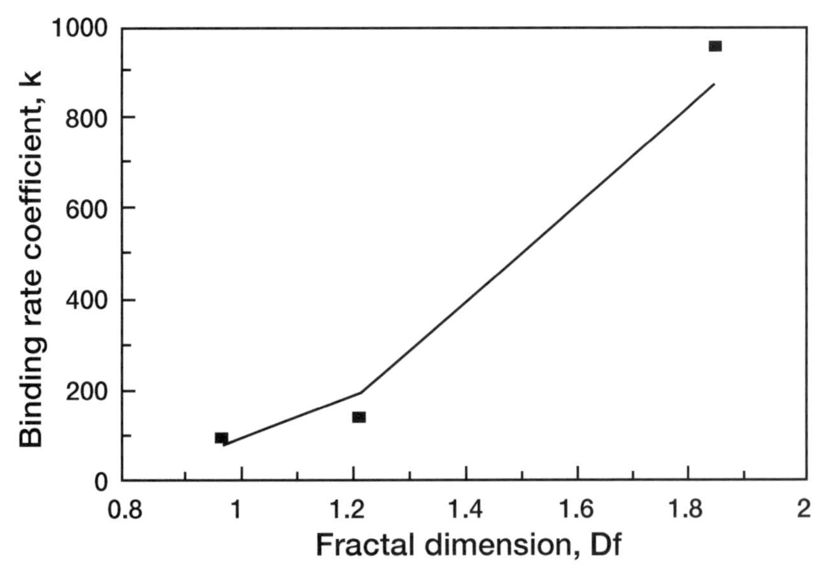

FIGURE 5.15 Influence of the fractal dimension, D_f, on the binding rate coefficient, k.

Loomans *et al.* (1997) have monitored peptide antibody binding using reflectometry. They were able to obtain the association and dissociation constants of the binding reaction between the antibody in solution and the immobilized antigen. The curve obtained using Eq. (5.1) for the binding between 100 μg/ml mouse antihuman chorionic gonadotrophin (hCG) monoclonal antibody OT-3A in solution to a derivative of peptide 3A (K-7 peptide) physically adsorbed to the surface is shown in Fig. 5.16a. A single-fractal analysis is adequate to describe the binding kinetics. The values of the binding coefficient, k, and the fractal dimension, D_f, are given in Table 5.6. Similarly, Fig. 5.16b shows the binding curve obtained between 100 μg/ml mouse anti-hCG monoclonal antibody OT-3A in solution to a derivative of peptide 3A (H-peptide) physically adsorbed to a surface. Once again, a single-fractal analysis is sufficient to adequately describe the binding kinetics. The values of k and D_f are given in Table 5.6.

Note that in both cases the fractal dimension is almost the same. D_f is equal to 2.5512 and 2.5974 for the K-7 peptide and the H-peptide, respectively. However, there is a significant difference in the binding rate coefficient for these two cases. The binding rate coefficient is larger by a factor of 2.28 for the K-7 peptide ($k = 1.6233$) when compared with the corresponding coefficient for the H-pepide ($k = 0.7103$). This indicates that the OT-3A–K-7 peptide binding reaction is more sensitive to the degree of heterogeneity (D_f) on the reflectometer surface than is the OT-3A–H peptide binding reaction. Also, since these two binding reactions are adequately described by a single-fractal analysis, one can say (with caution) that the binding mechanisms for both cases are similar. Note also (as previously mentioned) that the binding reaction of the antibody OT-3A in solution to these two peptides physically adsorbed to a surface leads to values of the fractal dimension (the degree of heterogeneity) that are within 1.8% of each other.

Figure 5.16c shows the binding curves obtained using a single-fractal analysis [Eq. (5.1)] and a dual-fractal analysis [Eq. (5.2)] for 100 μg/ml OT-3A in solution to Ata-peptide (a derivative of peptide 3A) physically adsorbed on a reflectometer surface. In this case, a single-fractal analysis does not provide an adequate fit, and thus a dual-fractal analysis is used. Table 5.6 shows the values of k and D_f for a single-fractal analysis and the values of $k_1$and k_2 and D_{f_1} and D_{f_2} for a dual-fractal analysis. Clearly, the dual-fractal analysis provides a much better fit. For the dual-fractal analysis, note that as the fractal dimension increases by about 48%—from $D_{f_1} = 2.0782$ to $D_{f_2} = 2.9264$—the binding rate coefficient increases by a factor of 4.36—from $k_1 = 0.3790$ to $k_2 = 1.6524$. Thus, the binding rate coefficient is quite sensitive to the degree of heterogeneity that exists on the reflectometer surface. Also, the changes in the fractal dimension and the binding rate coefficient observed in this case are in the same direction. It would be valuable

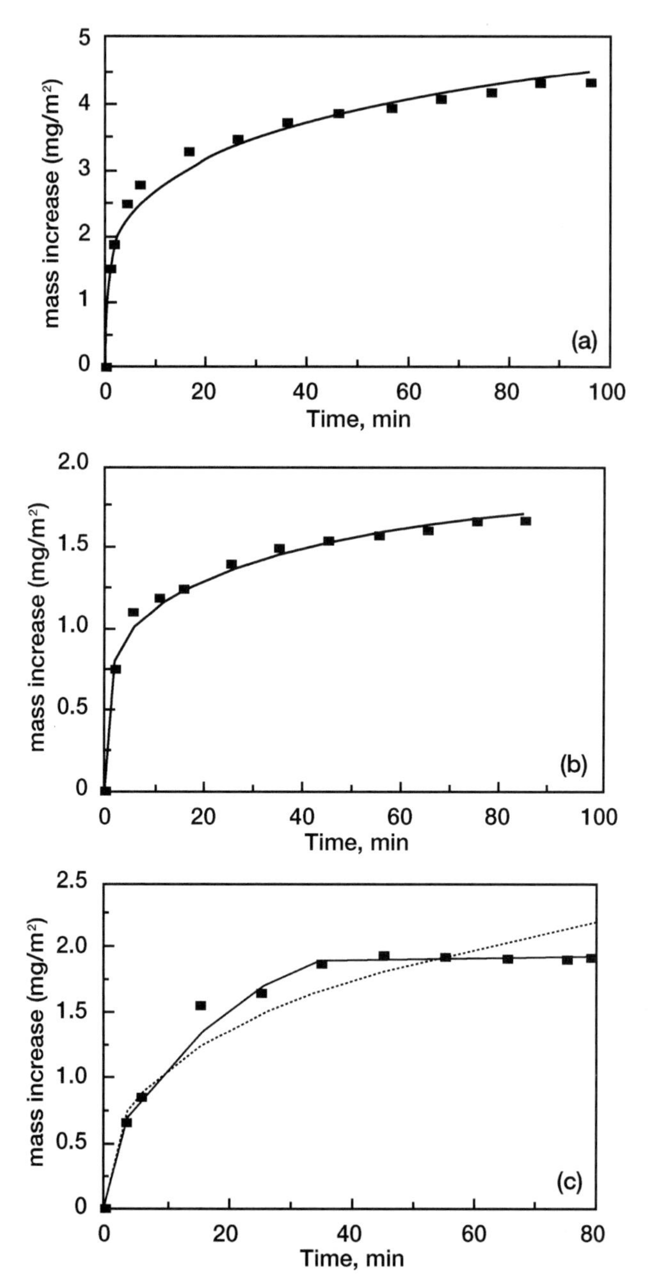

FIGURE 5.16 Binding of OT-3A antibody in solution to different physically adsorbed derivatives of peptide 3A by reflectometry (Loomans *et al.*, 1997). (a) K-7 peptide; (b) H-peptide; (c) Ata-peptide (- - -, single-fractal analysis; —, dual-fractal analysis).

TABLE 5.6 Influence of Different Peptides Adsorbed on a Surface on Fractal Dimensions and Binding Rate Coefficients: Single- and Dual-Fractal Analysis (Loomans *et al.*, 1997)

Antibody in solution/peptide on surface	k	D_f	k_1	k_2	D_{f_1}	D_{f_2}
OT-3A (100 μg/ml) in PBS/physically adsorbed derivative of peptide 3A (K-7 peptide)	1.623 ± 0.123	2.551 ± 0.028	na	na	na	na
OT-3A (100 μg/ml) in PBS/physically adsorbed derivative of peptide 3A (H-peptide)	0.710 ± 0.032	2.598 ± 0.023	na	na	na	na
OT-3A (100 μg/ml) in PBS/physically adsorbed derivative of peptide 3A (Ata-peptide)	0.487 ± 0.065	2.314 ± 0.076	0.379 ± 0.036	1.652 ± 0.018	2.078 ± 0.092	2.926 ± 0.023

to have some structural basis for this change in the binding mechanism of OT-3A to either the K-7 or the H-peptide (single-fractal mechanism applicable). However, no such structural basis or other possible explanation is offered at present.

5.3.3. Effect of Regeneration

Mauro *et al.* (1996) utilized fluorometric sensing to detect polymerase chain-reaction–amplified DNA using a DNA capture protein immobilized on a fiber-optic biosensor. The authors used amplified DNA labeled with the fluorophore tetramethylrhodamine and the AP-1 consensus nucleotide sequence recognized by GCN4. This DNA was noncovalently bound to IgG-modified fibers. Wanting to see if they could reuse the fiber, the authors performed regeneration studies. They focused their attention on conditions that would permit the release of the bound DNA while leaving the IgG-PG–GCN4 assembly in a functional state. Figure 5.17 shows the curves obtained using Eqs. (5.1) (single-fractal analysis) and (5.2) (dual-fractal analysis) for ten consecutive runs. The points are the experimental results obtained by Mauro *et al.* (1996). A dual-fractal analysis was required since the single-fractal analysis did not provide an adequate fit for the binding curves.

Table 5.7 shows the values of the binding rate coefficient, k, and the fractal dimension, D_f, obtained using Sigmaplot (1993) to fit the data. Since a dual-fractal analysis was used to model the binding curves, the results obtained from the single-fractal analysis will not be analyzed further. The D_{f_1} values reported for each of the ten runs were all equal to zero. This is due to the sigmoidal shape or concave nature of the curve (toward the origin) at very low values of time, t.

Figures 5.18a and 5.18b show the fluctuations in the binding rate coefficient, k_2, and in the fractal dimension, D_{f_2}, respectively, as the run number increases from one to ten. No pattern is easily discernible from the data presented in Figs. 5.18a and 5.18b. Table 5.7 and Fig. 5.18c indicate that an increase in D_{f_2} leads to a linear increase in k_2, but there is scatter in the data. An increase in D_{f_2} by about 16.9%—from a value of 1.9612 to 2.2938—leads to a 85.4% increase in k_2—from a value of 91.122 to 169. For the regeneration runs, k_2 may be given by

$$k_2 = (8.229 \pm 1.096) D_{f_2}^{3.3997 \pm 0.9319}. \tag{5.20}$$

Equation (5.20) predicts the k_2 values presented in Table 5.7 reasonably well. There is some deviation in the data. Note the high exponent dependence of k_2 on D_{f_2}. This underscores that k_2 is sensitive to the surface roughness or the

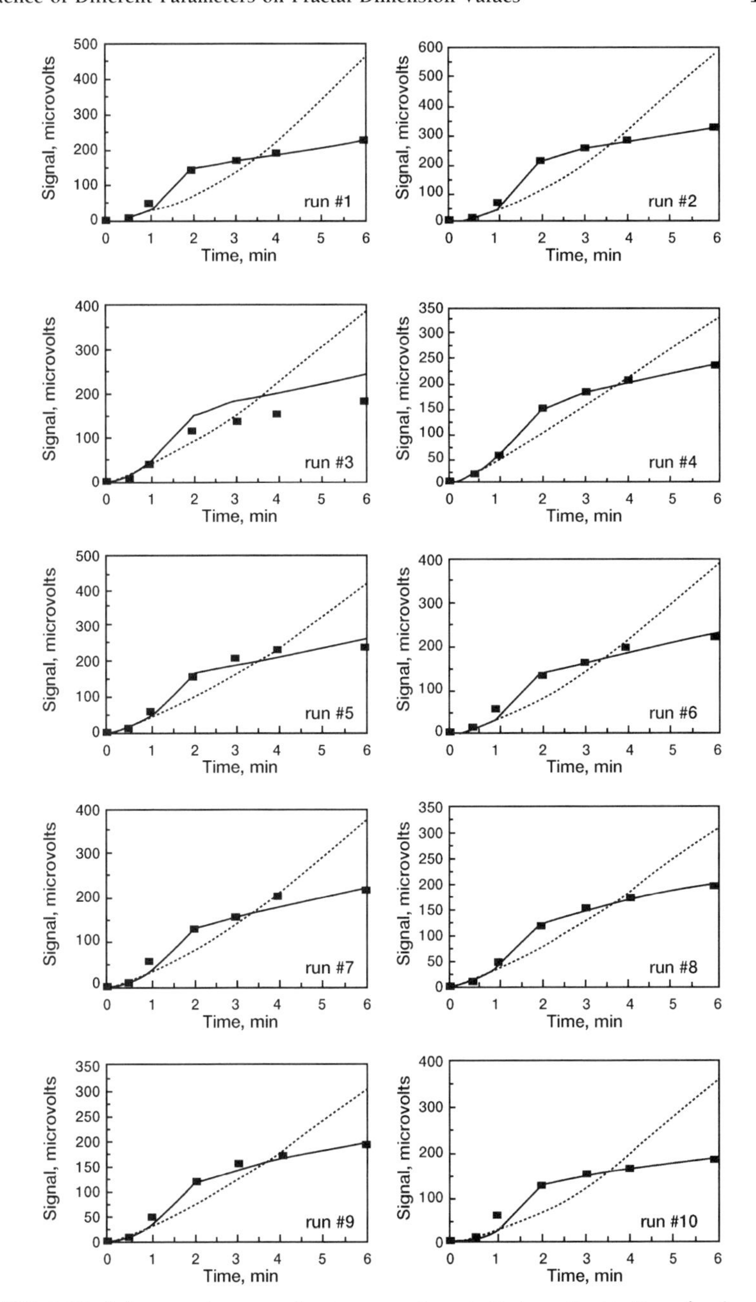

FIGURE 5.17 Influence of run number (regeneration studies) on the binding of polymerase chain reaction–amplified DNA in solution using a DNA capture protein immobilized on a fiber-optic biosensor (Mauro *et al.*, 1996). (- - -, single-fractal analysis; —, dual-fractal analysis).

TABLE 5.7 Influence of Run Number (Regeneration Cycle) on the Fractal Dimensions and on the Binding Rate Coefficients for Amplified DNA in Solution and a DNA Capture Protein Immobilized on a Fiber-Optic Biosensor (Mauro *et al.*, 1996)

Run	k	D_f	k_1	k_2	D_{f_1}	D_{f_2}
1	19.1 ± 26.4	0	$23.6 + 29.8$	108 ± 1.83	0	2.16 ± 0.042
2	40.3 ± 36.3	$0.054 + 0.62$	46.9 ± 29.6	169 ± 5.27	0	2.29 ± 0.076
3	35.9 ± 24.5	$0.357 + 0.50$	40.7 ± 19.5	115 ± 3.89	0	2.18 ± 0.08
4	48.2 ± 18.0	$0.848 + 0.31$	51.6 ± 2.90	116 ± 4.27	0	2.19 ± 0.09
5	38.9 ± 25.1	$0.330 + 0.48$	43.2 ± 16.2	128 ± 11.2	0	2.17 ± 0.21
6	29.5 ± 27.6	$0.142 + 0.64$	$34.4 + 34.9$	99.0 ± 4.9	0	2.06 ± 0.12
7	33.3 ± 31.1	0	33.2 ± 31.1	91.4 ± 6.5	0	1.96 ± 0.17
8	32.6 ± 29.4	0.474 ± 0.45	36.1 ± 16.4	91.1 ± 5.6	0	2.08 ± 0.15
9	33.1 ± 30.6	0.495 ± 0.46	36.8 ± 19.4	91.1 ± 5.4	0	2.08 ± 0.14
10	$23.4 + 29.5$	$0 + 0.791$	$28.3 + 39.9$	102 ± 3.8	0	2.23 ± 0.09

degree of heterogeneity that exists on the surface. No theoretical explanation is offered at present to explain the high exponent that occurs in the k–D_f correlation. (There are no suitable references available in the literature that mention this aspect.) There is an initial degree of heterogeneity that exists on the surface, and this determines the value of k. It is this heterogeneity that leads to the temporal nature of the binding rate coefficient. For a single-fractal analysis, it is assumed that this degree of heterogeneity remains constant during the reaction, exhibiting a single D_f value. When a dual-fractal analysis applies, there are two degrees of heterogeneity present in the reaction at different time intervals. These two different degrees of heterogeneity, D_{f_1} and D_{f_2}, lead to two different values of the binding rate coefficient, k_1 and k_2, respectively.

Abel *et al.* (1996) utilized a fiber-optic biosensor for the detection of 16-mer oligonucleotides in hybridization assays. The authors immobilized a biotinylated capture probe on the biosensor surface using either avidin or streptavidin. Fluorescence was utilized to monitor the hybridization with fluorescein-labeled complementary strands. These authors indicate that the capability of DNA and RNA fragments to bind selectively to complementary arranged nucleotides at other nucleic acids is the basis for *in vitro* tests. Therefore, hybridization methods that use nucleic acids as the biological recognition element may be utilized as an effective immunoassay.

Figure 5.19 shows the binding of 16^*CFl (complementary oligonuclotide) in a 10 nM solution to 16^*B (immobilized oligonucleoide) immobilized via sulfosuccinimidyl-6-(biotinamido)hexanoate (NHS-LC-biotin) and streptavidin to a biosensor. Abel *et al.* used both chemical and thermal regeneration. Figures 5.19a and 5.19b show the curves obtained using Eq. (5.1) for the

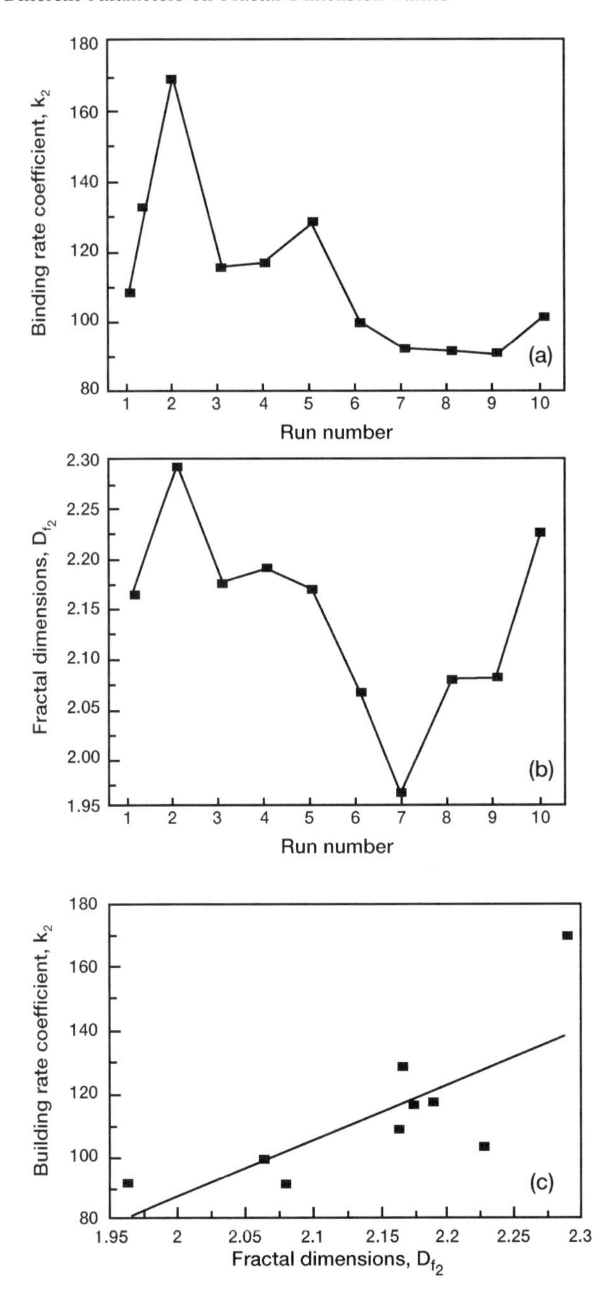

FIGURE 5.18 Influence of the run number on (a) the binding rate coefficient, k_2 and (b) the fractal dimension, D_{f_2}, during regeneration studies. (c) Increase in the binding rate coefficient, k_2, with an increase in the fractal dimension, D_{f_2}, during regeneration (Mauro *et al.*, 1996).

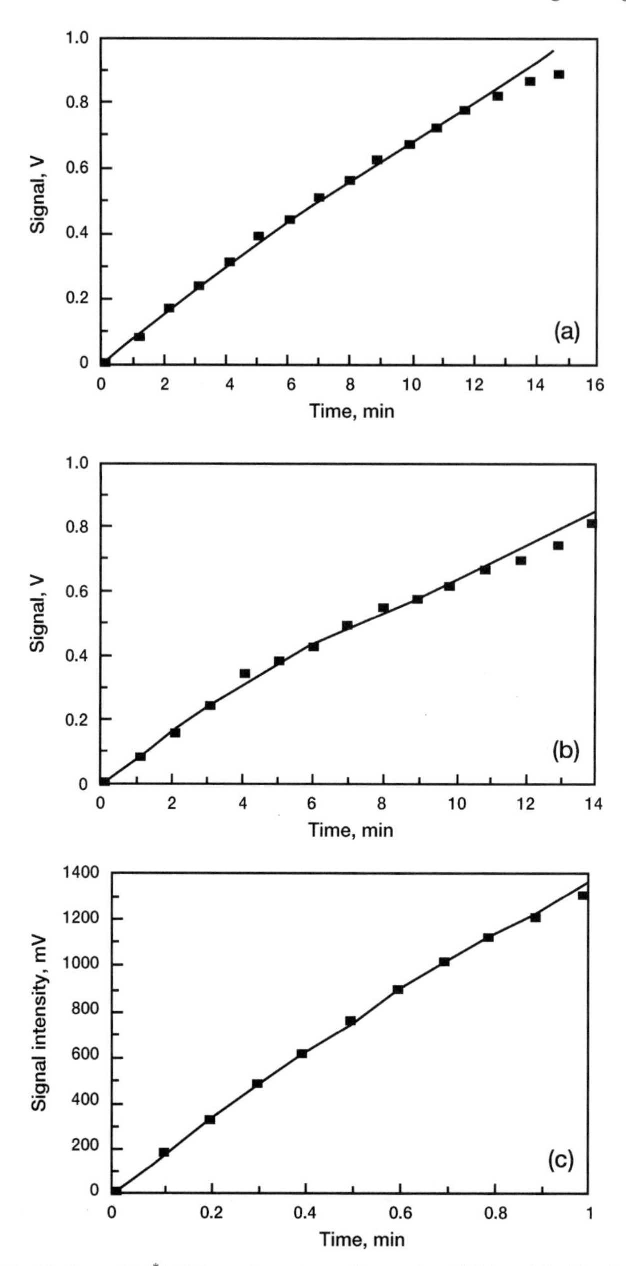

FIGURE 5.19 Binding of 16*CFl (complementary oligonucleotide) in a 10-nM solution to 16*B (immobilized oligonucleotide) immobilized via sulfosuccinimidyl-6-(biotinamido)hexanoate (NHS-LC-biotin) and streptavidin to a biosensor. (a) No regeneration; (b) chemical regeneration; (c) thermal regeneration.

TABLE 5.8 Influence of Chemical and Thermal Regeneration on the Fractal Dimensions and Binding Rate Coefficients for the Binding of 16^*CFl in Solution to 16^*B Immobilized on a Biosensor Surface (Abel *et al.*, 1996)

Analyte in solution/receptor on surface	k	D_f
10 nM 16^*CFl/16^*B immobilized via NHS-LC-biotin and streptavidin	1338 ± 31.3	1.272 ± 0.021
10 nM 16^*CFl/16^*B immobilized via NHS-LC-biotin and streptavidin (chemical regeneration)	86.53 ± 3.21	1.211 ± 0.026
10 nM 16^*CFl/16^*B immobilized via NHS-LC-biotin and streptavidin (thermal regeneration)	100.0 ± 6.70	1.394 ± 0.047

binding of 16^*CFl in solution to 16^*B immobilized on the biosensor surface, using chemical and thermal regeneration, respectively. In this case, a single-fractal analysis is sufficient to adequately describe the binding kinetics. Table 5.8 shows the values of the binding rate coefficients and the fractal dimensions obtained in these cases. Note that the temperature of hybridization was 26.7°C, and for thermal regeneration the fiber-optic surface was heated to 68.5°C. Note that as one goes from the chemical to the thermal regeneration cycle, k increases by about 15.5%—from a value of 86.53 to 100—and D_f increases by about 13.4%— from a value of 1.2112 to 1.3942.

Abel *et al.* also analyzed the binding kinetics of 10-nM 16^*CFl in solution to 16^*B immobilized to a biosensor surface. The only difference between this case and the previous two cases is that for the immobilization step the authors used avidin instead of streptavidin. A single-fractal analysis was employed here also to model the binding kinetics (Fig. 5.19c). The values of k and D_f are given in Table 5.8. In this case, the binding rate coefficient is much higher than in the previous two cases. The fractal dimension value of 1.2722 is between the two values obtained in the previous two cases. In this case, the fractal dimension analysis provides a quantitative estimate of the degree of heterogeneity that exists on the biosensor surface for each of the DNA hybridization assays. More data is required to establish trends or predictive equations for the binding rate coefficient in terms of either the analyte concentration in solution or the "estimated" fractal dimension of the biosensor surface under reaction conditions.

5.4. SUMMARY AND CONCLUSIONS

A fractal analysis of the binding of antigen (or antibody or, in general, analyte) in solution to the antibody (or antigen or, in general, receptor) immobilized on the biosensor surface provides a quantitative indication of the state of the disorder (fractal dimension, D_f) and the binding rate coefficient, k, on the surface. The fractal dimension provides a quantitative measure of the degree of heterogeneity that exists on the surface for the antibody–antigen (or, in general, analyte-receptor) systems. In our discussion, we gave examples wherein either a single- or a dual-fractal analysis was required to adequately describe the binding kinetics. The dual-fractal analysis was used only when the single-fractal analysis did not provide an adequate fit. This was done by the regression analysis provided by Sigmaplot (1993). The examples analyzed (1) the role of analyte concentration, (2) the effect of different surfaces, (3) the influence of regeneration, and (4) the effect of flow rate on binding rate coefficients and fractal dimensions during analyte-receptor binding in different biosensor systems.

In accord with the prefactor analysis for fractal aggregates (Sorenson and Roberts, 1997), quantitative (predictive) expressions were developed for the binding rate coefficient, k, as a function of the fractal dimension, D_f, for a single-fractal analysis and for the binding rate coefficients, k_1 and k_2, as a function of the fractal dimensions, D_{f_1} and D_{f_2}, for a dual-fractal analysis. Predictive expressions were also developed for the binding rate coefficient and the fractal dimension as a function of the analyte (antigen or antibody) concentration in solution.

The fractal dimension, D_f, is not a classical independent variable such as analyte (antigen or antibody) concentration. Nevertheless, the expressions obtained for the binding rate coefficient for a single-fractal analysis as a function of the fractal dimension indicate the high sensitivity of the binding rate coefficient on the fractal dimension. This is clearly brought out by the high order and fractional dependence of the binding rate coefficient on the fractal dimension. For example, in the case of the binding of premixed samples of 5 mM ATP and different concentrations of GroEL (13 to 129 nM) to GroES immobilized on a Ni^{++}-NTA surface, the order of dependence of the binding rate coefficient, k, on the fractal dimension, D_f, is 4.94. This emphasizes the importance of the extent of heterogeneity on the biosensor surface and its impact on the binding rate coefficient.

Note that the data analysis in itself does not provide any evidence for surface roughness or heterogeneity, and the existence of surface roughness or heterogeneity assumed may not be correct. Furthermore, there is deviation in the data that may be minimized by providing a correction for the depletion of the antigen (or antibody or analyte) in the vicinity of the surface (imperfect mixing) and by using a four-parameter equation. Other predictive expressions developed for the binding rate coefficient and for the fractal dimension as a function of the analyte (antigen or antibody) concentration in solution provide a means by which these parameters may be controlled.

In general, the binding rate coefficient increases as the fractal dimension increases. An increase in the binding rate coefficient value should lead to enhanced sensitivity and to a decrease in the response time of the biosensor. Both of these aspects would be beneficial in biosensor construction. If for a selective (or multiple) reaction system, an increase in the D_f value leads to an increase in the binding rate coefficient (of interest), this would enhance selectivity. Stability is a more complex issue, and one might intuitively anticipate that a distribution of heterogeneity of the receptor on the biosensor surface would lead to a more stable biosensor. This is especially true if the receptor has a tendency to inactivate or lose its binding capacity to the analyte in solution. Similar behavior has been observed for the deactivation of enzymes wherein a distribution of activation energies for deactivation (as compared to a single activation energy or deactivation) leads to a more stable enzyme (Malhotra and Sadana, 1987).

Whenever a distribution exists, it should be precisely determined, especially if different distributions are known to exist (Malhotra and Sadana, 1990). This would help characterize the distribution present on the surface and would influence the temporal nature of the binding rate coefficient on the surface. The present analysis only makes quantitative the extent of heterogeneity that exists on the surface, with no attempt at determining the qualitative nature of the distribution that exists on the surface. Much more detailed and precise data are required before any such attempt may be made. Finally, another parameter that is only rarely considered in the biosensor literature is robustness. This may be defined as insensitivity to measurement errors as far as biosensor performance is concerned. At this point, it is difficult to see how the binding rate coefficient and the fractal dimension would affect robustness.

REFERENCES

Abel, A. P., Weller, M. G., Duvencek, G. L., Ehrat, M., and Widmer, H. M., *Anal. Chem.* **68**, 2905–2912 (1996).

Bluestein, R. C., Diaco, R., Hutson, D. D., Miller, W. K., Neelkantan, N. V., Pankratz, T. J., Tseng, S. Y., and Vickery, E. K., *Clin. Chem.* **33**(9) 1543–1547 (1987).

Christensen, L. L. H., *Anal. Biochem.* **249**, 153–164 (1997).

Dabrowski, A. and Jaroniec, M., *J. Colloid & Interface Sci.* **73**, 475–482 (1979).

Eddowes, M. J., *Biosensors* **3**, 1–15 (1987/1988).

Edwards, P. R., Gill, A., Pollard-Knight, D. V., Hoare, M., Bucke, P. E., Lowe, P. A., and Leatherbarrow, R. J., *Biotechniques* **231**, 210 (1995).
Elbicki, J. M., Morgan, D. M., and Weber, S. G., *Anal. Chem.* **56**, 210 (1984).
Fagerstam, L. G., Frostell-Karlsson, A., Karlsson, R., Persson, B., and Ronnberg, I., *J. Chromatogr.* **597**, 397 (1997).
Fischer, R. J., Fivash, M., Casas-Finet, J., Bladen, S., and McNitt, K. L., *Methods* **6**, 121–133 (1994).
Giaver, I., *Immunol.* **116**, 766–771 (1976).
Glaser, R. W., *Anal. Biochem.* **213**, 152–158 (1993).
Havlin, S., in *The Fractal Approach to Heterogeneous Chemistry: Surfaces, Colloids, and Polymers* (D. Avnir, ed.), Wiley, New York, pp. 251–269, 1989.
Ikariyama, Y., Nishiguchi, S., Koyama, T., Kobatake, E., Aizawa, M., and Tsuda, M., *Anal. Chem.* **69**, 2600–2605 (1997).
Jaroniec, M. and Derylo, A., *Chem. Eng. Sci.* **36**, 1017–1019 (1981).
Jonsson, U., Fagerstam, L., Ivarsson, B., Johnsson, B., Karlsson, R., Lundh, K., Lofas, S., Persson, B., Roos, H., and Ronnberg, I., *Biotechniques* **11**, 620 (1991).
Koga, H., Rauchfuss, B., and Gunsalus, I. C., *Biochem. Biophys. Res. Commun.* **130**, 412–417 (1985).
Kopelman, R., *Science* **241**, 1620–1626 (1988).
Le Brecque, M., *Mosaic* **23**, 12–15 (1992).
Lindholm, J., Magnusson, K., Karlson, A. F., and Pluckhun, A., *Anal. Biochem.* **252**, 217–228 (1997).
Loomans, E. E. M. G., Beumer, T. A. M., Damen, K. C. S., Bakker, M. A., and Schielen, W. J. G., *J. Colloid & Interface Sci.*, **192**, 238–249 (1997).
Malhotra, A. and Sadana, A., *Biotechnol. Bioengg.* **30**, 108–112 (1987).
Malhotra, A. and Sadana, A., *J. Theor. Biol.* **145**, 143–152 (1990).
Matsuda, H., *J. Electroanal. Chem.* **179**, 107 (1976).
Mauro, J. M., Cao, L. K., Kondracki, L. M., Walz, S. E., and Campbell, J. R., *Analy. Biochem.* **235**, 61–72 (1996).
Milum, J. and Sadana, A., *J. Colloid & Interface Sci.* **187**, 128–138 (1997).
Morton, T. A., Myszka, D. G., and Chaiken, I. M., *Anal. Biochem.* **227**, 176 (1995).
Myszka, D. G., Morton, T. A., Doyle, M. L., and Chaiken, I. M., *Biophys. Chem.* **64**, 127–137 (1997).
Nath, N. Jain, S. R., and Anand, S., *Biosen. & Bioelectr.* **12**(6), 491–498 (1997).
Nellen, P. M. and Lukosz, W., *Sens. Actuators B* **B1**(1–6), 592–596 (1991).
Nieba, L., Nieba-Axmann, S. E., Persson, A., Hamalainen, M., Edebratt, F., Hansson, A., Lindholm, J., Magnusson, K., Karlson, A. F., and Pluckhun, A., *Anal. Biochem.* **252**, 217–228 (1997).
Nygren, H. and Stenberg, M., *J. Colloid & Interface Sci.* **107**, 560–566 (1985).
Nygren, H. A. *J. Immunol. Methods* **80**, 15–24 (1993).
Oscik J., Dabrowski, A., Jaroniec, M., and Rudzinski, W., *J. Colloid & Interface Sci.* **56**, 403–412 (1976).
Pisarchick , M. L., Gesty, D., and Thompson, N. L., *Biophys. J.* **63**, 215–223 (1992).
Place, J. F., Sutherland, R. M., Riley, A., and Mangan, S., in *Biosensors with Fiber Optics.* (D. Wise and L. B. Wingard Jr. eds.), Humana Press, New York, pp. 253–291, 1991.
Ramakrishnan, A. and Sadana, A., *J. Colloid & Interface Sci.* **224**, 219–230 (2000).
Richalet-Secordel, P. M., Rauffer-Bruyere, N., Christensen, L. L. H., Ofenloch-Haehnle, B., Seidel, C., and Van Regenmortel, M. H. V., *Anal. Biochem.* **249**, 165 (1997).
Rudzinski, W., Lattar, L., Zajac, J., Wofram, E., and Paszli, J., *J. Colloid & Interface Sci.* **96**, 339–359 (1983).

Sadana, A., *J. Colloid & Interface Sci.* **190**, 232–240 (1997).
Sadana, A. and Beelaram, A., *Biosen. & Bioelectr.* **10**, 1567–1574 (1995).
Sadana, A., *Biosen. & Bioelectr.* **14**, 515–531 (1999).
Sadana, A. and Madagula, A., *Biosen. & Bioelectr.* **9**, 43–55 (1994).
Sadana, A. and Sii, D., *J. Colloid & Interface Sci.* **151**(1), 166–177 (1992a).
Sadana, A. and Sii, D., *Biosen. & Bioelectr.* **7**, 559–568 (1992b).
Sadana, A. and Sutaria, M., *Biophys. Chem.* **65**, 29–44 (1997).
Sjolander, S. and Urbaniczky, C., *Anal. Chem.* **63**, 2338 (1991).
Sigmaplot, Scientific Graphing Software, Users's manual, Jandel Scientific, San Rafael, CA, 1993.
Sorenson, C. M. and Roberts, G. C., *J. Colloid & Interface Sci.* **186**, 447–452 (1997).
Stenberg, M. and Nygren, H. A., *Anal. Biochem.* **127**, 183–192 (1982).
Stenberg, M., Stiblert, L., and Nygren, H. A., *J. Theor. Biol.* **120**, 129–142 (1986).
Sutherland, R. M., Dahne, C., Place, J. F., and Ringrose, A. S., *Clin. Chem.* **30**, 1533–1538 (1984).
Viscek, T., *Fractal Growth Phenomena*, World Scientific Publishing, Pte, Singapore, 1989.
Williams, C. and Addona, T. A., *TIBTECH* **18**, 45 (2000).
Witten, T. A. and Sander, L. M., *Phys. Rev. Lett.* **47**, 1400 (1981).
Yen, K. M. and Gunsalus, I. C., *Proc. Natl. Acad. Sci.* **79**, 874–878 (1982).
Zhao, S. and Reichert, W. M., *Langmuir*, **8**, 2785 (1992).

CHAPTER 6

Fractal Dimension and the Binding Rate Coefficient

6.1. INTRODUCTION

Sensitive detection systems (or sensors) are required to distinguish a wide range of substances. Sensors find application in the areas of biotechnology, physics, chemistry, medicine, aviation, oceanography, and environmental control. These sensors, or biosensors, may be utilized to monitor the analyte-receptor reactions in real time (Myszka *et al.*, 1997). Scheller *et al.* (1991) have emphasized the importance of providing a better understanding of the mode of operation of biosensors to improve their sensitivity, stability, and specificity. A particular advantage of this method is that no reactant labeling is required. However, for the binding interaction to occur, one of the components has to be bound or immobilized onto a solid surface. This often leads to mass transfer limitations and subsequent complexities. Nevertheless, the solid-phase immunoassay technique represents a convenient method for the separation and/or detection of reactants (for example, antigen) in a solution since the binding of antigen to an antibody-coated surface (or vice versa) is sensed directly and rapidly. There is a need to characterize the reactions occurring at the biosensor surface in the presence of diffusional limitations that are inevitably present in these types of systems.

The details of the association of analyte (antibody or substrate) to a receptor (antigen or enzyme) immobilized on a surface is of tremendous significance for the development of immunodiagnostic devices as well as for biosensors (Pisarchick *et al.*, 1992). In essence, the analysis we will present is, in general, applicable to ligand-receptor and analyte-receptorless systems for biosensor and other applications (for example, membrane-surface reactions). External diffusional limitations play a role in the analysis of immunodiagnostic assays (Bluestein *et al.*, 1991; Eddowes, 1987/1988; Place *et al.*, 1991; Giaver *et al.*, 1976; Glaser, 1993; Fischer *et al.*, 1994). The influence of diffusion in such systems has been analyzed to some extent (Place *et al.*, 1991; Stenberg *et al.*, 1986; Nygren and Stenberg, 1985; Stenberg and Nygren, 1982; Morton *et al.*, 1995; Sadana and Sii, 1992a, 1992b; Sadana and Madagula, 1994; Sadana and Beelaram, 1995; Sjolander and Urbaniczky, 1991). The influence of partial (Christensen, 1997) and total (Matsuda, 1967; Elbicke *et al.*, 1984; Edwards *et al.*, 1995) mass transport limitations on analyte-receptor binding kinetics for biosensor applications is available. The analysis presented for partial mass transport limitation (Christensen, 1997) is applicable to simple one-to-one association as well as to the cases where there is heterogeneity of the analyte or the ligand. This applies to the different types of biosensors utilized for the detection of different analytes.

Chiu and Christpoulos (1996) emphasize that the strong and specific interaction of two complementary nucleic acid strands is the basis of hybridization assays. Syvanen *et al.* (1986) have analyzed the hybridization of nucleic acids by affinity-based hybrid collection. In their method, a probe pair is allowed to form hybrids with the nucleic acid in solution. They state that their procedure is quantitative and has a detection limit of 0.67 attamoles. Bier *et al.* (1997) analyzed the reversible binding of DNA oligonucleotides in solution to immobilized DNA targets using a grating coupler detector and surface plasmon resonance (SPR). These authors emphasize that the major fields of interest for hybridization analysis is for clinical diagnostics and for hygiene. The performance of these "genosensors" will be significantly enhanced if more physical insights are obtained into each step involved in the entire assay.

An optical technique that has gained increasing importance in recent years is the surface plasmon resonance (SPR) technique (Nylander *et al.*, 1991). This is particularly so due to the development and availability of the BIACORE biosensor, which is based on the SPR method and has found increasing industrial usage. Bowles *et al.* (1997) used the BIACORE biosensor to analyze the binding kinetics of Fab fragments of an antiparaquat antibody in solution to a paraquat analog (antigen) covalently attached at a sensor surface. Schmitt *et al.* (1997) also utilized a modified form of the BIACORE biosensor to analyze the binding of thrombin in solution to antithrombin covalently attached to a sensor surface. The performance of SPR and other biosensors will be enhanced as more physical insights are obtained into each of these analytical procedures.

Kopelman (1988) indicates that surface diffusion-controlled reactions that occur on clusters (or islands) are expected to exhibit anomalous and fractal-like kinetics. These fractal kinetics exhibit anomalous reaction orders and time-dependent rate (for example, binding) coefficients. As previously discussed, fractals are disordered systems in which the disorder is described by nonintegral dimensions (Pfeifer and Obert, 1989). These authors indicate that as long as surface irregularities show dilatational symmetry scale invariance such irregularities can be characterized by a single number, the fractal dimension. The fractal dimension is a global property and is insensitive to structural or morphological details (Pajkossy and Nyikos, 1989). Markel *et al.* (1991) indicate that fractals are scale self-similar mathematical objects that possess nontrivial geometrical properties. Furthermore, these authors indicate that rough surfaces, disordered layers on surfaces, and porous objects all possess fractal structure. A consequence of the fractal nature is a power-law dependence of a correlation function (in our case, the analyte-receptor complex on the surface) on a coordinate (for example, time).

Antibodies are heterogeneous, so their immobilization on a fiber-optic surface, for example, would exhibit some degree of heterogeneity. This is a good example of a disordered system, and a fractal analysis is appropriate for such systems. Furthermore, the antibody–antigen reaction on the surface is a good example of a low-dimension reaction system in which the distribution tends to be "less random" (Kopelman, 1988). A fractal analysis would provide novel physical insights into the diffusion-controlled reactions occurring at the surface.

Markel *et al.* (1991) indicate that fractals are widespread in nature. For example, dendrimers, a class of polymers with internal voids, possess unique properties. The stepwise buildup of six internal dendrimers into a dendrimer exhibits typical fractal (self-similar) characteristics (Gaillot *et al.*, 1997). Fractal kinetics also have been reported in biochemical reactions such as the gating of ion channels (Liebovitch and Sullivan, 1987; Liebovitch *et al.*, 1987), enzyme reactions (Li *et al.*, 1990), and protein dynamics (Dewey and Bann, 1992). Li *et al.* emphasize that the nonintegral dimensions of the Hill coefficient used to describe the allosteric effects of proteins and enzymes is a direct consequence of the fractal property of proteins.

Strong fluctuations in fractals have not been taken into account (Markel *et al.*, 1991). For example, strongly fluctuating fields bring about a great enhancement of Raman scattering from fractals. It would be beneficial to determine a fractal dimension for biosensor applications and to determine whether there is a change in the fractal dimension as the binding reaction proceeds on the biosensor surface. The final goal would be to determine how all of this affects the binding rate coefficient and subsequently biosensor performance. Fractal aggregate scaling relationships have been determined for both diffusion-limited processes and diffusion-limited scaling aggregation (DLCA) processes in spatial dimensions, 2, 3, 4, and 5, by Sorenson and Roberts (1997). Fractal dimension values for the kinetics of antigen–antibody binding (Sadana, 1997; Milum and Sadana, 1997) and for analyte-receptor binding (Sadana and Sutaria, 1997) for fiber-optic biosensor systems are available. In these studies the influence of the experimental parameters such as analyte concentration on the fractal dimension and on the binding rate coefficient (the prefactor in this case) were analyzed. We would like to delineate the role of surface roughness on the speed of response, specificity, stability, and sensitivity of fiber-optic and other biosensors. An initial attempt has been made to relate the influence of surface roughness (or fractal dimension) on the binding rate coefficient for fiber-optic and other biosensors (Sadana, 1998). High and fractional orders of dependence of the binding rate coefficient on the fractal dimension were obtained. We now extend these studies to other biosensor applications, including those where more than one fractal dimension is invloved at the biosensor surface—in other words, where complex binding mechanisms, as well as a change in the binding mechanism may be involved at the surface. Quantitative relationships for the binding rate coefficient as a function of the fractal dimension are obtained for different biosensor applications. The noninteger orders of dependence obtained for the binding rate coefficient on the fractal dimension further reinforces the fractal nature of these analyte-receptor binding systems.

6.2. THEORY

Milum and Sadana (1997) have analyzed the binding kinetics of antigen in solution to antibody immobilized on the biosensor surface. Sadana and Madagula (1993) have studied the influence of lateral interactions on the surface and variable rate coefficients. Here, we initially present a method of estimating actual fractal dimension values for analyte-receptor binding systems utilized in fiber-optic biosensors.

6.2.1. VARIABLE BINDING RATE COEFFICIENT

Kopelman (1988) has indicated that classical reaction kinetics is sometimes unsatisfactory when the reactants are spatially constrained on the microscopic level by either walls, phase boundaries, or force fields. Such heterogeneous reactions (for example, bioenzymatic reactions) that occur at interfaces of different phases exhibit fractal orders for elementary reactions and rate coefficients with temporal memories. In such reactions, the rate coefficient exhibits a form given by

$$k_1 = k' t^{-b}. \tag{6.1a}$$

In general, k_1 depends on time, whereas $k' = k_1$ $(t=1)$ does not. Kopelman indicates that in three dimensions (homogeneous space), $b=0$. This is in agreement with the results obtained in classical kinetics. Also, with vigorous stirring, the system is made homogeneous and, again, $b=0$. However, for diffusion-limited reactions occurring in fractal spaces, $b>0$; this yields a time-dependent rate coefficient.

The random fluctuations in a two-state process in ligand-binding kinetics has been analyzed (Di Cera, 1991). The stochastic approach can be used as a means to explain the variable binding rate coefficient. The simplest way to model these fluctuations is to assume that the binding rate coefficient, $k_1(t)$ is the sum of its deterministic value (invariant) and the fluctuation ($z(t)$). This $z(t)$ is a random function with a zero mean. The decreasing and increasing binding rate coefficients can be assumed to exhibit an exponential form (Cuypers *et al.*, 1987):

$$\begin{aligned} k_1 &= k_{1,0} \exp(-\beta t) \\ k_1 &= k_{1,0} \exp(\beta t). \end{aligned} \tag{6.1b}$$

Here, β and $k_{1,0}$ are constants.

Sadana and Madagula (1993) analyzed the influence of a decreasing and an increasing binding rate coefficient on the antigen concentration when the antibody is immobilized on the surface. These authors noted that for an increasing binding rate coefficient (after a brief time interval), as time increases, the concentration of the antigen near the surface decreases, as expected for the cases when lateral interactions are present or absent. The diffusion-limited binding kinetics of antigen (or antibody or substrate) in solution to antibody (or antigen or enzyme) immobilized on a biosensor surface has been analyzed within a fractal framework (Sadana, 1997; Milum and Sadana, 1997). Furthermore, experimental data presented for the binding of HIV virus (antigen) to the antibody anti-HIV immobilized on a surface displays a characteristic ordered "disorder" (Anderson, 1993). This indicates the possibility of a fractal-like surface. It is obvious that such a biosensor system (wherein either the antigen or the antibody is attached to the surface) along with its different complexities (heterogeneities on the surface and in solution, diffusion-coupled reaction, time-varying adsorption or binding rate coefficients, etc.) can be characterized as a fractal system. Sadana (1995) has analyzed the diffusion of reactants toward fractal surfaces and earlier Havlin (1989) has briefly reviewed and discussed these results.

Single-Fractal Analysis

Havlin (1989) indicates that the diffusion of a particle (antibody [Ab]) from a homogeneous solution to a solid surface (for example, antigen [Ag]-coated surface) where it reacts to form a product (antibody–antigen complex, $Ab \cdot Ag$) is given by

$$(Ab \cdot Ag) \sim \begin{cases} t^{(3-D_f)/2} & t \leq t_c \\ t^{1/2} & t > t_c \end{cases}. \tag{6.2a}$$

Here, D_f is the fractal dimension of the surface. Equation (6.2a) indicates that the concentration of the product $Ab \cdot Ag(t)$ in a reaction $Ab + Ag \rightarrow Ab \cdot Ag$ on a solid fractal surface scales at short and intermediate time frames as $Ab \cdot Ag \sim t^p$, with the coefficient $p = (3 - D_f)/2$ at short time frames and $p = \frac{1}{2}$ at intermediate time frames. This equation is associated with the short-term diffusional properties of a random walk on a fractal surface. Note that in a perfectly stirred kinetics on a regular (nonfractal) structure (or surface), k_1 is a constant; that is, it is independent of time. In other words, the limit of regular structures (or surfaces) and the absence of diffusion-limited kinetics leads to k_1 being independent of time. In all other situations, one would expect a scaling behavior given by $k_1 \sim k't^{-b}$ with $-b = p < 0$. Also, the appearance of the coefficient p different from $p = 0$ is the consequence of two different phenomena—the heterogeneity (fractality) of the surface and the imperfect mixing (diffusion-limited) condition.

Havlin indicates that the crossover value may be determined by $r_c^2 \sim t_c$. Above the characteristic length r_c, the self-similarity is lost. Above t_c, the surface may be considered homogeneous since the self-similarity property disappears and regular diffusion is now present. For the present analysis, we chose t_c arbitrarily. One may consider our analysis as an intermediate "heuristic" approach in that in the future one may also be able to develop an autonomous (and not time-dependent) model of diffusion-controlled kinetics.

Dual-Fractal Analysis

We can extend the preceding single-fractal analysis to include two fractal dimensions. At present, the time $(t = t_1)$ at which the first fractal dimension "changes" to the second fractal dimension is arbitrary and empirical. For the most part, it is dictated by the data analyzed and the experience gained by handling a single-fractal analysis. A smoother curve is obtained in the transition region if care is taken to select the correct number of points for the two regions. In this case, the product (antibody–antigen complex, $Ab \cdot Ag$)

concentration on the biosensor surface is given by

$$(\mathrm{Ab}\cdot\mathrm{Ag}) \sim \begin{cases} t^{(3-D_{f_1})/2} = t^{p_1} & t \leq t_1 \\ t^{(3-D_{f_2})/2} = t^{p_2} & t_1 < t \leq t_2 = t_c \\ t^{1/2} & t > t_c. \end{cases} \tag{6.2b}$$

6.3. RESULTS

In this discussion, a fractal analysis will be applied to data obtained for analyte-receptor binding data with different types of biosensors. This is one possible explanation for analyzing the diffusion-limited binding kinetics assumed to be present in all of the systems analyzed. The parameters thus obtained would provide a useful comparison of different situations. Alternate expressions involving saturation, first-order reaction, and no diffusion limitations are possible but seem to be deficient in describing the heterogeneity that inherently exists on the surface. The analyte-receptor binding reaction on the different types of biosensors analyzed is a complex reaction, and the fractal analysis via the fractal dimension and the binding rate coefficient provide a useful lumped-parameter analysis of the diffusion-limited situation.

6.3.1. Single-Fractal Analysis

Hirmo *et al.* (1998) have recently characterized *Helicobacter pylori* strains using sialic acid binding to a resonant mirror biosensor. These authors indicate that *Helicobacter pylori* is a gastric pathogen that causes type B gastritis and duodenal ulcer disease. In addition, *H. pylori* posseses a variety of cell-surface proteins (Clyne and Drumm, 1996). The binding of the cell-surface proteins (ligand-receptor interaction) using sialic acid binding specific for α-2,3-sialyllactose was characterized using this new optical biosensor technique. As the molecules bind to the sensing surface, there is a change in the refractive index. This results in a shift in the resonant angle (Cush *et al.*, 1993). Hirmo *et al.* emphasize that the advantage of their technique is that it is label-free and real-time monitoring of biomolecular events is possible.

Figure 6.1 shows the curves obtained using Eq. (6.2a) for the binding of hemagglutinating and poorly hemagglutinating *H. pylori* cells to sialyl (α-2,3) lactose-conjugated (20 mol%) polyacrylamide (3′SL-PAA, MW ∼ 30 kDa) immobilized on a resonant mirror biosensor (RMB) using bacterial cell suspensions. In these two cases, a single-fractal analysis is adequate to

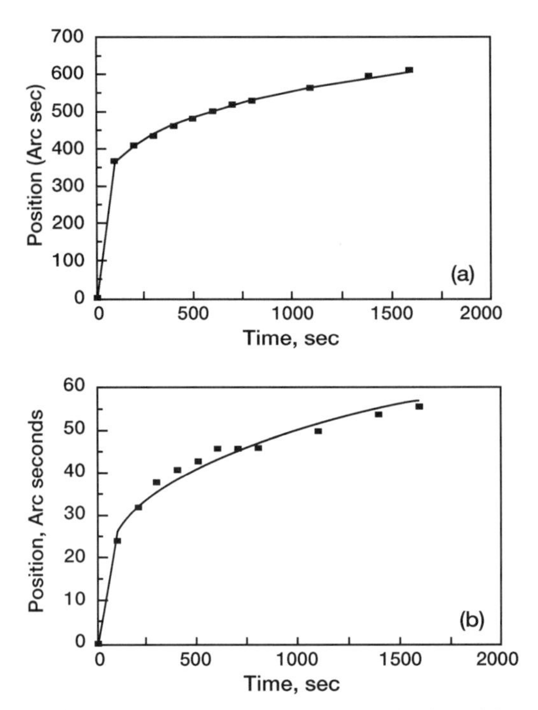

FIGURE 6.1 Binding of hemagglutinating *H. pylori* cells to sialyl (α-2,3) lactose-conjugated (20 mol%) polyacrylamide (3′SL-PAA) immobilized on a resonant mirror biosensor (Hirmo *et al.*, 1998). (a) Hemagglutinating; (b) poorly hemagglutinating.

describe the binding kinetics. Table 6.1a shows the values of the binding rate coefficient, k, and the fractal dimension, D_f. The values of k presented in Table 6.1 were obtained from a regression analysis using Sigmaplot (1993) to model the experimental data using Eq. (6.2a), wherein $(\text{Ab} \cdot \text{Ag}) = kt^p$. The k and D_f values presented in Table 6.1 are within 95% confidence limits. For example, for the binding of hemaglluttinating *H. pylori* cells in solution to silayl (α-2,3) lactose immobilized on the RMB, the value of k reported is 150.022 ± 1.432. The 95% confidence limits indicates that 95% of the k values will lie between 148.590 and 151.454. This indicates that the Table 6.1 values are precise and significant. The curves presented in the figures are theoretical curves. Note that as one goes from the poorly hemagglutinating to hemagglutinating cells there is about a 7.8% increase in D_f—from 2.4298 to 2.620, and an increase in k by a factor of 21.3—from 7.043 to 150.0. Note that the change in D_f and k are in the same direction. These two results indicate that k is rather sensitive to the fractal dimension or the degree of heterogeneity that exists on the surface.

TABLE 6.1 Influence of Different Parameters on Fractal Dimensions and Binding Rate Coefficients for the Binding of *Helicobacter pylori* to Sialylglycoconjugates Using a Resonant Mirror Biosensor (Hirmo *et al.*, 1998)

Analyte concentration in solution/receptor on surface	Binding rate coefficient, k	Fractal dimension, D_f
(a) Hemagglutinating *H. pylori* cells/immobilized sialyl ($\alpha - 2,3$) lactose-conjugated (20 mol%) polyacrylamide (3′SL-PAA, MW ~ 30 Da) on a resonant mirror biosensor (RMB)	150.022 ± 1.432	2.620 ± 0.00712
Poorly hemagglutinating *H. pylori* cells/immobilized 3′SL-PAA on RMB	7.0433 ± 0.3579	2.4298 ± 0.03720
(b) 250 μg/ml cell-surface proteins extracted from *H. pylori* strain 52/immobilized 3′SL-PAA on RMB	7.4460 ± 0.0425	2.2424 ± 0.00695
250 μg/ml cell-surface proteins extracted from *H. pylori* strain 33/immobilized 3′SL-PAA on RMB	51.274 ± 1.016	2.8501 ± 0.0152
(c) 250 μg/ml cell-surface proteins from *H. pylori* strain 52 (in the absence of free sialyl (α-2,3)lactose/immobilized 3′SL-PAA on RMB	82.995 ± 2.699	2.4836 ± 0.0204
250 μg/ml cell-surface proteins from *H. pylori* strain 52 in the presence of 5 mM free sialyl (α-2,3) lactose/immobilized 3′SL-PAA on RMB	178.36 ± 6.31	2.6356 ± 0.0202
250 μg/ml cell-surface proteins from *H. pylori* strain 52 in the presence of 10 mM free sialyl (α-2,3) lactose/immobilized 3′SL-PAA on RMB	193.438 ± 4.655	2.6152 ± 0.0152
(d) Two-times diluted *H. pylori* cells of strain 52/immobilized 3′SL-PAA on RMB	0.1774 ± 0.0109	1.2240 ± 0.0430
Nondiluted *H. pylori* cells of strain 52/immobilized 3′SL-PAA on RMB	2.2641 ± 0.1078	1.7980 ± 0.0226
Two-times concentrated *H. pylori* cells of strain 52/immobilized 3′SL-PAA on RMB	8.0477 ± 0.6249	1.9884 ± 0.0352
Five-times concentrated *H. pylori* cells of strain 52/immobilized 3′SL-PAA on RMB	163.486 ± 2.822	2.6484 ± 0.0080

Figures 6.2a and 6.2b show the binding curves obtained using Eq. (6.2a) for the binding of 250 μg/ml cell-surface proteins extracted from *H. pylori* strain 52 and 33 in solution, respectively, to 3′SL-PAA immobilized on an RMB. Once again, a single-fractal analysis is adequate to describe the binding kinetics. Table 6.1b shows the values of k and D_f. Once again, note that higher values of the fractal dimension and the binding rate coefficient are exhibited

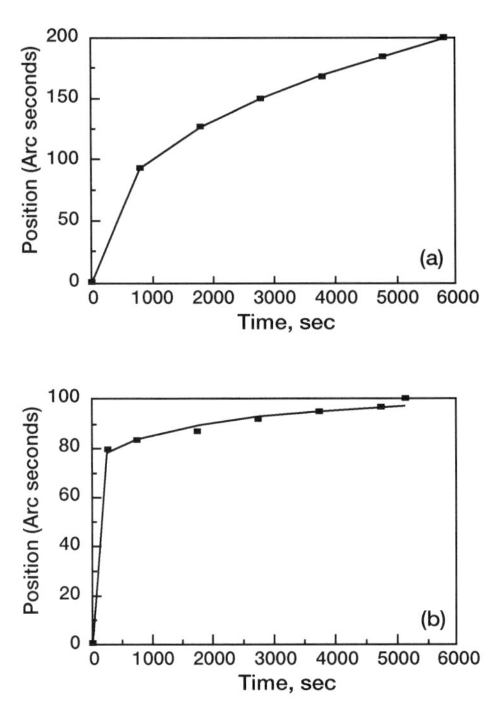

FIGURE 6.2 Binding of 250 μg/ml cell-surface proteins extracted from *H. pylori* strain (a) 52 and (b) 33 in solution to 3′SL-PAA immobilized on a resonant mirror biosensor (Hirmo *et al.*, 1998).

by strain 52 when compared to strain 33. A 27.2% increase in D_f—from 2.24 to 2.85—leads to an increase in k by a factor of 7.19—from 7.446 to 51.27.

Figures 6.3a–6.3c show the curves obtained using Eq. (6.2a) for the binding of cell-surface proteins from *H. pylori* strain 52 in solution in the absence (Fig. 6.3a) and in the presence of 5 and 10 mM free sialyl (α-2,3 lactose) (Figs. 6.3b and 6.3c, respectively) to 3′SL-PAA inmmobilized on an RMB. Again, a single-fractal analysis is adequate to describe the binding kinetics. Table 6.1c shows the values of k and D_f. In the free sialyl concentration range of 0 to 10 mM, the binding rate coefficient k is given by

$$k = (161.34 \pm 3.579)[\text{sialyl lactose}]^{0.0727 \pm 0.0024}. \qquad (6.3a)$$

Figure 6.4a shows that this predictive equation fits the values of k presented in Table 6.1c reasonably well. Note that only three data points are available, and one data point is available in the absence of silayl lactose concentration. Due to the lack of experimental data points, this point was also used in the

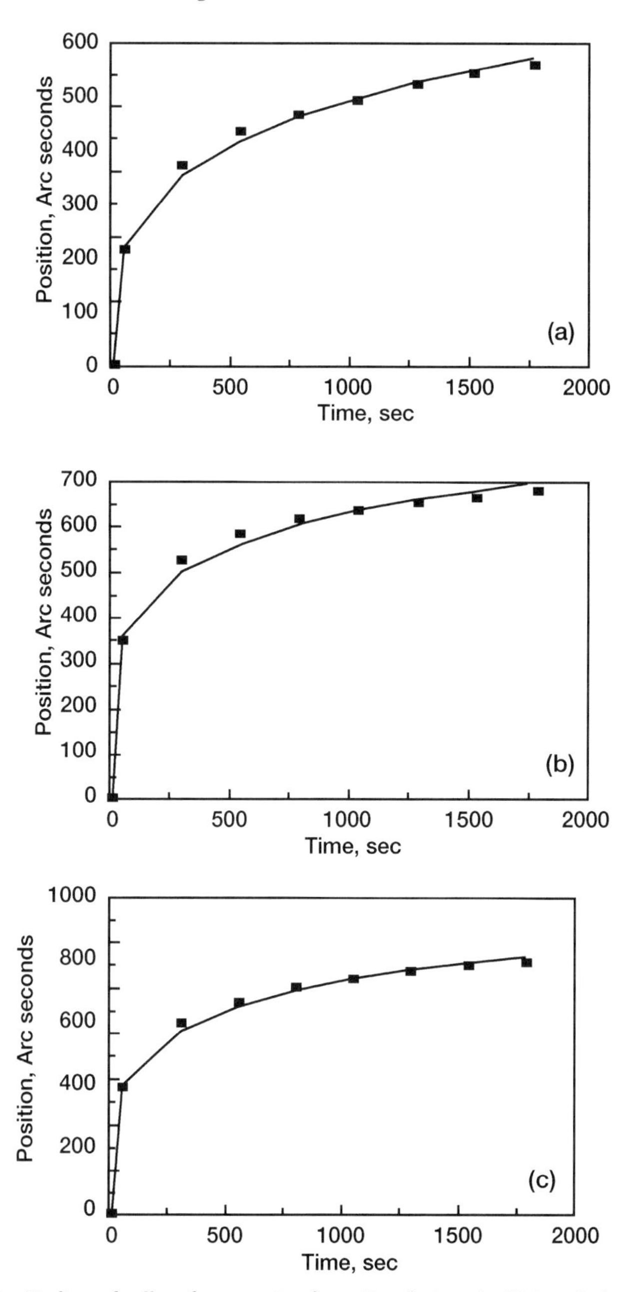

FIGURE 6.3 Binding of cell-surface proteins from *H. pylori* strain 52 in solution in the (a) absence of and in the presence of (b) 5 mM and (c) 10 mM free sialyl (α-2,3 lactose) to 3′SL-PAA immobilized on a resonant mirror biosensor (Hirmo *et al.*, 1998).

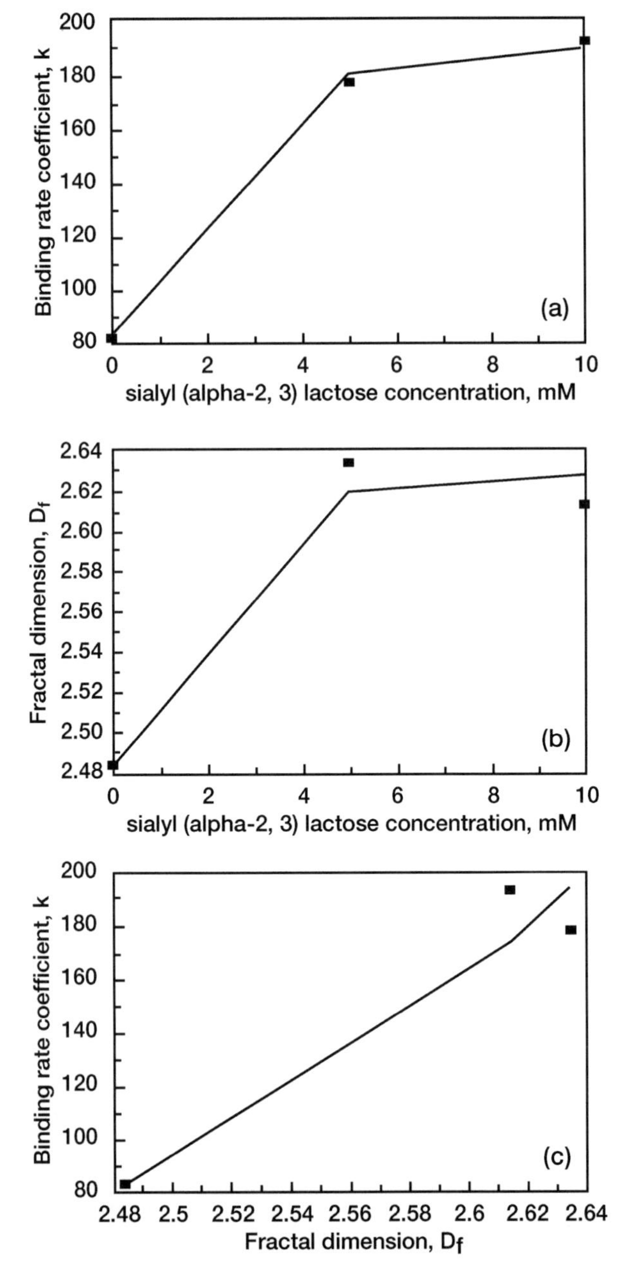

FIGURE 6.4 Influence of (a) sialyl (α-2,3 lactose) concentration (in mM) on the binding rate coefficient, k; (b) sialyl (α-2,3 lactose) concentration (in mM) on the fractal dimension, D_f; (c) the fractal dimension, D_f, on the binding rate coefficient, k.

modeling. For this case, a very small (0.0001 mM) silayl concentration was used. Using this value, a predicted k value of 82.91 was obtained. This compares very favorably with the experimental k value of 82.995 obtained in the absence of free sialyl concentration in solution. More data points would have provided a better fit. Note the very low value of the exponent.

In the free sialyl lactose concentration range (0 to 10 mM) utilized in solution, the fractal dimension is given by

$$D_{\mathrm{f}} = (2.5997 \pm 0.0207)[\text{sialyl lactose}]^{0.00492 \pm 0.00087}. \tag{6.3b}$$

Figure 6.4b shows that this predictive equation fits the values of the fractal dimension presented in Table 6.1c reasonably well. Once again, the data point for the absence of free sialyl lactose concentration in solution was used in the model. Here too, a very low concentration (0.0001 mM) of sialyl lactose was used. In this case too, a predicted value of 2.4846 was obtained for 0.0001 mM of sialyl lactose. This compares very favorably with the estimated D_{f} value of 2.4836 obtained in the absence of free sialyl lactose concentration. Once again, note the very low value of the exponent. This indicates that the fractal dimension is not very sensitive to the free sialyl lactose concentration in solution.

Figure 6.4c shows that the binding rate coefficient, k, increases as the fractal dimension, D_{f}, increases. This is in accord with the prefactor analysis of fractal aggregates (Sorenson and Roberts, 1997) and with the analyte-receptor binding kinetics for biosensor applications (Sadana, 1998). For the data presented in Table 6.1c, the binding rate coefficient is given by

$$k = (0.000216 \pm 0.000032) D_{\mathrm{f}}^{14.149 \pm 2.980}. \tag{6.3c}$$

This predictive equation fits the values of k presented in Table 6.1c reasonably well. The very high exponent dependence indicates that the binding rate coefficient is very sensitive to the degree of heterogeneity that exists on the surface. Some of the deviation may be attributed to the depletion of the analyte in the vicinity of the surface (imperfect mixing). At this time however, no correction is presented to account for the imperfect mixing. More data points are required to more firmly establish a quantitative relationship in this case. Perhaps one may require a functional form for k that involves more parameters as a function of the fractal dimension. One possible form could be

$$k = aD_{\mathrm{f}}^{b} + cD_{\mathrm{f}}^{d}, \tag{6.3d}$$

where a, b, c, and d are coefficients to be determined by regression. This functional form may describe the data better, but one would need more points to justify the use of a four-parameter equation.

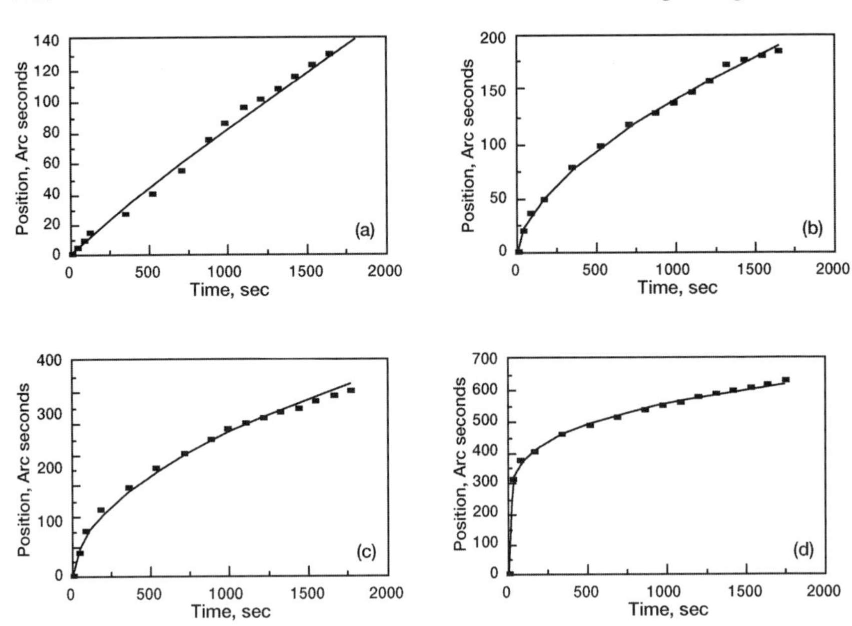

FIGURE 6.5 Binding of different concentrations of *H. pylori* of strain 52 in solution to 3′SL-PAA immobilized on a resonant biosensor (Hirmo *et al.*, 1998). (a) Two-times diluted; (b) nondiluted; (c) two-times concentrated; (d) five-times concentrated.

Figures 6.5a–6.5d show the binding curves obtained using Eq. (6.2a) for the binding of *H. pylori* cells of strain 52 in solution to 3′SL-PAA immobilized on an RMB. In this case, different cell concentrations were used. Once again, a single-fractal analysis is adequate to describe the binding kinetics. Table 6.1d shows the values of k and D_f. Diluted and concentrated cell concentrations were used. For the diluted and concentrated cell concentrations used, the binding rate coefficient is given by

$$k = (1.512 \pm 2.421)[\text{normalized cell concentration}]^{2.8666 \pm 0.2239}. \qquad (6.4a)$$

The normalized cell concentration is the cell concentration divided by the nondiluted cell concentration. Figure 6.6a shows that the predictive equation fits the values of k presented in Table 6.1d reasonably well. This is in spite of the high error in the estimate of the coefficient in the predictive equation. There is some scatter in the data at the higher normalized cell concentration. More data points would more firmly establish this equation. The binding rate coefficient is quite sensitive to the normalized cell concentration, as indicated by the value of the exponent.

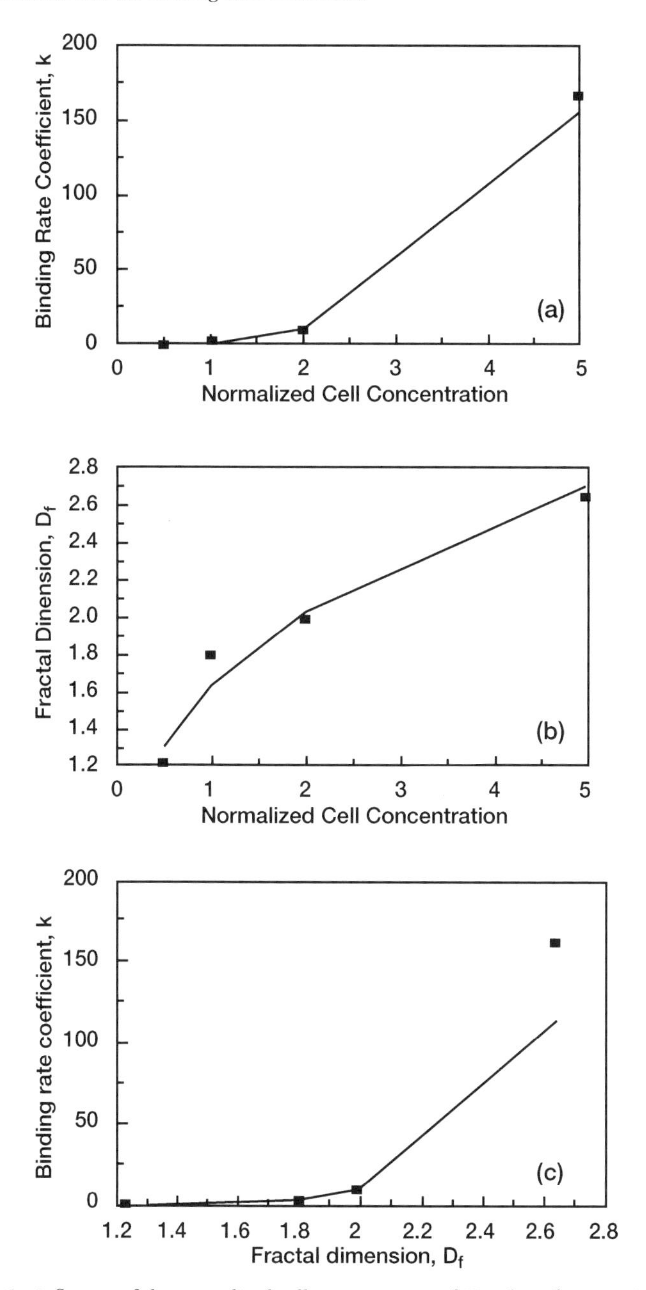

FIGURE 6.6 Influence of the normalized cell concentration of *H. pylori* of strain 52 in solution on (a) the binding rate coefficient, k; (b) the fractal dimension, D_f. (c) Influence of the fractal dimension, D_f, on the binding rate coefficient, k.

Figure 6.6b indicates that the fractal dimension, D_f, increases as the normalized cell concentration increases. In the normalized cell concentration analyzed, D_f is given by

$$D_f = (1.6246 \pm 0.1474)[\text{normalized cell concentration}]^{0.3161 \pm 0.0509}. \quad (6.4b)$$

Figure 6.6b shows that this predictive equation fits the values of D_f presented in Table 6.1d reasonably well. Note the low value of the exponent, which indicates that the fractal dimension is not very sensitive to the normalized cell concentration.

An increase in D_f leads to an increase in k. For this cell concentration range, k is given by

$$k = (0.0218 \pm 0.0145)D_f^{8.7795 \pm 0.9171}. \quad (6.4c)$$

There is some scatter in the data at the highest value of D_f utilized. Nevertheless, the predictive equation fits the values of k presented in Table 6.1d reasonably well (Fig. 6.6c). More data points, especially at the higher fractal dimension, for the D_f values would more firmly establish this equation. Once again, some of this deviation may be attributed to the depletion of the analyte in the vicinity of the surface (imperfect mixing). Again, no correction is presented to account for imperfect mixing. Furthermore, and as indicated earlier, one may require a functional form for k that involves more parameters as a function of the fractal dimension. The possible form suggested is given in Eq. (6.3d), where the coefficients a, b, c, and d are to be determined by regression. Once again, as observed previously, the high exponent dependence indicates that the binding rate coefficient, k, is very sensitive to the degree of heterogeneity that exists on the surface.

Wink *et al.* (1998) have utilized surface plasmon resonance (SPR) to analyze liposome-mediated enhancement of immunoassay sensitivity of proteins and peptides. They developed a sandwich immunoassay for interferon-γ (IFN-γ). A 16 kDa cytokine was used as the capture monoclonal antibody, which was physically adsorbed on a polystyrene surface. These authors were careful to point out the advantage and the disadvantage of this technique. The advantage is that no chemical labeling is required. The disadvantage is that SPR detects an aspecific parameter, a refractive index. Figure 6.7a shows the binding of 20 ng/ml interferon-γ in solution to cytokine physically adsorbed on a polystyrene surface. A single fractal analysis is adequate to describe the binding kinetics. Table 6.2a shows the values of k and D_f.

A high-affinity RNA was used by Kleinjung *et al.* (1998) as a recognition element in a biosensor. Taking advantage of recent developments in the

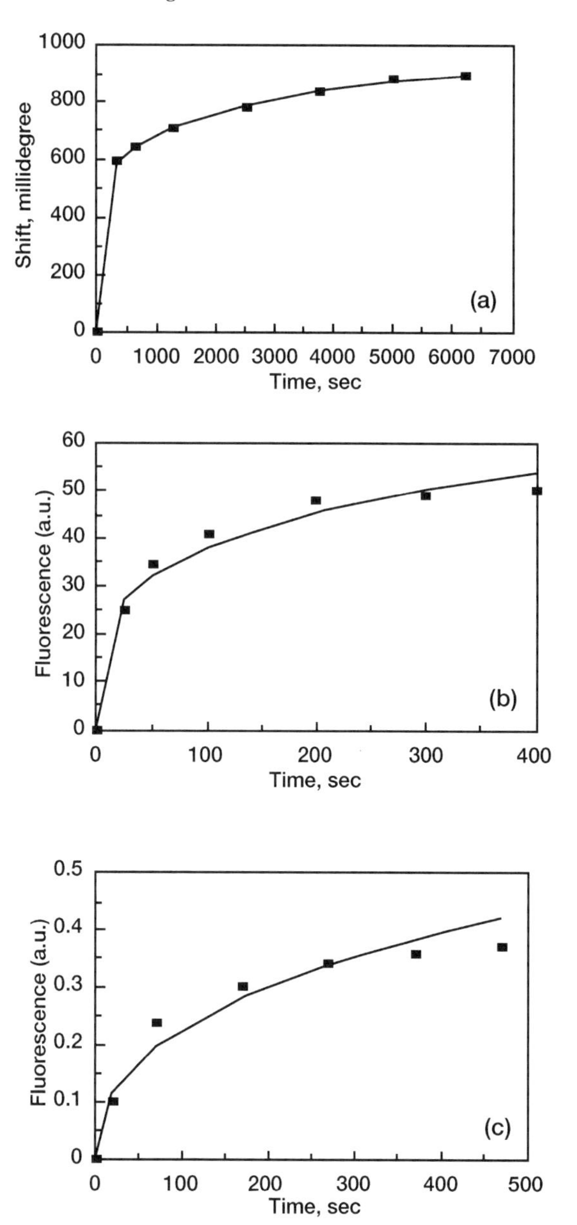

FIGURE 6.7 Binding of (a) 20 ng/ml interferon (IFN)-γ in solution to 16 kDa cytokine (monoclonal antibody) adsorbed on a polystyrene surface (with liposome amplificaton) (Wink *et al.*, 1998); (b) L-adenosine^{FITC} in solution to high-affinity RNA attached to a fiber-optic biosensor via an biotin–avidin bridge (Kleinjung *et al.*, 1998); (c) 0.02 mg/ml antibiotin in solution to biotinylated indium-tin oxide electode (Asanov *et al.*, 1998).

TABLE 6.2 Influence of Different Parameters on Fractal Dimensions and Binding Rate Coefficients for Different Antibody-Antigen and Analyte-Receptor Systems

Analyte concentration in solution/receptor on surface	k	D_f	Reference
(a) 20 ng/ml interferon (IFN)-γ/16-kDa cytokine (monoclonal antibody) adsorbed on polystyrene surface (with liposome amplification)	266.61 ± 2.832	2.7216 ± 0.007	Wink *et al.*, 1998
(b) L-adenosine^{FITC}/high-affinity RNA attached to a fiber-optic biosensor via an avidin-biotin bridge	12.707 ± 1.096	2.5192 ± 0.068	Kleinjung *et al.*, 1998
(c) 0.02 mg/ml antibody (antibiotin)-FITC/ biotinylated indium-tin oxide (ITO) electrode	0.0350 ± 0.005	2.1950 ± 0.115	Asanov *et al.*, 1998

understanding of the basic nature of the nucleic acids, these authors indicate that the inherent nature of some nucleic acids to combine the geneotype (nucleotide sequence) and a phenotype (ligand binding or catalytic activity) permits one to identify molecular targets. The authors immobilized an L-adenosine-specific RNA via a biotin–avidin bridge to an optical fiber core. They measured the binding using total internal reflection fluorescence of L-adenosine conjugated to fluorescein isothiocyanate. Figure 6.7b shows the curve obtained using Eq. (6.2a) for the binding of L-adenosine^{FITC} to high-affinity RNA attached to a fiber-optic biosensor via the biotin–avidin bridge. Once again, a single-fractal analysis is sufficient to adequately describe the binding kinetics. Table 6.2b shows the values of the binding rate coefficient, k and the fractal dimension, D_f.

Asanov *et al.* (1998) have developed a new biosensor platform on which to analyze antibody–antigen and streptavidin–biotin interactions. They analyzed the binding of 0.02 ng/ml antibody (antibiotin) to biotin immobilized to a transparent indium-tin oxide (ITO) working electrode. They used total internal reflection fluorescence (TIRF) to monitor biospecific interactions and electrochemical polarization to control the interactions between biotin and antibiotin. Figure 6.7c shows the curve obtained for the binding of 0.02 mg/ml antibody (antibiotin)-FITC in solution to biotinylated indium-tin (ITO) oxide. Again, a single-fractal analysis is sufficient to adequately describe the binding kinetics. Table 6.2c shows the values of k and D_f.

Rogers and Edelfrawi (1991) have analyzed the pharmacological activity of a nicotinic acetylcholine receptor (nAChR) optical biosensor. They utilized three fluorescein isothiocyanate (FITC)-labeled neurotoxic peptides: α-bungarotoxin (α-BGT), α-naja toxin (α-NT), and α-conotoxin (GI, α-CNTX). These peptides vary in the reversibility of their receptor inhibition. Nonspecific binding is a problem. Thus, these authors measured by evanescent fluorescence the nonspecific binding of α-BGT, α-NT, and α-CNTX on two sets of quartz fibers. Figure 6.8 shows the curves obtained using Eq. (6.2a) for the binding of the neurotoxic peptides. In each case, a single fractal analysis is adequate to describe the binding kinetics. Table 6.3 shows the values of the binding rate coefficient, k, and the fractal dimension, D_f. The binding rate coefficient presented in Table 6.3 was obtained from a regression analysis using Sigmaplot (1993) to model the experimental data using Eq. (6.2a), wherein $(\mathrm{Ab} \cdot \mathrm{Ag}) = kt^p$. The k and D_f values presented in the table are within 95% confidence limits. For example, the value of k reported for the binding of FITC-α-BGT to the quartz fiber is 45.025 ± 2.389. The 95% confidence limits indicates that 95% of the k values will fall between 45.025 and 47.414. This indicates that the Table 6.3 values are precise and significant. The curves presented in the figures are theoretical curves.

Note that as one goes from α-BGT to α-NT to α-CNTX the binding rate coefficient increases from 45.025 to 85.649 to 1454.7, respectively. Also, the fractal dimension values exhibit an increase from 1.2336 to 1.7432 to 2.4938, respectively, as one goes from α-BGT to α-NT to α-CNTX. Note that an increase in the fractal dimension leads to an increase in the binding rate coefficient, even though different neurotoxic peptides are utilized. The nonspecific binding kinetics of 5 and 10 nM FITC-α-NT to another set of quartz fibers are available in Sadana (1997). This data analysis is not given here but the binding rate coefficient and the fractal dimension values are given in Table 6.3.

Figure 6.9 shows that k increases as D_f increases for the nonspecific binding of these neurotoxic peptides to quartz fibers. This is in accord with the prefactor analysis of fractal aggregates (Sorenson and Roberts, 1997). Note that the fractal dimension is not an actual variable such as temperature or concentration that may be directly controlled. It is evaluated and estimated from Eqs. (6.2a) and (6.2b). One may consider it as a derived variable. In any case, it provides a quantitative estimate of the degree of heterogeneity or surface roughness. For the data presented in Table 6.3a, the binding rate coefficient is given by

$$k = (14.743 \pm 12.220) D_f^{4.6716 \pm 1.1513}. \tag{6.5}$$

This predictive equation fits the values of k presented in Table 6.3a reasonably well. There is some deviation at the high fractal dimension value. Considering that the data is plotted from two different sets of experiments, that is rather

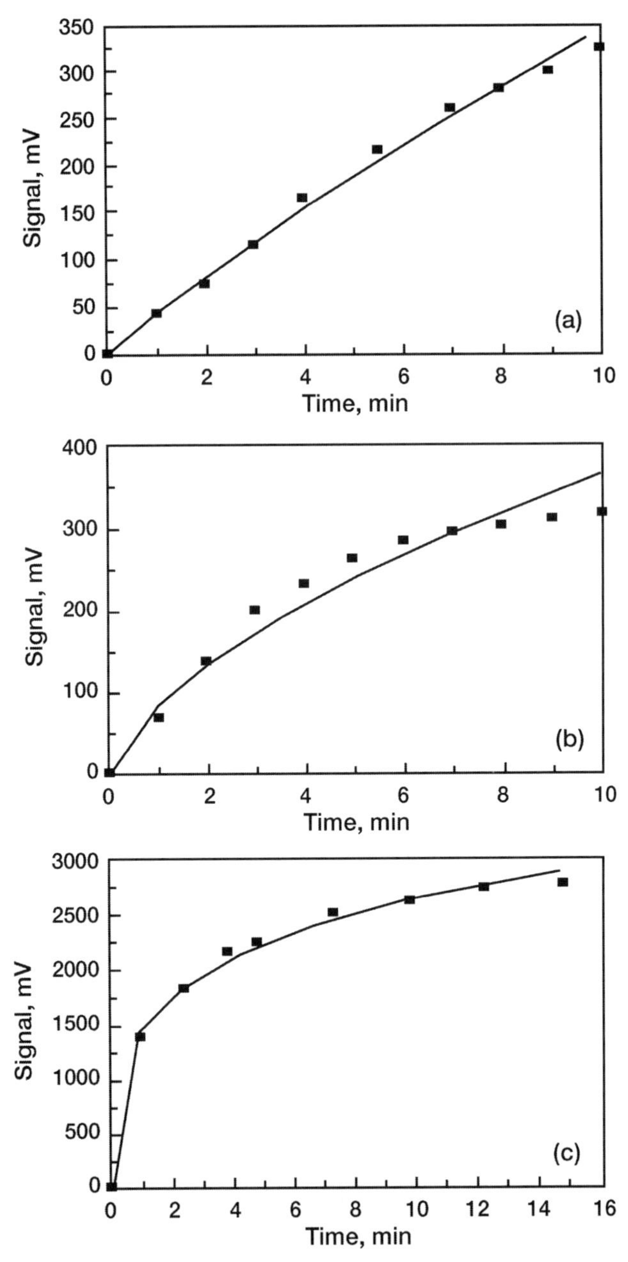

FIGURE 6.8 Nonspecific binding of (FITC)-labeled neurotoxic peptides to quartz fibers (Rogers and Edelfrawi, 1991). (a) α-bungarotoxin (α-BGT); (b) α-naja toxin (α-NT), (c) α-conotoxin (α-CNTX) (single-fractal analysis).

TABLE 6.3 Influence of Different Parameters on Fractal Dimensions and Binding Rate Coefficients for Different Antibody–Antigen and Analyte-Receptor Reaction Kinetics (Rogers and Edelfrawi, 1991)

Analyte concentration in solution/receptor on surface	k	D_f
FITC-α-BGT	45.025 ± 2.389	1.2336 ± 0.0472
FITC-α-NT	85.649 ± 11.52	1.7432 ± 0.1148
FITC-α-CNTX	1454.7 ± 48.64	2.4938 ± 0.0273
5 nM FITC-α-BGT	150.6 ± 9.60	1.47 ± 0.074
10 nM FITC-α-NT	99.1 ± 2.13	1.54 ± 0.029

reasonable. Note the high exponent dependence of k on D_f. This underscores that the binding rate coefficient is sensitive to the surface roughness or heterogeneity that exists on the surface. However, the data analysis in itself does not provide any evidence for surface roughness or heterogeneity, so the assumed existence of surface roughness or heterogeneity may not be correct.

Richalet-Secordel *et al.* (1997) utilized the BIACORE biosensor to analyze the binding of ligand–analyte pairs under conditions where the mass transport

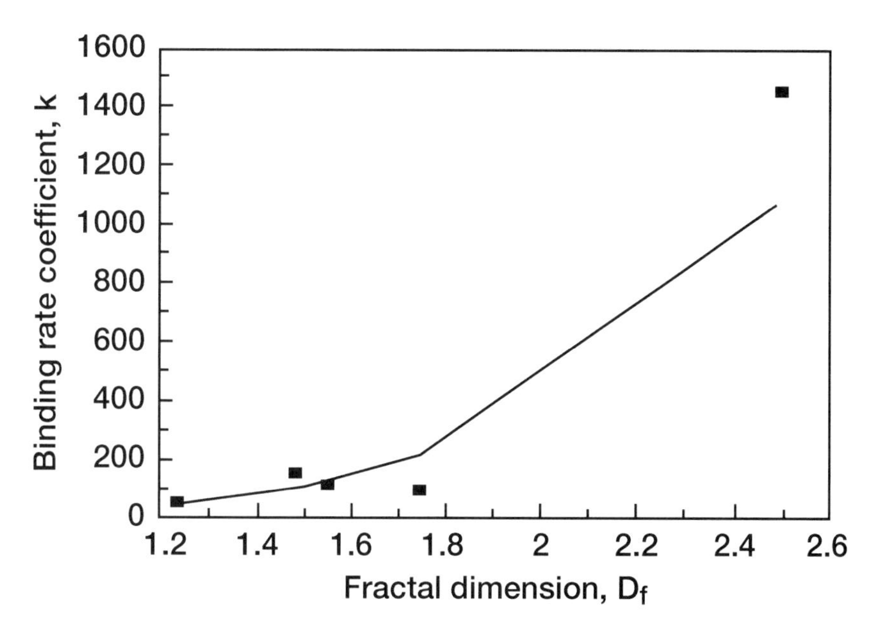

FIGURE 6.9 Influence of the fractal dimension, D_f, on the binding rate coefficient, k, for the nonspecific binding of neurotoxic peptides (α-BGT, α-NT, α-CNTX) to quartz fibers.

is only partially rate limiting. They used this method successfully to measure the concentration of monoclonal antibodies and monoclonal antibody fragments (Fab). The authors used different flow rates of the analyte, which interacted with a ligand that was covalently linked to a dextran matrix bound to a biosensor surface. The authors indicate that the binding of the protein leads to an increase in the refractive index. This increase is monitored in real time by the change in the resonance angle. For these set of experiments, these authors immobilized 5860 RU of streptavidin on the dextran matrix. This was followed by 582 RU of the biotinylated gp120 peptide. Figure 6.10 shows the

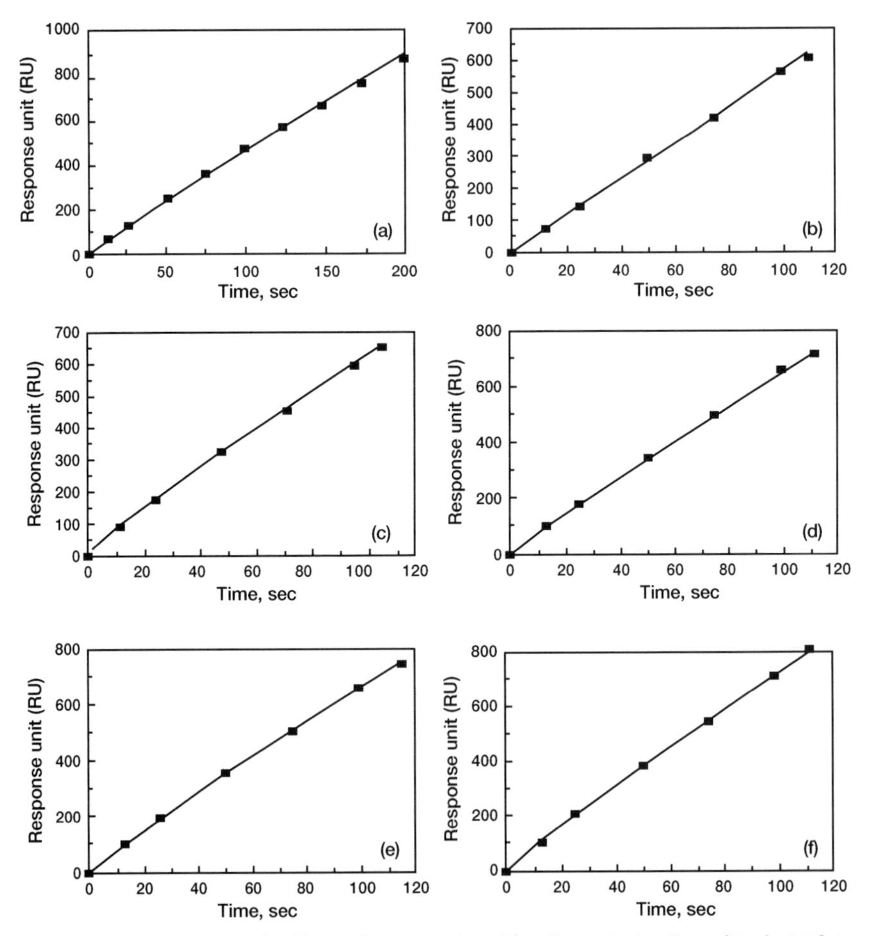

FIGURE 6.10 Influence of different flow rates (in μl/min) on the binding of Mab 0.5β in solution to biotinylated gp120 peptide immobilized on a BIACORE biosensor (Richalet-Secordel *et al.*, 1997).

binding of Mab 0.5β in solution to the gp120 peptide. The Mab was diluted 1/20,000 and injected for 2 min at different flow rates (ranging from 2 to 100 μl/min) over the gp120 peptide.

Note that in each case presented in Figure 6.10, a single-fractal analysis is adequate to describe the binding kinetics. Table 6.4a shows the values of the binding rate coefficients and the fractal dimensions obtained for the different flow rates ranging from 2 to 100 μl/min. The table indicates that an increase in D_f leads to an increase in k. See also Fig. 6.11, where the increase in k with an increase D_f is more obvious. In the fractal dimension range presented in Table 6.4, k may be represented by

$$k = (5.1395 \pm 0.1149) D_f^{4.0926 \pm 0.1576}. \tag{6.6}$$

Table 6.4 Influence of Different Parameters on Fractal Dimensions and Binding Rate Coefficients for Different Antibody-Antigen and Analyte-Receptor Reaction Kinetics (Richalet-Secordel *et al.*, 1997)

Analyte concentration in solution/receptor on surface	k	D_f
(a) Mab 0.5β diluted 1/20,000; flow rate 2 μl/min/582 RU biotin-ylated gp120 peptide	5.7255 ± 0.110	1.0962 ± 0.0142
Mab 0.5β; 5 μl/min/582 RU gp120 peptide	5.9264 ± 0.197	1.0322 ± 0.0340
Mab 0.5β;10 μl/min/582 RU gp120 peptide	8.0141 ± 0.161	1.1232 ± 0.0206
Mab 0.5β;25 μl/min/582 RU gp120 peptide	9.3402 ± 0.074	1.1572 ± 0.0170
Mab 0.5β;50 μl/min/582 RU gp120 peptide	10.606 ± 0.227	1.1884 ± 0.0218
Mab 0.5β; 100 μl/min/582 RU gp120 peptide	12.625 ± 0.141	1.2454 ± 0.0114
(b) mFab 20.5.3 diluted 1/4000; flow rate 2 μl/min/598 RU biotin-ylated gp32 peptide	1.7941 ± 0.038	1.3584 ± 0.019
mFab 20.5.3; 5 μl/min/598 RU gp32 peptide	2.7268 ± 0.153	1.4536 ± 0.0589
mFab 20.5.3;10 μl/min/gp32 peptide	1.3728 ± 0.028	1.6272 ± 0.0286
mFab 20.5.3; 25 μl/min/gp32 peptide	2.1789 ± 0.071	1.1746 ± 0.0345
mFab 20.5.3; 50 μl/min/gp32 peptide	2.9564 ± 0.196	1.2914 ± 0.0690
mFab 20.5.3; 100 μl/min/gp32 peptide	2.7408 ± 0.138	1.2260 ± 0.0666

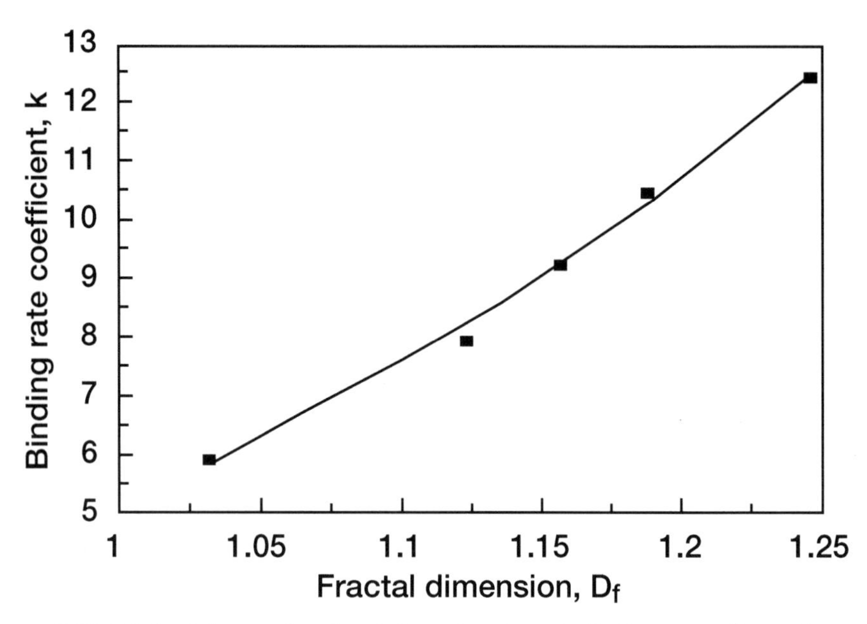

FIGURE 6.11 Influence of the fractal dimension, D_f, on the binding rate coefficient, k.

Note that Eq. (6.6) predicts the binding rate coefficient values presented in Table 6.4 reasonably well. The high exponent dependence of k on D_f once again indicates that the k value is very sensitive to the degree of heterogeneity that exists on the BIACORE surface. Note that the range of the fractal dimension presented in Fig. 6.11 and Table 6.4 is rather narrow. More data points are required over a wider range of D_f values to further establish Eq. (6.6). As indicated earlier, the data analysis in itself does not provide any evidence for surface roughness or heterogeneity, so the assumed existence of surface roughness or heterogeneity may not be correct.

Richalet-Secordel *et al.* (1997) also utilized the BIACORE biosensor to analyze the binding of mFab 20.5.3 in solution to biotinylated gp32 peptide (598 RU). Once again, different flow rates were used. Figure 6.12 shows that for the range of 2 to 100 μl/min flow rates a single-fractal analysis is adequate to describe the binding kinetics. Table 6.4b shows the values of k and D_f for the flow rates used. For the different flow rates presented in Table 6.4, the binding rate coefficient may be represented by

$$k = (2.331 \pm 0.541) D_f^{0.2110 \pm 1.1720}. \tag{6.7}$$

Ingersoll and Bright (1997) utilized fluorescence to probe biosensor interfacial dynamics. These authors immobilized active dansylated IgG at the

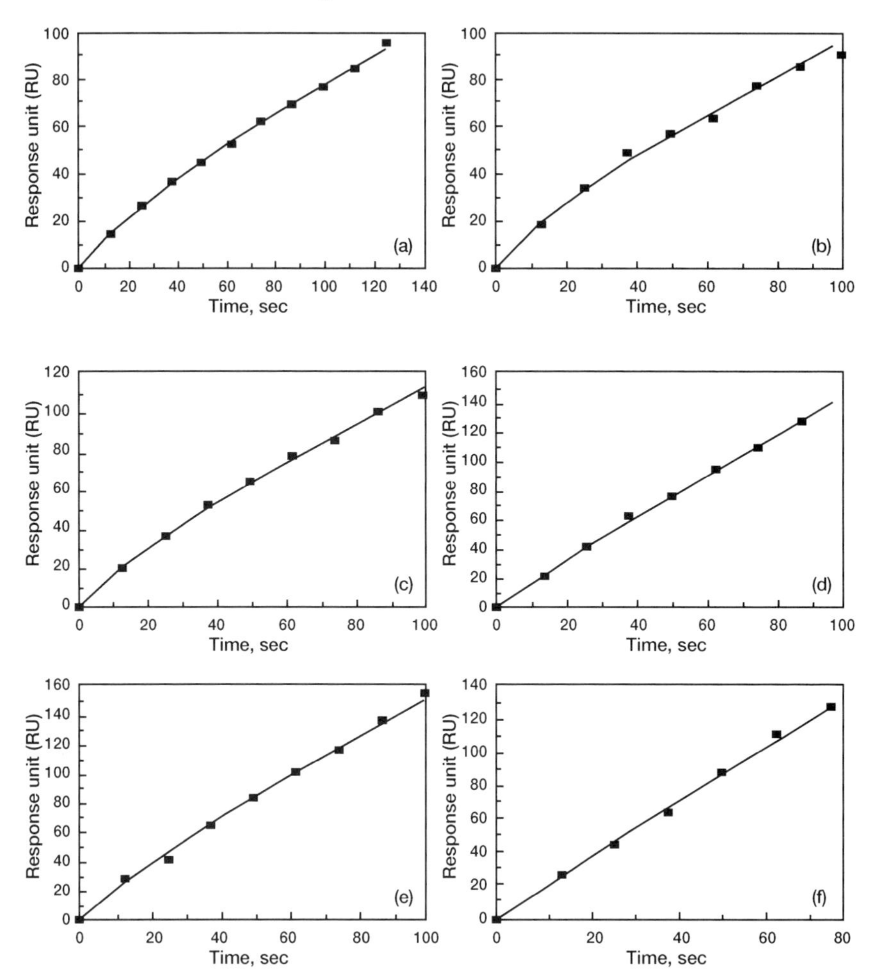

FIGURE 6.12 Influence of different flow rates (in μl/min) on the binding of mFab 20.5.3 in solution to biotinylated gp32 peptide immobilized on a BIACORE biosensor (Richalet-Secordel *et al.*, 1997). (a) 2; (b) 5; (c) 10; (d) 25; (e) 50; (f) 100.

interface. They controlled the active IgG by mixing active anti-BSA IgG with the Fc segment of the same IgG. Figure 6.13 shows the curves obtained using Eq. (6.2a) for the binding of 1 μM BSA in solution to the anti-BSA-protein G fused to the biosensor surface for three different IgG/Fc ratios. A single-fractal analysis is again sufficient to adequately model the binding kinetics.

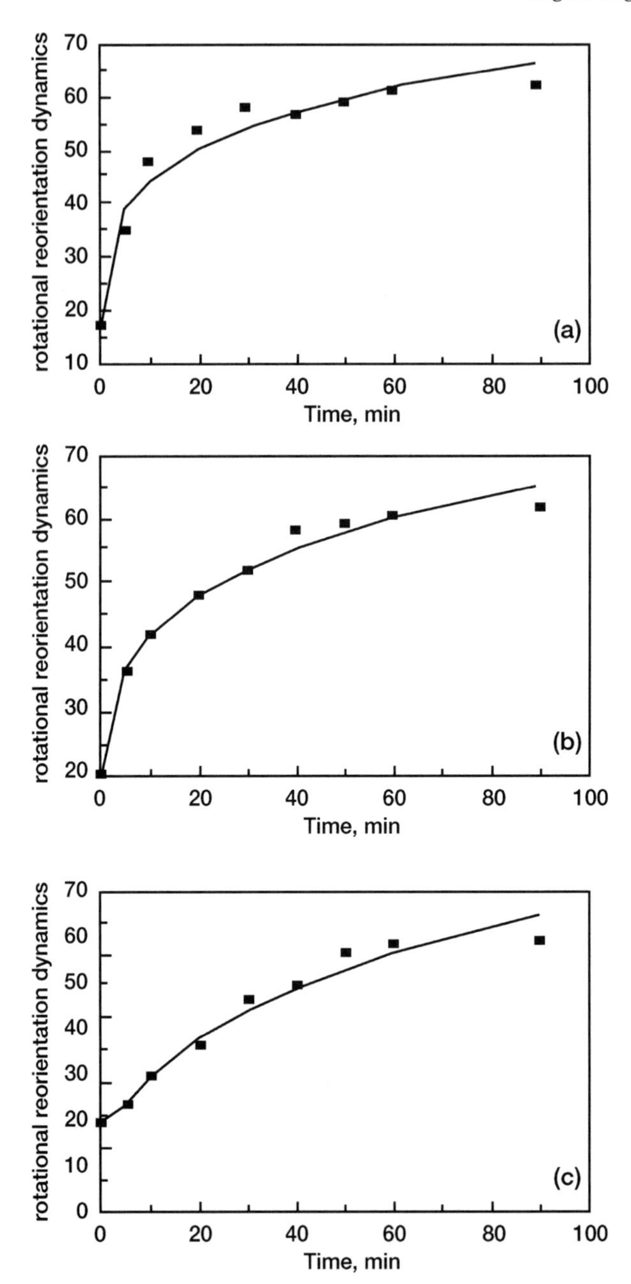

FIGURE 6.13 Influence of the IgG/Fc ratio on the binding of 1 μM BSA in solution to the anti-BSA-protein G fused to the biosensor surface (Ingersoll and Bright, 1997). (a) IgG/Fc = 0.6; (b) 0.8; (c) 1.0.

TABLE 6.5 Influence of Different Parameters on the Fractal Dimension and on the Binding Rate Coefficient for Different Antigen–Antibody Binding Kinetics

Analyte concentration in solution/receptor on surface	k	D_f	Reference
(a) 1 μM BSA/BSA IgG with the F_c segment of the same IgG; $IgG/F_c = 0.6$	29.199 ± 2.226	2.6348 ± 0.0574	Ingersoll and Bright, 1997
1 μM BSA/BSA IgG with F_c; $IgG/F_c = 0.8$	26.726 ± 0.920	2.6076 ± 0.0264	Ingersoll and Bright, 1997
1 μM BSA/BSA IgG with F_c; $IgG/F_c = 1.0$	13.700 ± 0.783	2.3038 ± 0.0434	Ingersoll and Bright, 1997
(b) 10 nM 16*CFl (oligonucleotide)/16*B immobilized via NHS-LC-biotin and streptavidin (chemical regeneration)	86.53 ± 3.21	1.2112 ± 0.0260	Abel *et al.*, 1996
10 nM 16*CFl/16*B immobilized via NHS-LC-biotin and streptavidin (thermal regeneration)	100.00 ± 6.70	1.3942 ± 0.0466	Abel *et al.*, 1996
10 nM 16*CFl/16*B immobilized via NHS-LC-biotin and avidin	1338 ± 31.3	1.2722 ± 0.021	Abel *et al.*, 1996

Table 6.5a shows that an increase in the IgG/Fc ratio leads to a decrease in the binding rate coefficient, k, and to a decrease in the fractal dimension, D_f. As the IgG/Fc ratio increases from a value of 0.6 to 1.0, k decreases from 29.199 to 13.700. Figure 6.14a also shows this decrease in k with an increase in the IgG/Fc ratio. In the IgG/Fc ratio range of 0.6 to 1.0 analyzed, the binding rate coefficient is given by

$$k = (15.541 \pm 4.919)(\text{IgG/Fc})^{-1.4260 \pm 0.7592}. \tag{6.8}$$

This predictive equation fits the values of k presented in Table 6.5a reasonably well. More data points are required over a wider range of IgG/Fc ratios to more clearly define this equation. The exponent (negative in this case) dependence of k on the IgG/Fc ratio lends further support to the fractal nature of the system.

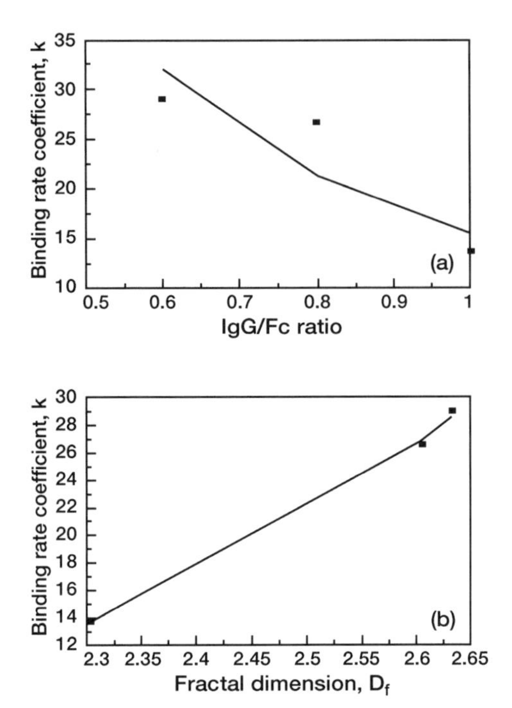

FIGURE 6.14 Influence of (a) the IgG/Fc ratio on the binding rate coefficient, k; (b) the fractal dimension, D_f, on the binding rate coefficient, k.

Figure 6.14b shows the increase in k with an increase in D_f. For the IgG/Fc ratio range of 0.6 to 1.0 analyzed, the binding rate coefficient is given by

$$k = (0.1348 \pm 0.003) D_f^{5.5351 \pm 0.2084}. \tag{6.9}$$

This predictive equation fits the values of k presented in Table 6.5a well. More data points are required over a wider range of D_f values to further define this equation. The fractional high exponent dependence of k on the fractal dimension further reinforces the fractal nature of the system. The high exponent dependence of k on D_f further emphasizes that k is very sensitive to the degree of heterogeneity that exists on the surface.

Abel *et al.* (1996) utilized a fiber-optic biosensor for the detection of 16-mer oligonucleotides in hybridization assays. The authors immobilized a biotinylated capture probe on the biosensor surface using either avidin or streptavidin. Fluorescence was utilized to monitor the hybridization with fluorescein-labeled complementary strands. These authors indicate that the capability of DNA and RNA fragments to bind selectively to complementary arranged nucleotides at other nucleic acids is the basis for *in vitro* tests. Therefore, hybridization methods that use nucleic acids as the biological recognition element may be utilized as an effective immunoassay.

Figure 6.15 shows the binding of 16*CFl (complementary oligonuclotide) in a 10 nM solution to 16*B (immobilized oligonucleoide) immobilized via sulfosuccinimidyl-6-(biotinamido)hexanoate (NHS-LC-biotin) and streptavidin to a biosensor. Abel *et al.* used both chemical and thermal regeneration. Figures 6.15a and 6.15b show the curves obtained using Eq. (6.2a) for the binding of 16*CFl in solution to 16*B immobilized on the biosensor surface, using chemical and thermal regeneration, respectively. A single-fractal analysis is once again sufficient to adequately describe the binding kinetics. Table 6.5b shows the values of the binding rate coefficients and the fractal dimensions obtained in these cases. Note that the temperature of hybridization was 26.7°C, and for thermal regeneration the fiber-optic surface was heated to 68.5°C. Note that as one goes from the chemical to the thermal regeneration cycle, k increases by about 15.5%—from a value of 86.53 to 100—and D_f increases by about 13.4%—from a value of 1.2112 to 1.3942.

Abel *et al.* also analyzed the binding kinetics of 10 nM 16*CFl in solution to 16*B immobilized to a biosensor surface. The only difference between this case and the previous two cases is that for the immobilization step the authors used avidin instead of streptavidin. A single-fractal analysis was employed here also to model the binding kinetics (Fig. 6.15c). The values of k and D_f are given in Table 6.5b. In this case, the binding rate coefficient is much higher than in the previous two cases. The fractal dimension value of 1.2722 is between the two values obtained in the previous two cases. In this case, the fractal dimension analysis provides a quantitative estimate of the degree of heterogeneity that exists on the biosensor surface for each of the DNA hybridization assays. More data is required to establish trends or predictive equations for the binding rate coefficient in terms of either the analyte concentration in solution or the "estimated" fractal dimension of the biosensor surface under reaction conditions.

Schmitt *et al.* (1997) have analyzed the binding of thrombin in solution to antithrombin immobilized on a transducer surface using a BIACORE biosensor. The authors utilized reflectometric interference spectroscopy (RIfS) as a transducer. The inhibitor group was immobilized on the transducer surface with carboxymethyldextran (CMD) by diisopropylcarbidiimide (DIC) activation. Figure 6.16a shows the curves obtained using Eq. (6.2a) for the binding of 1 μg/ml thrombin in solution to 2 mg antithrombin immobilized on the transducer surface. A single-fractal analysis is sufficient to describe the binding kinetics. The entire binding curve is used to obtain the fractal dimension and the binding rate coefficient.

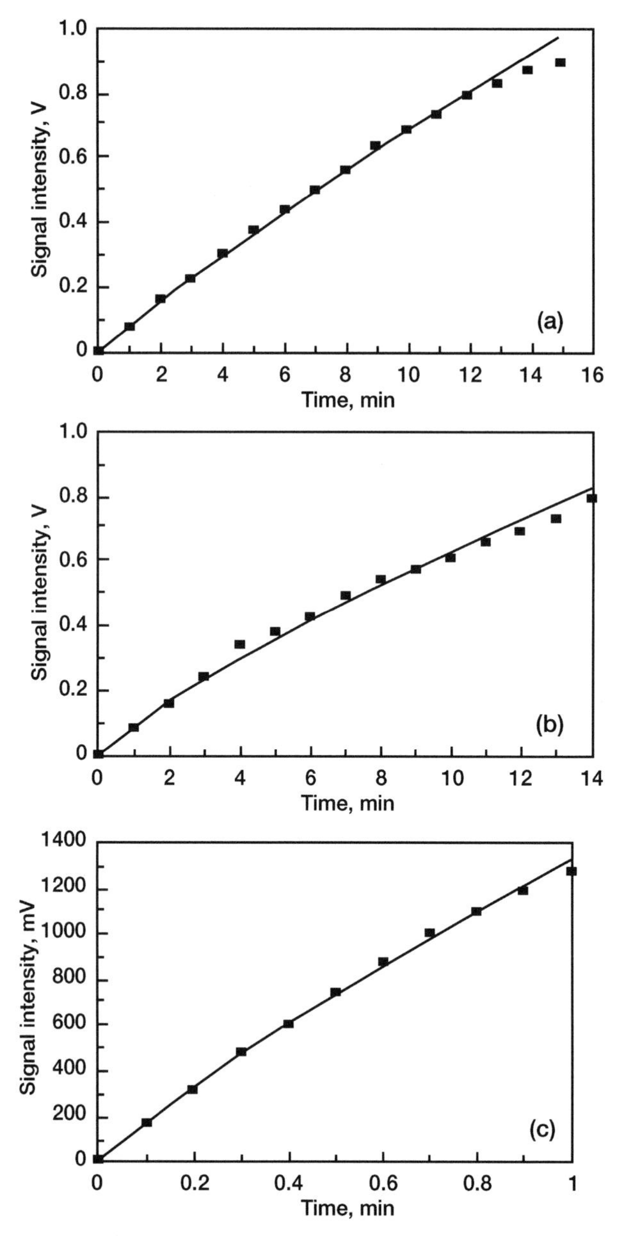

FIGURE 6.15 Binding of 16*CFl (complementary oligonucleotide) in a 10 nM solution to 16*B (immobilized oligonucleotide) immobilized via sulfosuccinimidyl-6-(biotinamido)hexanoate (NHS-LC-biotin) and streptavidin to a biosensor. Influence of (a) chemical and (b) thermal regeneration (Abel *et al.*, 1996). (c) Binding of 16*CFI (complementary oligonucleotide) in a 10 nM solution to 16*B (immobilized oligonucleotide) immobilized via sulfosuccinimidyl-6-(biotinamido)hexanote (NHS-LC-biotin) and avidin to a biosensor (Abel *et al.*, 1996).

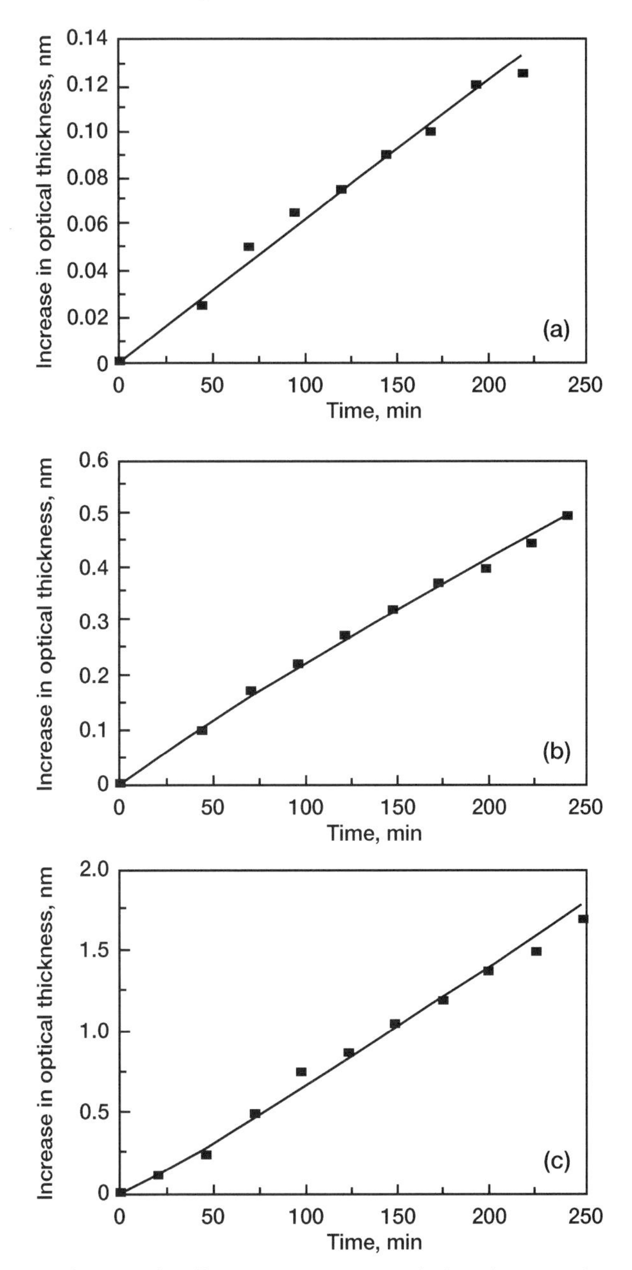

FIGURE 6.16 Influence of different concentrations of thrombin in solution to 2 mg antithrombin immobilized on a transducer surface (Schmitt *et al.*, 1997). (a) 1.0 μg/ml; (b) 2.0 μg/ml; (c) 5.0 μg/ml.

TABLE 6.6 Influence of Thrombin Concentration in Solution on Binding Rate Coefficients and Fractal Dimensions for Its Binding to Antithrombin Immobilized on a Transducer Surface Using a BIACORE Biosensor (Schmitt *et al.*, 1997)

Thrombin concentration in solution (μg/ml)/2 mg antithrombin immobilized on transducer surface modified with carboxymethyldextran (CMD) by diisopropyl-carbidiimide (DIC) activation	Binding rate coefficient, k (nm t^{-p}), from Eq. (6.2a)*	Fractal dimension, D_f
1.0	0.00072 ± 0.000069	1.06 ± 0.13
2.0	0.0033 ± 0.0007	1.16 ± 0.06
5.0	0.0048 ± 0.0004	0.85 ± 0.07

* 1 nm in optical thickness is approximately equal to 1 ng protein bound per mm^2 of transducer surface

Table 6.6 shows the values of the binding rate coefficient, k, and the fractal dimension, D_f. The values of k and D_f presented in the table were obtained from a regression analysis using Sigmaplot (1993) to model the experimental data using Eq. (6.2a), wherein (analyte-receptor) $= kt^p$. The k and D_f values presented are within 95% confidence limits. For example, for the binding of 1 μg/ml thrombin in solution to 2 mg antithrombin immobilized on the transducer surface shown in Fig. 6.16a, the value of k reported is 0.000722 ± 0.000069. The 95% confidence limit indicates that 95% of the k values will lie between 0.000653 and 0.000791. This indicates that the Table 6.6 values are precise and significant. The curves presented in the figures are theoretical curves.

Figures 6.16b and 6.16c show the binding of 2.0 and 5.0 μg/ml thrombin in solution to antithrombin immobilized on a sensor surface, respectively. In both cases, a single-fractal analysis is sufficient to adequately describe the binding kinetics. The values of k and D_f are given in Table 6.6. Note that as one increases the thrombin concentration in solution from 1 to 5 μg/ml the binding rate coefficient increases from a value of 0.000722 to 0.004839 and the fractal dimension decreases from a value of 1.0644 to 0.8498.

There is no nonselective adsorption of the thrombin. Our analysis, at present, does not include this nonselective adsorption. We do recognize that in some cases this may be a significant component of the adsorbed material and that this rate of association, which is of a temporal nature, would depend on surface availability. Accommodating the nonselective adsorption into the model would lead to an increase in the degree of heterogeneity on the surface since, by its very nature, nonspecific adsorption is more heterogeneous than specific adsorption. This would lead to higher fractal dimension values since the fractal dimension is a direct measure of the degree of heterogeneity that exists on the surface. Future analyses of analyte-receptor binding data may include this aspect in the analysis, which would be exacerbated by the presence of inherent external diffusion limitations.

Furthermore, we do not present any independent proof or physical evidence for the existence of fractals in the analyses of these analyte-receptor binding systems except by indicating that it has been applied in other areas and that it is a convenient means of making more quantitative the degree of heterogeneity that exists on the surface. Thus, in all fairness, this is just one possible way by which to analyze this analyte-receptor binding data. One might justifiably argue that appropriate modeling may be achieved by using a Langmuir or other approach. However, a major drawback of the Langmuir approach is that it does not allow for the heterogeneity that exists on the surface.

Table 6.6 indicates that k increases as the thrombin concentration in solution increases. Figure 6.17a shows the increase in k with an increase in the thrombin concentration in solution. In the thrombin concentration range (1 to 5 μg/ml) analyzed, k is given by

$$k = (0.00094 \pm 0.00072)[\text{thrombin}]^{1.1420 \pm 0.4992}. \qquad (6.10a)$$

There is scatter in the data, which is clearly indicated in the error estimates of the values of the exponent and the constant. More data points are required to more firmly establish this equation. Nevertheless, the equation is of value since it provides an indication of the change in k as the thrombin concentration in solution changes.

Table 6.6 also indicates that D_f decreases as the thrombin concentration in solution increases. Figure 6.17b shows the increase in D_f with an increase in the thrombin concentration in solution. In the thrombin concentration range (1 to 5 μg/ml) analyzed, D_f is given by

$$D_f = (1.1416 \pm 0.1863)[\text{thrombin}]^{-0.1505 \pm 0.1324}. \qquad (6.10b)$$

Again, there is scatter in the data, which is clearly indicated in the value of the exponent. More data points would more firmly establish this equation. Nevertheless, the equation is of value since it provides an indication of the change in D_f as the thrombin concentration in solution changes.

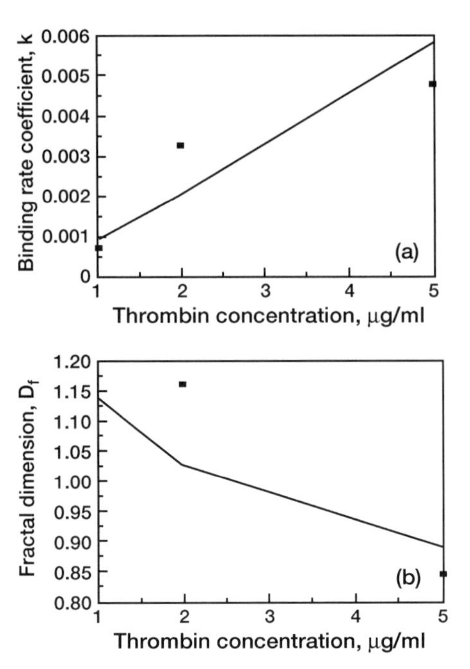

FIGURE 6.17 Influence of the thrombin concentration (in μg/ml) in solution on (a) the binding rate coefficient, k; (b) the fractal dimension, D_f.

6.3.2. Dual-Fractal Analysis

Kyono *et al.* (1998) have used the scintillation proximity assay (SPA) to detect hepatitis C virus helicase activity. These authors indicate that hepatitis C virus (HCV) is a major etiologic agent of non-A and non-B viral hepatitis (Choo *et al.*, 1989; Kuo *et al.*, 1989). Kyono *et al.* indicate that at the C-terminal two-thirds of the nonstructural protein 3 (NS3) of hepatitis C virus possesses RNA helicase activity. This enzyme is expected to be one of the target molecules of anti-HCV drugs. The authors utilized the SPA system to detect the helicase activity of NS3 protein purified by an immunoaffinity column. A polyclonal antibody to HCV was adsorbed on the immunoaffinity column. Figure 6.18 shows the curves obtained using Eqs. (6.2a) and (6.2b) for the time course of helicase activity by purified HCV NS3 protein. Note that in this case a dual-fractal analysis is required to provide a reasonable fit. Table 6.7 shows the values of the binding rate coefficient, k, and the fractal dimension, D_f, obtained using a single-fractal analysis as well as the binding

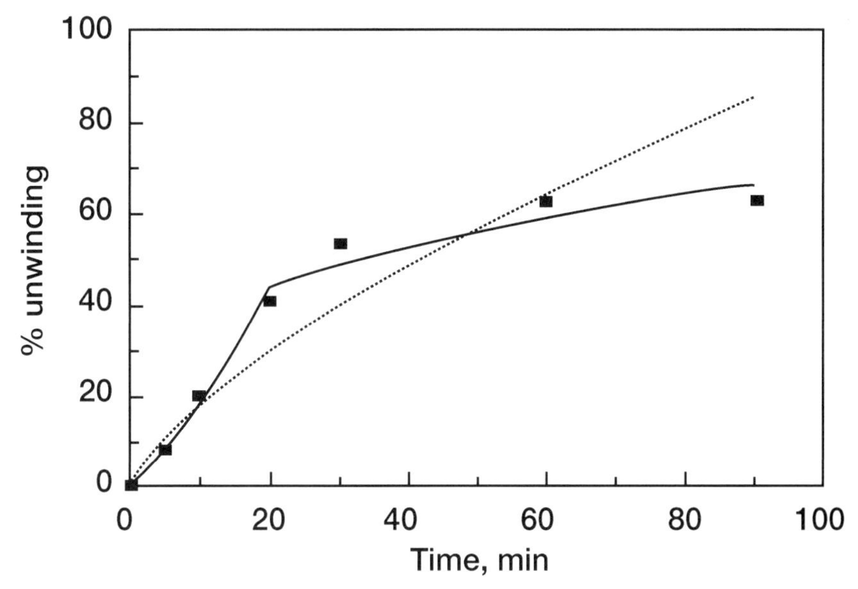

FIGURE 6.18 Binding of 20 ng/ml per well helicase in solution to immobilized nonstructural protein 3 (NS3) of hepatitis virus (HCV) (Kyono *et al.*, 1998) (- - - single-fractal analysis; — dual-fractal analysis).

rate coefficients, k_1 and k_2, and the fractal dimensions, D_{f_1} and D_{f_2} using a dual-fractal analysis.

Asanov *et al.* (1998) have analyzed the nonspecific and biospecific adsorption of IgG-FITC and γ-IgG-FITC in solution to a biotinylated ITO surface. They analyzed this in the absence and in the presence of bovine serum albumin (BSA). Figures 6.19a and 6.19b show the curves obtained using Eqs. (6.2a) and (6.2b) for the binding of the analyte in solution to the receptor immobilized on the ITO surface. Figure 6.19a shows the binding of 0.03 mg/ml IgG-FITC in solution and in the presence of 0.01 mg/ml BSA to the biotinylated ITO surface. Note that a single-fractal analysis does not provide an adequate fit, so a dual-fractal analysis is required. Table 6.8 shows the values of k and D_f obtained using a single-fractal analysis as well as k_1 and k_2, and the fractal dimensions, D_{f_1} and D_{f_2} using a dual-fractal analysis.

Figure 6.19b shows the binding of 0.03 mg/ml IgG-FITC in solution in the absence of BSA to the biotinylated ITO surface. In this case, a single-fractal analysis is sufficient to adequately describe the binding kinetics. Table 6.8 shows the values of k and D_f. On comparing this with the result in Fig. 6.19a, one notes that the presence of BSA in solution leads to complexities in the binding.

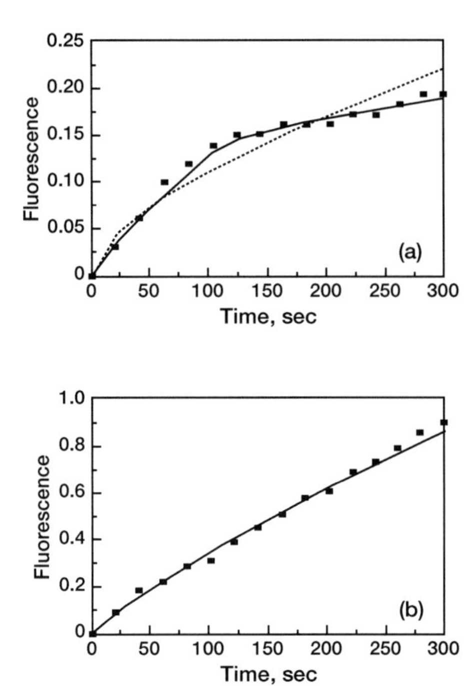

FIGURE 6.19 Binding of 0.03 mg/ml IgG-FITC in solution (a) in the presence of 0.01 mg/ml BSA (- - -, single-fractal analysis; —, dual-fractal analysis) and (b) in the absence of BSA to the biotinylated ITO surface (Asanov *et al.*, 1998).

TABLE 6.7 Influence of Different Parameters on Fractal Dimensions and Binding Rate Coefficients for Analyte-Receptor Reaction Kinetics (Kyono *et al.*, 1998)

Analyte concentration in solution/receptor on surface	k	D_f	k_1	k_2	D_{f_1}	D_{f_2}
20 ng per well helicase/ nonstructural protein 3 (NS3) of hepatitis virus (HCV) [HCV NS3 protein] using a scintillation proximity assay (SPA) system	3.6058 $\pm$1.344	1.5936 $\pm$0.271	1.5671 $\pm$0.197	18.710 $\pm$2.001	0.8782 $\pm$0.1736	2.4402 $\pm$0.173

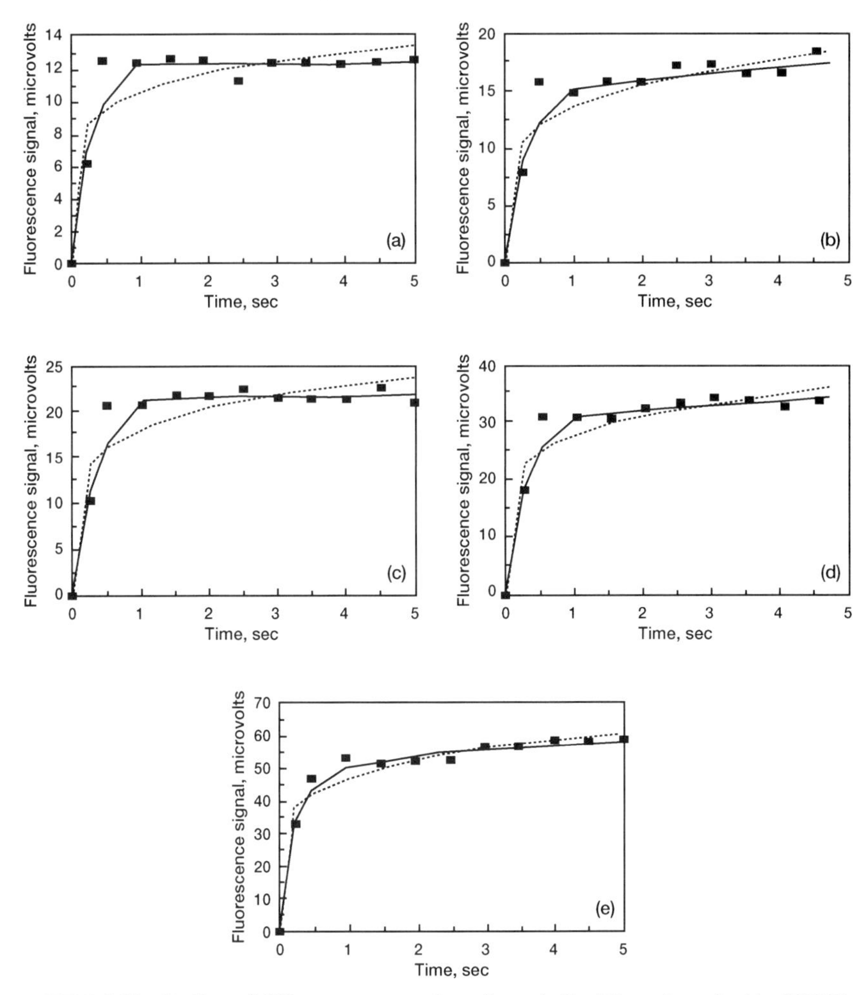

FIGURE 6.20 Binding of different concentrations (in ng/ml) of lipopolysaccharide (TRITC-LPS) in solution to polymyxin B immobilized on a fiber-optic biosensor (James *et al.*, 1996). (a) 10; (b) 25; (c) 50; (d) 100; (e) 200.

An evanescent fiber-optic biosensor has been developed to detect lipopolysaccharide (LPS) found in the outer cell membrane of gram-negative bacteria by James *et al.* (1996). These authors analyzed the kinetics and stability of the binding of 25 to 200 ng/ml TRITC-LPS (also known as endotoxin) in solution to polymyxin B immobilized on the fiber-optic biosensor. Figure 6.20 shows the curves obtained using Eqs. (6.2a) and (6.2b) for the binding of LPS in solution to the polymyxin immobilized on the

TABLE 6.8 Influence of Nonspecific and Specific Adsorption on the Fractal Dimensions for the Binding of Anti-biotin Antibody in Solution to Biotin Immobilized on a Transparent Indium-Tin Oxide (ITO) Electrode (Asanov *et al.*, 1998)

Analyte concentration in solution/ receptor on surface	k	D_f	k_1	k_2	D_{f_1}	D_{f_2}
0.03 mg/ml IgG-FITC adsorption in the presence of 0.01 mg/ml BSA/biotinylated ITO surface	0.00657 ± 0.0012	1.7812 ± 0.1132	0.00251 ± 0.00023	0.03962 ± 0.0011	1.2766 ± 0.1463	2.4570 ± 0.061
0.03 mg/ml IgG-FITC in the absence of BSA/biotinylated ITO surface	0.00844 ± 0.0004	1.3688 ± 0.00048	na	na	na	na

TABLE 6.9 Influence of Different Parameters on Fractal Dimensions and Binding Rate Coefficients for Analyte-Receptor Reaction Kinetics (James *et al.*, 1996)

Analyte concentration in solution/receptor on surface	k	D_f	k_1	k_2	D_{f_1}	D_{f_2}
10 ng/ml LPS/polymyxin B	10.451 ± 1.836	2.7190 ± 0.1064	13.731 ± 4.612	12.094 ± 0.469	2.0236 ± 0.5908	$3.00 - 0.050$
25 ng/ml LPS/polymyxin B	13.900 ± 2.207	2.6218 ± 0.0968	17.019 ± 6.169	15.384 ± 0.700	2.0932 ± 0.6310	2.6218 ± 0.0968
50 ng/ml LPS/polymyxin B	18.433 ± 3.148	2.6528 ± 0.1036	23.571 ± 7.710	21.519 ± 0.604	2.000 ± 0.5772	2.9508 ± 0.0364
100 ng/ml LPS/polymyxin B	28.34 ± 3.443	2.6792 ± 0.0754	33.939 ± 9.434	31.136 ± 0.825	2.2158 ± 0.4528	2.8633 ± 0.0344
200 ng/ml LPS/polymyxin B	47.289 ± 3.665	2.7102 ± 0.0484	56.082 ± 5.447	51.391 ± 1.711	2.3326 ± 0.1890	2.7102 ± 0.0484

biosensor. Table 6.9 shows the values of the binding rate coefficients and the fractal dimensions obtained using a single- and a dual-fractal analysis. Figure 6.20 shows that for each of the 25 to 200 ng/ml TRITC-LPS concentrations used, a dual-fractal analysis provides a better fit. Therefore, only the dual-fractal analysis is analyzed further.

Table 6.9 indicates that for a dual-fractal analysis an increase in the TRITC-LPS concentration in solution leads to an increase in the binding rate coefficients, k_1 and k_2. Figure 6.21 also shows this increase. In the 25 to 200 ng/ml TRITC-LPS concentration, k_1 is given by

$$k_1 = (4.120 \pm 0.573)[\text{TRITC-LPS}]^{0.4696 \pm 0.0556} \tag{6.11a}$$

and k_2 is given by

$$k_2 = (3.5515 \pm 0.4574)[\text{TRITC-LPS}]^{0.4827 \pm 0.0518}. \tag{6.11b}$$

These predictive equations fit the values of the k_1 and k_2 presented in Table 6.9 reasonably well. The exponent dependence of the binding rate coefficients

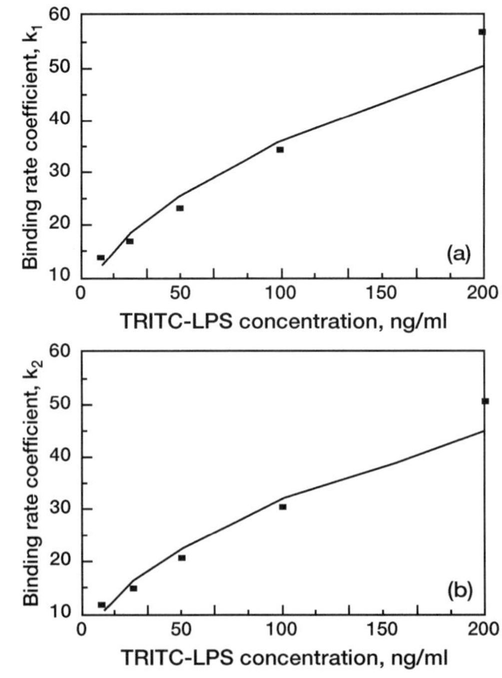

FIGURE 6.21 Influence of the TRITC-LPS concentration in solution on the binding rate coefficients. (a) k_1; (b) k_2.

on the TRITC-LPS concentration in solution lends support to the fractal nature of the system.

Figure 6.22a shows the influence of the TRITC-LPS concentration on the fractal dimension, D_{f_1}. An increase in the TRITC-LPS concentration in solution leads to an increase in D_{f_1}. In the 25 to 200 ng/ml TRITC-LPS concentration, D_{f_1} is given by

$$D_{f_1} = (1.7882 \pm 0.0778)[\text{TRITC-LPS}]^{0.04501 \pm 0.0182}. \tag{6.11c}$$

The fit of this predictive equation is reasonable. Since there is some scatter in the data, more data points are required to more firmly establish the equation. The fractal dimension, D_{f_1}, is not very sensitive to the TRITC-LPS concentration in solution, as noted by the very low exponent dependence of D_{f_1} on TRITC-LPS concentration.

Figure 6.22b shows the influence of the TRITC-LPS concentration in solution on the fractal dimension D_{f_2}. In the 25 to 200 ng/ml TRITC-LPS

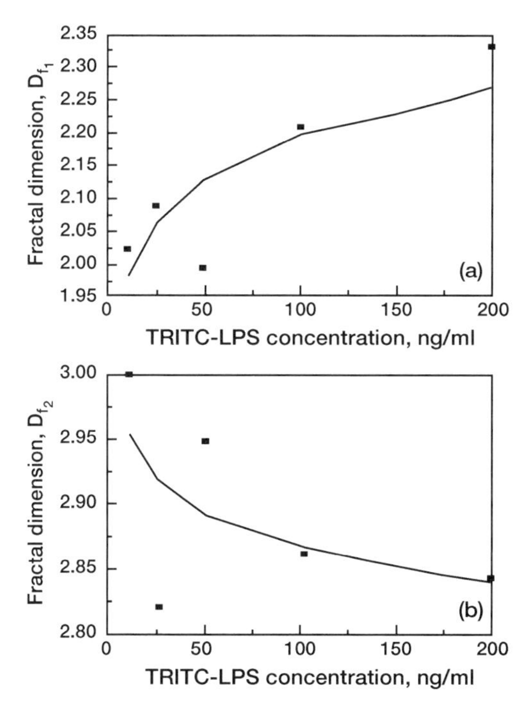

FIGURE 6.22 Influence of the TRITC-LPS concentration in solution on the fractal dimensions. (a) D_{f_1}; (b) D_{f_2}.

concentration, D_{f_2} is given by

$$D_{f_2} = (3.0445 \pm 0.0746)[\text{TRITC-LPS}]^{-0.01294 \pm 0.0103}. \tag{6.11d}$$

Note that an increase in the TRITC-LPS concentration in solution leads to a decrease in D_{f_2}. There is scatter in the data at the lower TRITC-LPS concentrations used. This scatter is also indicated by the "error" in estimating the exponent dependence in Eq. (6.11d). Once again, the fractal dimension, D_{f_2} is rather insensitive to the TRITC-LPS concentrations in solution, as noted by the low exponent dependence on [TRITC-LPS] in Eq. (6.11d). More data points are required, especially at the lower TRITC-LPS concentrations, to more firmly establish the predictive equation.

Figure 6.23a shows that the binding rate coefficient, k_1, increases as the fractal dimension, D_{f_1}, increases. For the data presented in Table 6.9, k_1 is given by

$$k_1 = (0.0793 \pm 0.0292) D_{f_1}^{7.6335 \pm 2.4367}. \tag{6.11e}$$

The fit of the above predictive equation is reasonable. The binding rate

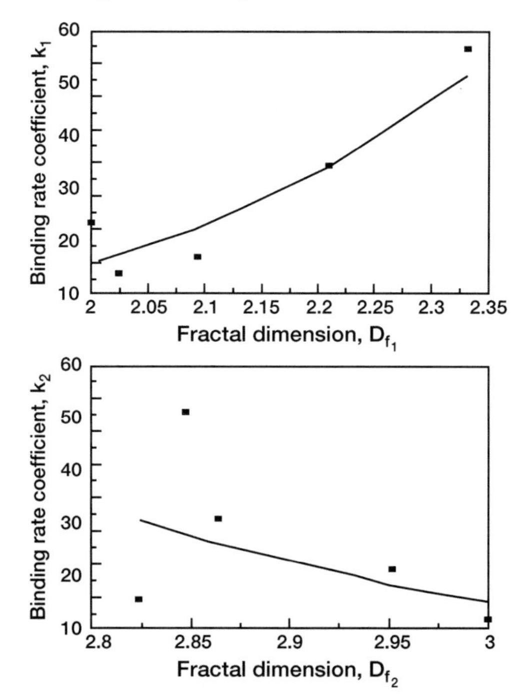

FIGURE 6.23 Influence of (a) the fractal dimension, D_{f_1}, on the binding rate coefficient, k_1; (b) the fractal dimension, D_{f_2} on the binding rate coefficient, k_2.

coefficient, k_1, is very sensitive to the fractal dimension, D_{f_1}, as noted by the high exponent dependence of k_1 on D_{f_1}. Once again, this underscores that the binding rate coefficient, k_1, is very sensitive to the surface roughness or heterogeneity that exists on the biosensor surface.

Figure 6.23b shows that the binding rate coefficient, k_2, decreases as the fractal dimension, D_{f_2}, increases. For the data presented in Table 6.9, k_2 is given by

$$k_2 = (4.9640 \pm 3.7788) D_{f_2}^{-11.55 \pm 10.9525}. \tag{6.11f}$$

There is scatter in the data, especially at the lower values of D_{f_2} presented. This is reflected in the both of the coefficients shown in Eq. (6.11f). More data is required at the lower D_{f_2} values to more firmly establish the predictive equation. However, the k_2 presented is still very sensitive to D_{f_2} or the heterogeneity that exists on the biosensor surface. Once again, the data analysis in itself does not provide any evidence for surface roughness or heterogeneity, and the assumed existence of surface roughness or heterogeneity may not be correct.

Mauro *et al.* (1996) used fluorometric sensing to detect polymerase chain-reaction–amplified DNA using a DNA capture protein immobilized on a fiber-optic biosensor. These authors utilized amplified DNA labeled with the fluorophore tetramethylrhodamine and the AP-1 consensus nucleotide sequence recognized by GCN4. This DNA was noncovalently bound to IgG-modified fibers. Wanting to see if they could reuse the fiber, the authors performed regeneration studies. They focused their attention on conditions that would permit the release of the bound DNA while leaving the IgG-PG/GCN4 assembly in a functional state. Figure 6.24 shows the curves obtained using Eqs. (6.2a) (single-fractal analysis) and (6.2b) (dual-fractal analysis) for ten consecutive runs. The points are the experimental results obtained by Mauro *et al.* A dual-fractal analysis was required since the single-fractal analysis did not provide an adequate fit for the binding curves.

Table 6.10 shows the values of k and D_f obtained using Sigmaplot (1993) to fit the data. The values of the parameters presented are within 95% confidence limits. For example, the value of k reported for run 2 (single-fractal analysis) is 40.248 ± 36.3232. The 95% confidence limit indicates that 95% of the k values will lie between 3.925 and 76.571. Since a dual-fractal analysis was needed to adequately model the binding curves, the results obtained from the single-fractal analysis will not be analyzed further. The D_{f_1} values reported for each of the ten runs were all equal to zero. This is due to the sigmoidal shape or concave nature of the curve (toward the origin) at very low values of time, t.

Figures 6.25a and 6.25b show the fluctuations in the binding rate coefficient, k_2, and in the fractal dimension, D_{f_2}, respectively, as the run

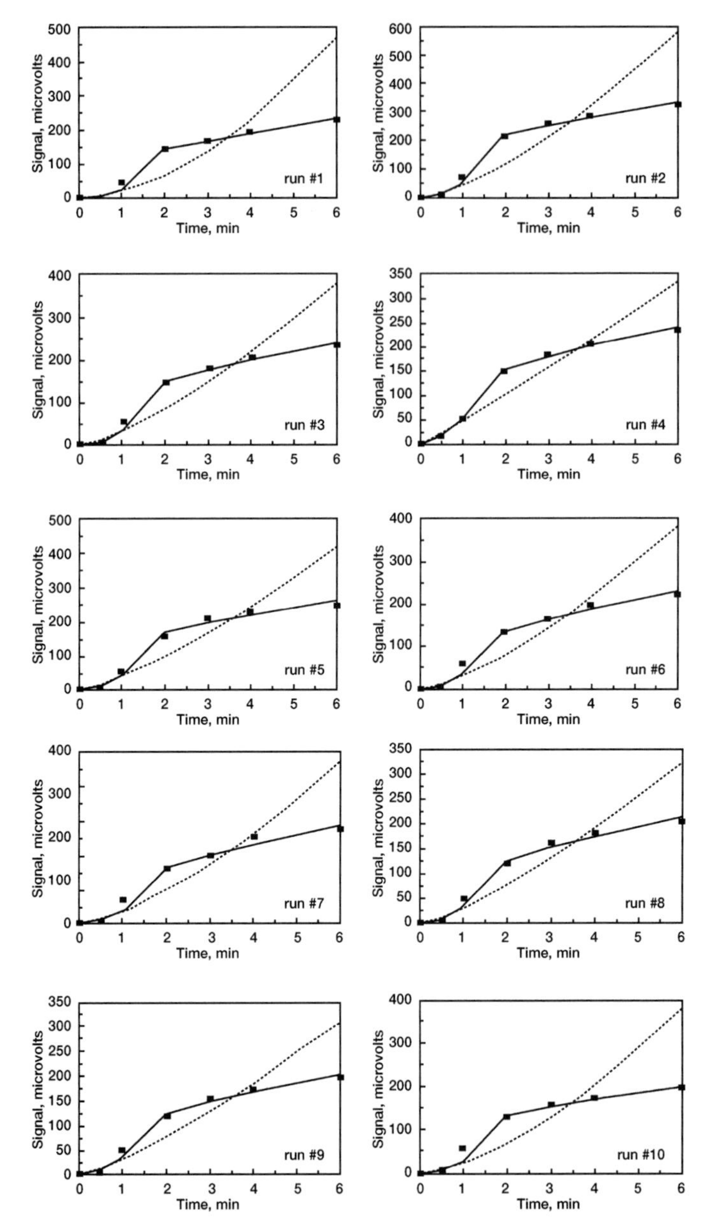

FIGURE 6.24 Influence of the run number (regeneration study) on the binding of polymerase chain-reaction–amplified DNA in solution using a DNA capture protein immobilized on a fiber-optic biosensor (Mauro *et al.*, 1996) (- - -, single-fractal analysis; —, dual-fractal analysis).

TABLE 6.10 Influence of Run Number (Regeneration Cycle) on the Fractal Dimensions and Binding Rate Coefficients for Amplified DNA in Solution and a DNA Capture Protein Immobilized on a Fiber-Optic Biosensor[a] (Mauro *et al.*, 1996)

Run number	k	D_f	k_1	k_2	D_{f_1}	D_{f_2}
One (fresh)	$19.076 + 26.414$	0	$23.607 + 29.833$	107.669 ± 1.828	0	2.1644 ± 0.0418
Two	40.248 ± 36.323	$0.0544 + 0.6238$	46.934 ± 29.645	169.00 ± 5.273	0	2.2938 ± 0.0764
Three	35.967 ± 24.530	0.3570 ± 0.5042	40.6558 ± 19.523	115.432 ± 3.886	0	2.1760 ± 0.0824
Four	48.212 ± 17.995	$0.8478 + 0.3076$	51.639 ± 2.9063	116.229 ± 4.273	0	2.1904 ± 0.0898
Five	38.945 ± 25.110	$0.3300 + 0.4826$	43.203 ± 16.159	127.613 ± 11.169	0	2.1694 ± 0.2088
Six	29.514 ± 27.640	$0.1418 + 0.6404$	$34.445 + 34.915$	99.029 ± 4.918	0	2.0646 ± 0.1206
Seven	33.251 ± 31.088	0	33.251 ± 31.088	91.417 ± 6.526	0	1.9612 ± 0.1716
Eight	32.596 ± 29.405	0.4736 ± 0.4530	36.098 ± 16.385	91.122 ± 5.554	0	2.0794 ± 0.1472
Nine	33.145 ± 30.602	0.4954 ± 0.4662	30.820 ± 19.374	91.136 ± 5.417	0	2.0820 ± 0.1438
Ten	$23.406 + 29.513$	$0. + 0.7912$	$28.281 + 39.879$	102.200 ± 3.804	0	2.2266 ± 0.0910

[a]4.8 ng/ml amplified DNA with fluorophone tetramethylrhodamine and the AP-1 consensus nucleotide sequence recognized by GCN4 in solution/PG/GCN4 protein non-covalently bound to IgG-modified fibers.

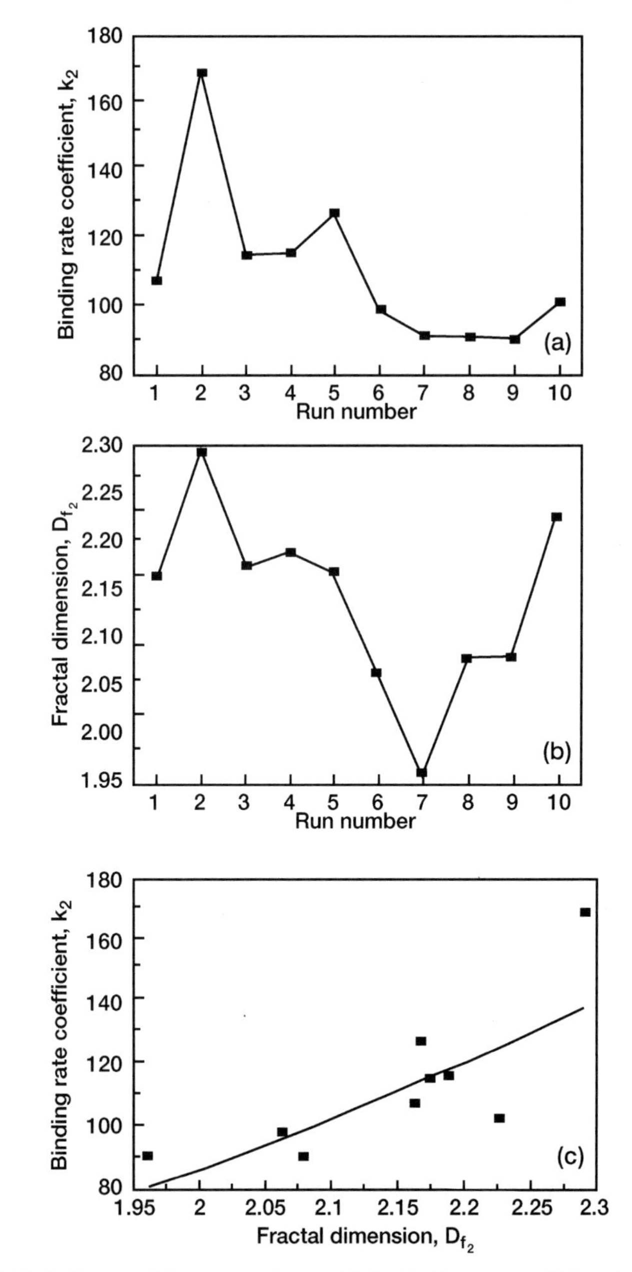

FIGURE 6.25 Influence of the run number on (a) the binding rate coefficient, k_2, and (b) the fractal dimension, D_{f_2}, during regeneration studies. (c) Increase in the binding rate coefficient, k_2, with an increase in the fractal dimension, D_{f_2}, during regeneration.

number increases from one to ten. No pattern is easily discernible from the data presented in Figs. 6.25a and 6.25b. Table 6.10 and Fig. 6.25c indicate that an increase in D_{f_2} leads to a linear increase in k_2, but there is scatter in the data. An increase in D_{f_2} by about 16.9%—from a value of 1.9612 to 2.2938—leads to a 85.4% increase in k_2—from a value of 91.122 to 169. For the regeneration runs, k_2 may be given by

$$k_2 = (8.229 \pm 1.096) D_{f_2}^{3.3997 \pm 0.9319}. \tag{6.12}$$

Equation (6.12) predicts the k_2 values presented in Table 6.10 reasonably well. There is some deviation in the data. Note the high exponent dependence of k_2 on D_{f_2}. This underscores that k_2 is sensitive to the surface roughness or the degree of heterogeneity that exists on the surface.

Blair *et al.* (1997) utilized a fiber-optic biosensor to monitor volatile organic compounds in water. The authors analyzed the transport of trichloroethylene, 1,1,1-trichloroethane, and toluene in aqueous solution using a Fickian diffusion model. They utilized an evanescent fiber-optic chemical sensor (EFOCS) and presumed that the transport occurs through a polydimethylsiloxane film. A hydrophobic polymer material was utilized as the cladding material for the fiber-optic biosensor. We now reanalyze their data using a fractal analysis, since that assists in describing the heterogeneity that exists on the biosensor surface. Figures 6.26a–6.26c show the curves obtained using Eqs (6.2a) and (6.2b) for the binding of three different concentrations (74, 148, and 370 ppm) of trichloroethylene in solution to the EFOCS biosensor. Clearly, a dual-fractal analysis provides a better fit for all three concentrations. The binding rate coefficient and fractal dimension values are presented in Table 6.11.

Table 6.11 indicates that both k_1 and k_2 increase as the trichloroethylene concentration in solution increases from 74 to 370 ppm. In this concentration range, k_1 is given by

$$k_1 = (0.6625 \pm 0.4276)[\text{trichloroethylene}]^{1.0711 \pm 0.4512}.$$

Figure 6.27a shows that this predictive equation fits the values of k_1 presented in Table 6.11 reasonably well. More data points are required to more firmly establish this equation. The dependence of k_1 on concentration is close to first order. The fractional order of dependence of k_1 on the tricloroethylene concentration reinforces the fractal nature of the system.

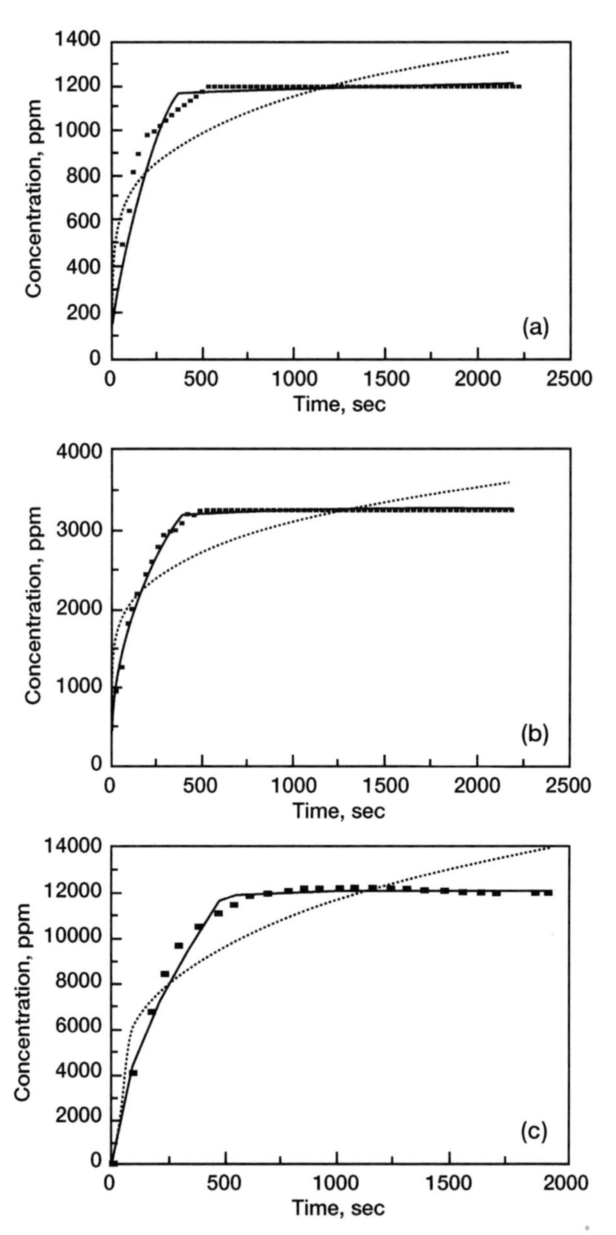

FIGURE 6.26 Binding rate curves for three different trichloroethylene concentrations in solution (in ppm) to the evanescent fiber-optic chemical sensor (EFOCS) (Blair *et al.*, 1997). (- - -, single-fractal analysis; —, dual-fractal analysis) (a) 74; (b) 148; (c) 370.

TABLE 6.11 Influence Of Different Parameters on Fractal Dimensions and Binding Rate Coefficients for Analyte-Receptor Binding Kinetics (Blair *et al.*, 1997)

Analyte concentration in solution/receptor on surface	k	D_f	k_1	k_2	D_{f_1}	D_{f_2}
74 ppm trichloroethylene/hydrophobic polymer	270.818 ± 46.152	2.580 ± 0.042	52.447 ± 9.118	1038.83 ± 14.16	1.9454 ± 0.1106	2.9604 ± 0.0736
148 ppm trichloroethylene/hydrophobic polymer	828.719 ± 108.039	2.6152 ± 0.0328	212.79 ± 12.16	2964.32 ± 32.726	2.0924 ± 0.0380	2.9747 ± 0.0059
370 ppm trichloroethylene/hydrophobic polymer	1796.971 ± 264.143	2.4584 ± 0.0709	311.646 ± 27.342	10523.66 ± 130.12	1.8266 ± 0.1093	2.9616 ± 0.0155

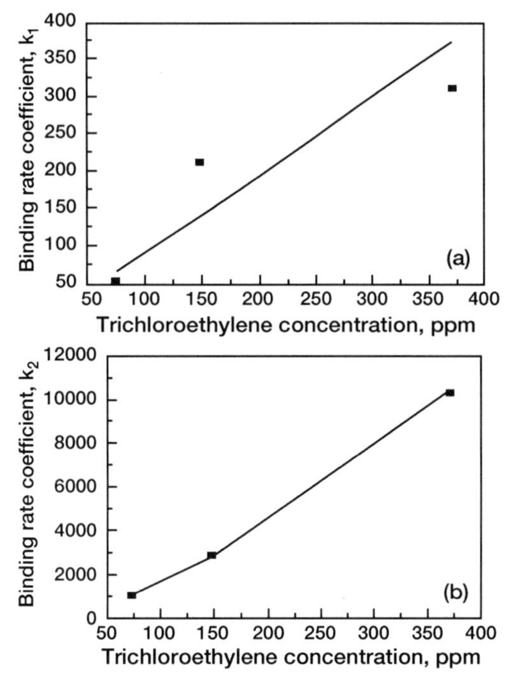

FIGURE 6.27 Increase in the binding rate coefficients (a) k_1 and (b) k_2 with an increase in the trichloroethylene concentration (in ppm) in solution (Blair *et al.*, 1997).

In the 74 to 370 ppm trichloroethylene concentration in solution utilized, k_2 is given by

$$k_2 = (2.1878 \pm 0.0940)[\text{trichloroethylene}]^{1.4632 \pm 0.03684}.$$

Figure 6.27b shows that this predictive equation fits the values of k_2 presented in Table 6.11 well. More data points are required to more firmly establish this equation. Once again, the fractional order of dependence of k_2 on the trichloroethylene concentration reinforces the fractal nature of the system.

Table 6.12 summarizes some of the binding rate expressions obtained as a function of both the analyte concentration in solution and the fractal dimension. Expressions obtained for the single- and the dual-fractal analysis are presented. Even though only a few examples are presented, the table does begin to provide us with an overall perspective of the nature of the binding rate coefficient dependence on the analyte concentration in solution and on the fractal dimension.

TABLE 6.12 Binding Rate Coefficient Expressions as a Function of the Fractal Dimension and the Analyte Concentration in Solution

Analyte/receptor system	Binding rate coefficient expression	Reference
Polymerase chain-reaction–amplified DNA/DNA capture protein (regeneration)	$k_2 = (8.279 \pm 1.096)D_{f_2}^{3.3997 \pm 0.9319}$	Mauro *et al.*, 1996
1 μM BSA/anti-BSA-protein G	$k = (15.541 \pm 4.919)(\mathrm{IgG/Fc})^{-1.4260 \pm 0.7592}$ $k = (0.1348 \pm 0.003)D_f^{5.5351 \pm 0.2084}$	Ingersoll and Bright, 1997
74 to 370 ppm trichloroethylene/hydrophobic polymer	$k_1 = (0.6625 \pm 0.4276)[\mathrm{trichloroethylene}]^{1.0711 \pm 0.4512}$	Blair *et al.*, 1997

Su *et al.* (1997) analyzed the hybridization kinetics of interfacial RNA homopolymer using a thickness-shear mode acoustic wave biosensor. The authors indicate that the binding or hybridization mechanism involves the diffusion of the RNA probe molecules in solution followed by duplex formation at the surface. Their analysis included the influence of temperature, buffer solutions, and blocking agents on the hybridization kinetics.

Figure 6.28a shows the binding of RNA homopolymer in solution to polycytidylic acid (5′) (poly C) immobilized on an electrode surface. The rough electrode surface was treated with Denhardt's reagent (stock solution containing 10 g of Ficoll, 10 g of polyvinylpyrrolidone, and 10 g of bovine serum albumin in 500 mL of water.) This was run 2, carried out at 24°C. A single-fractal analysis is sufficient to adequately describe the binding kinetics. The values of k and D_f are given in Table 6.13. Figure 6.28b shows the binding of RNA homopolymer in solution to poly C immobilized on an electrode surface. In this case, run 1, the rough electrode surface was untreated. This time a dual-fractal analysis is required to adequately describe the binding kinetics. The values of k and D_f for a single-fractal analysis and k_1 and k_2 and D_{f_1} and D_{f_2} for a dual-fractal analysis are given in Table 6.13.

Figure 6.28c shows the binding of RNA homopolymer in solution to poly C immobilized on an electrode surface. In this case, the rough electrode surface was treated with Denhardt's reagent and ss salmon test DNA together. This is run 3, performed at 24°C. Once again, a dual-fractal analysis is required to adequately describe the binding kinetics. The values of k and D_f, k_1 and k_2, and D_{f_1} and D_{f_2} are given in Table 6.13. Note that when one goes from a single-fractal analysis to a dual-fractal analysis to describe the binding curves,

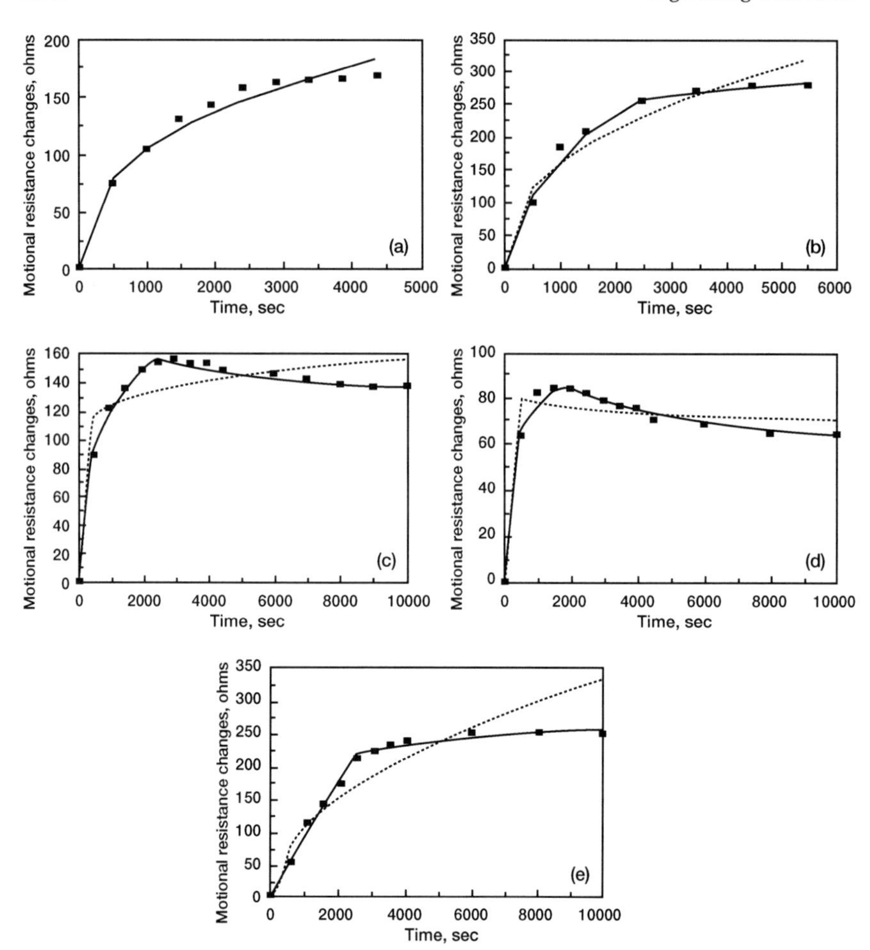

FIGURE 6.28 RNA homopolymerization on a electrode-solution surface (Su *et al.*, 1997). (a) Run 2; (b) run 1; (c) run 3; (d) run 4; (e) run 5.

TABLE 6.13 Influence of Different Conditions on the Fractal Dimensions and Binding Rate Coefficients for RNA Homopolymer Hybridization on an Electrode-Solution Interface (Su *et al.*, 1997)

Run number	k	D_f	k_1	k_2	D_{f_1}	D_{f_2}
Two	8.66 ± 0.59	2.27 ± 0.06	na	na	na	na
One	9.87 ± 1.49	2.19 ± 0.13	2.99 ± 0.39	102.46 ± 1.63	1.84 ± 0.21	2.76 ± 0.05
Three	62.34 ± 7.92	2.80 ± 0.07	11.19 ± 0.44	358.82 ± 5.52	2.32 ± 0.06	$3.0 - 0.02$
Four	101.81 ± 11.28	$3.0 - 0.07$	18.04 ± 1.29	322.69 ± 5.76	2.58 ± 0.13	$3.0 - 0.02$
Five	3.98 ± 0.89	2.03 ± 0.14	0.36 ± 0.03	87.54 ± 2.37	1.36 ± 0.13	2.76 ± 0.04

there is an inherent change in the binding mechanism taking place. For example, consider the curves presented in Fig. 6.28a versus Figs. 6.28b and 6.28c. Figure 6.28a requires a single-fractal analysis to describe the binding curve, whereas in Figs. 6.28b and 6.28c a dual-fractal analysis is required.

Figure 6.28d (run 4, carried out at 24°C) shows the binding of RNA homopolymer in solution to poly C immobilized on an electrode surface. In this case, 5 X SSC (NaCl and sodium citrate) hybridization buffer solution was used. A dual-fractal analysis is required to adequately describe the binding kinetics, and the values of k and D_f, k_1 and k_2, and D_{f_1} and D_{f_2} are given in Table 6.13. Figure 6.28e (run 5) shows the binding of RNA homopolymer in solution to poly C immobilized on an electrode surface where the electrode surface was smooth. Once again, a dual-fractal analysis is required to adequately describe the binding kinetics, and the values of k and D_f, k_1 and k_2, and D_{f_1} and D_{f_2} are given in Table 6.13.

Compare the results presented in Figs. 6.28b (run 1) and 6.28e (run 5), where a dual-fractal analysis was used to describe the binding kinetics. Both are untreated electrode surfaces. In run 1, the electrode surface was rough, and in run 5, the electrode surface was smooth. One might anticipate that the fractal dimension in run 1 would be higher than that observed in run 5. This should be true at time t close to zero, which correctly describes the nature of the surface and has not been influenced by the reaction occurring at the surface. Table 6.13 indicates that, as expected, the fractal dimension for run 1 ($D_{f_1} = 1.8438$) is higher than that for run 5 ($D_{f_1} = 1.3602$). The second fractal dimensions, D_{f_2}, obtained for run 1 (2.7637) and run 5 (2.7616) are very close to each other. In this case, the reaction surface has now been made complex due to the reaction taking place on the surface. Besides, the maximum value that the fractal dimension can have is 3. Thus, no comments are made with regard to a comparison of the D_{f_2} values observed.

Finally, concerning the value of the D_{f_1} and k_1 obtained in Table 6.13, note that an increase in D_{f_1} leads to an increase in k_1. This is also seen in Fig. 6.29. For run 1, 3, 4, and 5, where a dual-fractal analysis is applicable, the binding rate coefficient, k_1, is given by

$$k_1 = (0.059 \pm 0.010) D_{f_1}^{6.17 \pm 0.32}. \tag{6.14}$$

Bowles *et al.* (1997) analyzed the binding of a Fab fragment of an antiparaquat monoclonal antibody in solution and an immobilized antigen in the form of a paraquat analog immobilized on a sensor surface. One of their aims was to develop a method of obtaining kinetic constants from data (sensorgrams) that exhibited other than first-order behavior or deviated from pseudo-first-order behavior. Figure 6.30a shows the binding of a 4 μM Fab fragment of an antiparaquat antibody in solution to an antigen (paraquat

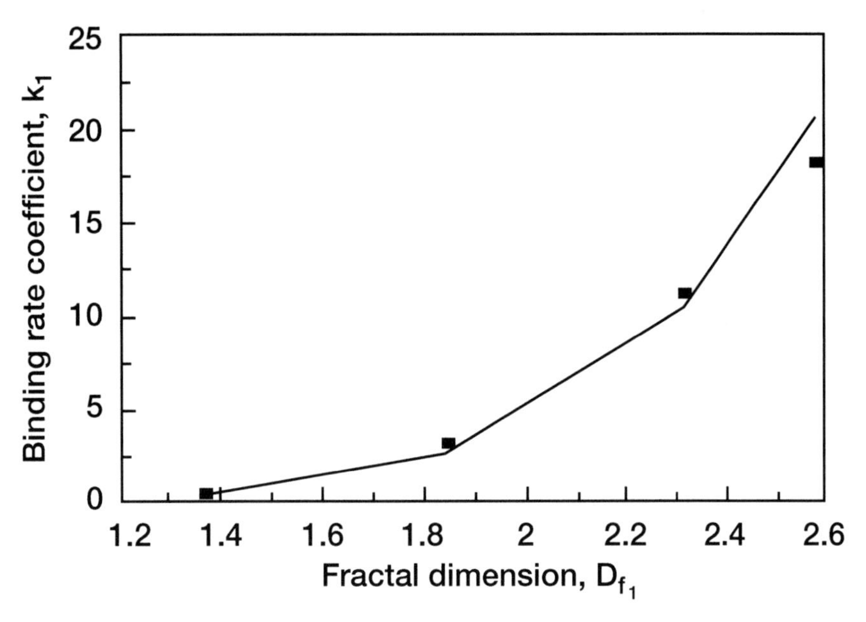

FIGURE 6.29 Influence of the fractal dimension, D_{f_1}, on the binding rate coefficient, k_1.

analog) immobilized on a sensor chip. A dual-fractal analysis is required to adequately describe the binding kinetics. The values of k and D_f, k_1 and k_2, and D_{f_1} and D_{f_2} are given in Table 6.14. Figure 6.30 shows the binding of 1.25 μM Fab fragment of an antiparaquat antibody in solution to a paraquat (analog) sites immobilized on a sensor chip. In this case too, a dual-fractal analysis is required to adequately describe the binding kinetics. The values of k and D_f, k_1 and k_2, and D_{f_1} and D_{f_2} are given in Table 6.14. Figure 6.30c shows the binding of 10 μM Fab fragment of an antiparaquat antibody in solution to paraquat (analog) sites immobilized on a sensor chip. Again, a dual-fractal analysis is required to adequately describe the binding kinetics, and the values of k and D_f, k_1 and k_2, and D_{f_1} and D_{f_2} are given in Table 6.14. Note that an increase in the Fab fragment concentration in solution by a factor of 8—from 1.25 to 10 μM—leads to increases in the binding rate coefficients, k_1 and k_2, and in the fractal dimensions, D_{f_1} and D_{f_2}. For example, k_1 increases by a factor 11.85—from 650.418 to 7705.94.

For the three runs presented in Table 6.14, note that an increase in the fractal dimension, D_{f_1}, leads to an increase in the binding rate coefficient, k_1. For these runs, k_1 is given by

$$k_1 = (13.1512 \pm 10.6046) D_{f_1}^{7.2450 \pm 1.6553}. \qquad (6.15)$$

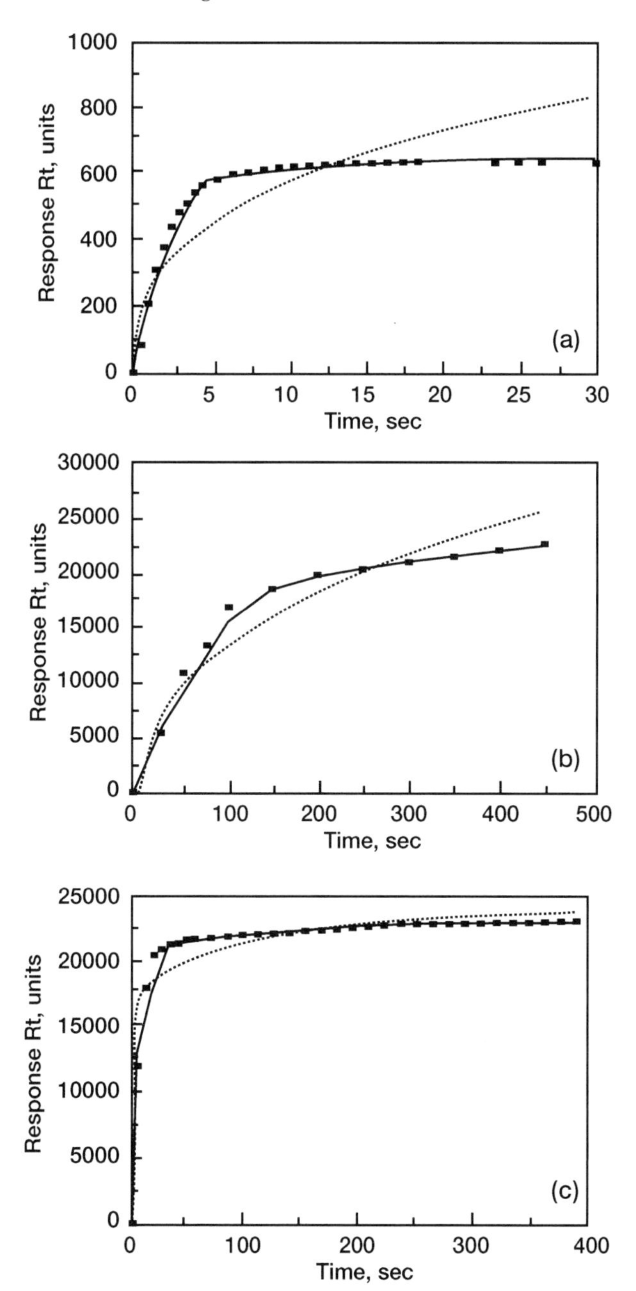

FIGURE 6.30 Binding of the Fab fragment of an antiparaquat antibody in solution to a paraquat analog immobilized on a sensor chip (Bowles *et al.*, 1997). (a) 4 μM; (b) 1.25 μM; (c) 10 μM.

TABLE 6.14 Fractal Dimensions and Binding Rate Coefficients for the Binding of Fab Fragment of an Antiparaquat Antibody in Solution to Paraquat Analog Immobilized on a Sensor Chip (Bowles *et al.*, 1997)

Fab fragment in solution/receptor on surface	k	D_f	k_1	k_2	D_{f_1}	D_{f_2}
4 μM Fab fragment of an antiparaquat antibody/paraquate analog immobilized on sensor chip	256.52 ± 72.64	2.30 ± 0.09	193.78 ± 38.26	527.55 ± 10.38	1.53 ± 0.15	2.87 ± 0.02
1.25 μM Fab fragment of an antiparaquat antibody/paraquat analog immobilized on sensor chip	1867.78 ± 319.43	2.14 ± 0.11	650.42 ± 82.27	7623.69 ± 46.90	1.61 ± 0.17	2.64 ± 0.10
10 μM Fab fragment of an antiparaquat antibody/paraquat analog immobilized on sensor chip	14074.87 ± 1097.87	2.82 ± 0.03	7705.94 ± 757.34	18275.05 ± 75.08	2.43 ± 0.11	2.92 ± 0.00

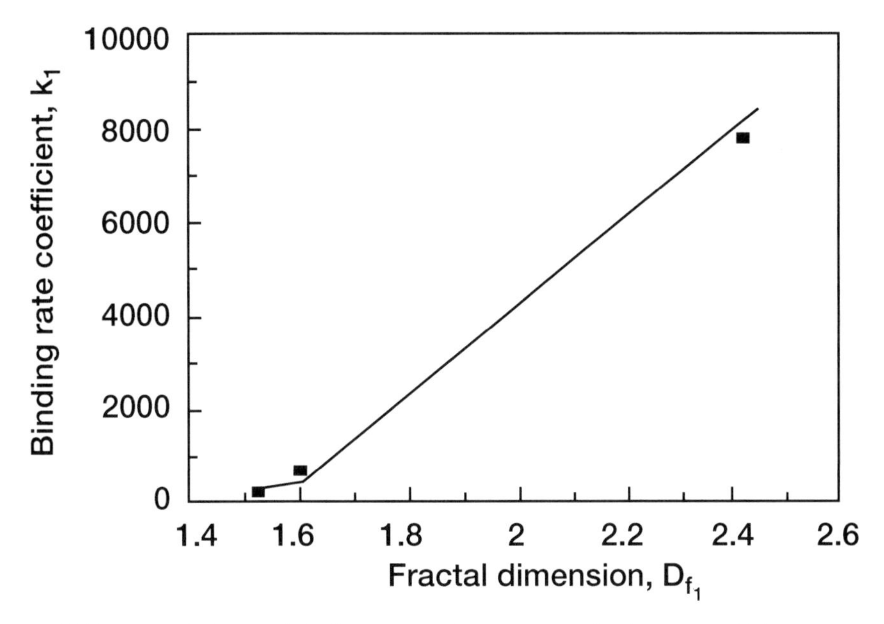

FIGURE 6.31 Influence of the fractal dimension, D_{f_1}, on the binding rate coefficient, k_1.

Figure 6.31 shows this equation fits the data reasonably well. There is some deviation in the estimated value of the exponent in spite of the fact that the fit is very reasonable. Since only three data points are available, more data points would more firmly establish this equation. The binding rate coefficient is very sensitive to the degree of heterogeneity on the surface, as may be noted by the very high value of the exponent. The fractional order of dependence further reinforces the fractal nature of the system.

Note that an increase in the degree of heterogeneity on the surface leads to an increase in the binding rate coefficient. One possible explanation for this effect is that due to the rough surfaces additional interactions arise; for example, there is a generation of turbulent flow by the protruding surfaces and entrapment of fluid in small "holes" on the surface (Martin *et al.*, 1991). This turbulence would enhance the localized mixing, which would lead to a decrease in the diffusional limitations. This decrease should then lead to an increase in the binding rate coefficient. Other suitable (and perhaps more appropriate) explanations to elucidate this effect are possible.

6.4. CONCLUSIONS

A fractal analysis of analyte-receptor binding kinetics presented provides examples of cases when (1) a single-fractal analysis is applicable, (2) a single-as well as a dual-fractal analysis is applicable, and (3) only a dual-fractal analysis is applicable. The fractal dimension, D_f, provides a quantitative measure of the degree of heterogeneity that exists on the surface. Also, when there is a change from a single- to a dual-fractal analysis or when a dual-fractal analysis is applicable throughout the reaction, this indicates that there is change in the binding mechanism as the reaction proceeds on the biosensor surface.

The predictive relationships developed for the binding rate coefficient as a function of a reactant concentration are of particular value since they provide an avenue by which one may manipulate the binding rate coefficient. These relationships have been developed for all the three cases mentioned. In each case, the noninteger dependence of the binding rate coefficient on the reactant (or analyte) concentration lends further support for the fractal nature of the system.

The fractal dimension is not a typical independent variable such as is analyte concentration, that may be directly manipulated. It is estimated from Eqs. (6.2a) and (6.2b) and may be considered as a derived variable. The predictive relationships developed for the binding rate coefficient as a function of the fractal dimension when any of the three cases are prevalent are of considerable value since (1) they directly link the binding rate coefficient to the degree of heterogeneity that exists on the surface, and (2) they provide a means by which the binding rate coefficient may be manipulated—by changing the degree of heterogeneity that exists on the surface. In some cases, the binding rate coefficient is rather sensitive to D_f or the degree of heterogeneity that exists on the biosensor surface. This may be seen in the high order of dependence (and in some cases very high order) of the binding rate coefficient on the fractal dimension.

More such studies are required to determine whether the binding rate coefficient is sensitive to the fractal dimension or the degree of heterogeneity that exists on the biosensor surface. If this is indeed so, then this would encourage experimentalists to pay more attention to the nature of the surface, as well as to how it may be manipulated to control the relevant parameters and biosensor performance in desired directions. Finally, in a more general sense, the treatment should also be more or less applicable to nonbiosensor applications wherein further physical insights could be obtained.

REFERENCES

Abel, A. P., Weller, M. G., Duveneck, G. L., Ehrat, M., and Widmer, H. M., *Anal. Chem.* **68**, 2905 (1996).

Anderson, J., NIH Panel Review Meeting, Case Western Reserve University, Cleveland, OH, July 1993.

Asanov, N., Wilson, W. W., and Oldham, P. B., *Anal. Chem.* **70**, 1156 (1998).
Bier, F. F., Kleinjung, F., and Scheller, F. W., *Sensors & Actuators* **38–39**, 78–82 (1997).
Blair, D. S., Burgess, L. W., and Brodsky, A. M., *Anal. Chem.* **69**(3), 2238 (1997).
Bluestein, B. I., Craig, M., Slovacek, R., Stundtner, L., Uricouli, C., Walziak, I., and Luderer, A., in *Biosensors with Fiberoptics* (D. Wise and L. B. Wingard Jr., eds.), Humana Press, Clifton, NJ, pp. 181–223, 1991.
Bowles, M. R., Hall, D. R., Pond, S. M., and Winzor, D. J., *Anal. Biochem.* **244**, 133–143 (1997).
Chiu, N. H. L. and Christpoulos, T. K., *Anal. Chem.* **68**, 2304–2308 (1996).
Choo, Q. L., Kuo, G., Weiner, A. J., Overby, L. R., Bradley, D. W., and Houghton, M., *Science* **244**, 359–362 (1989).
Christensen, L. L. H., *Anal. Biochem.* **249**, 153–164 (1997).
Clyne, M. and Drumm, B., *FEMS Immunol. Med. Microbiol.* **16**, 141 (1996).
Cush, R., Crinin, J. M., Stewart, W. J., Maule, C. H., Molloy, J., and Goddard, N. J., *Biosen. & Bioelectr.* **8**, 347 (1993).
Cuypers, P. A., Willems, G. M., Hemker, H. C., and Hermans, W. T. in "Blood in contact with natural and artificial surfaces," E. F. Leonard, V. T. Turitto, and C. Vroman, eds., *Annals N.Y. Acad. Sci.* **516**, 244–252 (1987).
Dewey, T. G. and Bann, J. H., *Biophys. J.* **63**, 594–598 (1992).
Di Cera, E., *J. Chem. Phys.* **95**(2), 5082–5086 (1991).
Eddowes, M. J., *Biosensors* **3**, 1–15 (1987/1988).
Edwards, P. R., Gill, A., Pollard-Knight, D. V., Hoare, M., Buckle, P. E., Lowe, P. A., and Leatherbarrow, R. J., *Anal. Biochem.* **231**, 210–217 (1995).
Elbicke, J. M., Morgan, D. M., and Weber, S. G., *Anal. Chem.* **56**, 978–985 (1984).
Fischer, R. J., Fivash, M., Casas-Finet, J., Bladen, S., and McNitt, K. L., *Methods: Companion to Methods in Enzymol.* **6**, 121–133 (1994).
Gaillot, C., Larre, C., Caminade, A-M., and Majoral, J-P., *Science* **277**, 1981–1984 (1997).
Giaver, I., *J. Immunol.* **116**, 766–771 (1976).
Glaser, R. W., *Anal. Biochem.* **213**, 152–161 (1993).
Havlin, S., in *The Fractal Approach to Heterogeneous Chemistry: Surfaces, Colloids, and Polymers* (D. Avnir, ed.), Wiley, New York pp. 251–269 (1989).
Hirmo, S., Artusson, E., Puu, G., Wadstrom, T., and Nilsson, B., *Anal. Biochem.* **257**, 63 (1998).
Ingersoll, C. M. and Bright, F. V., *Anal. Chem.* 403A–408A (1997).
James, E. A., Schmeltzer, K., and Ligler, F. S., *Appl. Biochem. & Biotech.* **60**, 180 (1996).
Kopelman, R., *Science* **241**, 1620–1626 (1988).
Kleinjung, F., Klussman, S., Erdmann, V. A., Scheller, F. W., Furste, J. P., and Bier, F. F., *Anal. Chem.* **70**, 328–331 (1998).
Kuo, G., Chou, Q. L., Alter, H. H., Gitnick, G. L., Redeker, A. G., Purcell, R. H., Miyamura, T., Dienstag, J. L., Alter, M. J., and Stevens, C. E., *Science* **244**, 362–364 (1989).
Kyono, K., Miyashiro, M., and Taguchi, I., *Anal. Biochem.* **257**, 120 (1998).
Liebovitch, L. S. and Sullivan, J. M., *Biophys. J.* **52**, 979–988 (1987).
Liebovitch, L. S., Fischbarg, J., Koniarek, J. P., Todorova, I., and Wang, M., *Math. Biosci.* **84**, 37–68 (1987).
Li, H., Chen, S. and Zhao, H., *Biophys. J.* **58**, 1313–1320 (1990).
Markel, V. A., Muratov, L. S., Stockman, M. I., and George, T. F., *Phys. Rev. B* **43**(10), 8183–8195 (1991).
Martin, S. J., Granstaff, V. E., and Frye, G. C., *Anal. Chem.* **65**, 2910 (1991).
Matsuda, H., *J. Electroanal. Chem.* **179**, 107–117 (1967).
Mauro, J. M., Cao, L. K., Kondracki, L. M., Walz, S. E. and Campbell, J. R., *Anal. Biochem.* **235**, 61 (1996).
Milum, J. and Sadana, A., *J. Colloid & Interface Sci.* **187**, 128–138 (1997).

Morton, T. A., Myszka, D. G., and Chaiken, I. M., *Anal. Biochem.* **227**, 176–185 (1995).
Myszka, D. G., Morton, T. A., Doyle, M. L., and Chaiken, I. M., *Biophys. Chem.* **64**, 127–137 (1997).
Nygren, H. A. and Stenberg, M., *J. Colloid & Interface Sci.* **107**, 560–566 (1985).
Nylander, C., Liedberg, B., and Lind, T., *Sensors & Actuators* **3**, 79 (1991).
Pajkossy, T. and Nyikos, L., *Electrochim. Acta* **34**(2), 171–177 (1989).
Pfeifer, P. and Obert, M., in *The Fractal Approach to Heterogeneous Chemistry: Surfaces, Colloids, and Polymers* (D. Avnir, ed.), Wiley, New York, pp. 11–43, 1989.
Pisarchick, M. L., Gesty, D., and Thompson, N. L., *Biophys. J.* **63**, 215–223 (1992).
Place, J. F., Sutherland, R. M., Riley, A., and Mangan, C., in *Biosensors with Fiberoptics* (D. Wise and L. B. Wingard Jr., eds.), Humana Press, Clifton, NJ, pp. 253–291, 1991.
Richalet-Secordel, P. M., Rauffer-Bruyere, N., Christensen, L. L. H., Ofenloch-Haehnle, B., Seidel, C., and van Regenmortel, M. H. V., *Anal. Biochem.* **249**, 165–173 (1997).
Rogers, J. and Edelfrawi, M., *Biosen. & Bioelectr.* **6** (1991).
Sadana, A., *Biotechnol. Progr.* **11**, 50–57 (1995).
Sadana, A., *J. Colloid & Interface Sci.* **190**, 232–240 (1997).
Sadana, A., *J. Colloid & Interface Sci.* **198**, 164–178 (1998).
Sadana, A. and Beelaram, A., *Biosen. & Bioelectr.* **10**(3–4), 301–316 (1995).
Sadana, A. and Madagula, A., *Biotechnol. Progr.* **9**, 259–266 (1993).
Sadana, A. and Madagula, A., *Biosen. & Bioelectr.* **9**, 45–55 (1994).
Sadana, A. and Sii, D., *J. Colloid & Interface Sci.* **151**(1), 166–177 (1992a).
Sadana, A. and Sii, D., *Biosen. & Bioelectr.* **7**, 559–568 (1992b).
Sadana, A. and Sutaria, M., *Biophys. Chem.* **65**, 29–44 (1997).
Schmitt, H. M., Brecht, A., Piehler, J., and Gauglitz, G., *Biosen. & Bioelectr.* **12**(8), 809–816 (1997).
Schmitt, H. M., Brecht, A., Piehler, J., and Gauglitz, G., *Biosen. & Bioelectr.* **12**(8), 809 (1997).
Scheller, F. W., Hintsche, R., Pfeifer, D., Schubert, D., Reidel, K., and Kindevater, R., *Sensors & Actuators* **4**, 197–206 (1991).
Sigmaplot, Scientific Graphing Software, User's manual, Jandel Scientific, San Rafael, CA, 1993.
Sjolander, S. and Urbaniczky, C., *Anal. Chem.* **63**, 2338–2345 (1991).
Sorenson, C. M. and Roberts, G. C. *J. Colloid & Interface Sci.* **186**, 447–456 (1997).
Stenberg, M. and Nygren, H. A., *Anal. Biochem.* **127**, 183–192 (1982).
Stenberg, M., Stiblert, L., and Nygren, H. A., *J. Theor. Biol.* **120**, 129–142 (1986).
Syvanen, A. C., Laaksonen, M., and Sonderlund, H., *Nucleic Acids Res.* **14**(2), 5037–5048 (1986).
Su, H., Chong, S., and Thompson, M., *Biosen. & Bioelectr.* **12**(3), 161 (1997).
Wink, T., Zuillen, S. J., Bult, A., and Bennekom, W. P., *Anal. Chem.* **70**, 827 (1998).

CHAPTER 7

Fractal Dimension and the Dissociation Rate Coefficient

7.1. INTRODUCTION

A promising area in the investigation of biomolecular interactions is the development of biosensors, which are finding application in the areas of biotechnology, physics, chemistry, medicine, aviation, oceanography, and environmental control. One advantage of these biosensors is that they can be used to monitor the analyte-receptor reactions in real time (Myszka *et al.*, 1997). In addition, some techniques—like the surface plasmon resonance (SPR) biosensor—do not require radio labeling or biochemical tagging (Jonsson *et al.*, 1991), are reusable, have a flexible experimental design, provide a rapid and automated analysis, and have a completely integrated system. Moreover, the SPR combined with mass spectrometry (MS) exhibits the potential to provide a proteomic analysis (Williams and Addona, 2000). In addition to evaluating affinities and interactions, the SPR can be used to determine unknown concentrations, to determine specificity, for kinetic analysis, to check for allosteric effects, and to compare binding patterns of different species.

Nilsson *et al.* (1995) utilized the BIACORE (biospecific interaction analysis core) surface plasmon resonance (SPR) biosensor to monitor DNA manipulations. DNA fragments were immobilized on to the SPR biosensor surface and used to monitor DNA hybridization kinetics, enzymatic modifications, and DNA strand separation. Houshmand *et al.* (1999) utilized the SPR biosensor to analyze the competitive binding of peptides (linear NH_3-CPNSLTPADPTMDY-COOH and NH_3-NSLTPCNNKPSNRC-COOH with an intramolecular S-S bridge) and a large T-antigen to the corresponding antibodies (LT1 and F4). Using these same peptides, these authors also analyzed the gentle and specific dissociation of the large T-antigen–antibody complexes. Loomans *et al.* (1997) utilized the SPR biosensor to monitor peptide–surface and peptide–antibody interactions. These authors noted that antibody binding activity as well as affinity could be improved or even restored by both the chemical modification of the peptides and an increase in the molecular size of the peptides.

There is a need to characterize the reactions occurring at the biosensor surface in the presence of diffusional limitations that are inevitably present in these types of systems. It is essential to characterize not only the binding, or associative, reaction (by a binding rate coefficient, k_{bind} or k_{ads}) but also the desorption, or dissociation, reaction (by a desorption rate coefficient, k_{des} or k_{diss}). This significantly assists in enhancing the biosensor performance parameters—such as reusability, multiple usage for the same analyte, and stability—as well as providing further insights into the sensitivity, reproducibility, and specificity of the biosensor. The ratio of k_{diss} to k_{bind} (equal to K) may be used to help further characterize the biosensor-analyte-receptor system. In essence, the analysis of just the binding step is incomplete, and the analysis of the binding and the dissociation step provides a more complete picture of the analyte-receptor reaction on the surface. Moreover, the analysis of dissociation kinetics alone also provides fresh physical insights into the reaction occurring on the biosensor surface.

The details of the association/dissociation of the analyte (antibody or substrate) to a receptor (antigen or enzyme) immobilized on a surface is of tremendous significance for the development of immunodiagnostic devices as well as for biosensors (Pisarchick *et al.*, 1992). The analysis we will present is, in general, applicable to ligand-receptor and analyte-receptorless systems for biosensor and other applications (e.g., membrane-surface reactions). External diffusional limitations play a role in the analysis of immunodiagnostic assays (Bluestein *et al.*, 1991; Eddowes, 1987/1988; Place *et al.*, 1991; Giaver, 1976; Glaser, 1993; Fischer *et al.*, 1994). The influence of diffusion in such systems has been analyzed to some extent (Place *et al.*, 1991; Stenberg *et al.*, 1986; Nygren and Stenberg, 1985; Stenberg and Nygren 1982; Morton *et al.*, 1995; Sjolander and Urbaniczky, 1991; Sadana and Sii, 1992a, 1992b; Sadana and Madagula, 1994; Sadana and Beelaram, 1995). The influence of partial (Christensen, 1997) and total (Matsuda, 1967; Elbicki *et al.*, 1984; Edwards *et al.*, 1995) mass transport limitations on analyte-receptor binding kinetics for biosensor applications is also available. The analysis presented for partial mass transport limitation (Christensen, 1997) is applicable to simple one-to-one association as well as to cases in which there is heterogeneity of the analyte or the ligand. This applies to the different types of biosensors utilized for the detection of different analytes.

In all of these analyses, only the association or the binding of the analyte to the receptor is analyzed. Apparently, up until now, the dissociation kinetics (of the analyte-receptor complex on the surface) has not been studied or reviewed in great detail, except in some isolated studies (Loomans *et al.*, 1997; Ramakrishnan and Sadana, 2000). This book attempts to address this issue further by analyzing both the association and the dissociation phases of the analyte-receptor kinetics on the biosensor surface. In addition, an example will be presented wherein the kinetics of dissociation alone is presented. In general, the analysis should be applicable to analyte-receptor reactions occurring on different surfaces (for example, cellular surfaces). This provides a more complete picture for the analyte-receptor biosensor system, just as an analysis of the unfolding/folding of an enzyme provides a better picture of the mechanistic reactions involved in converting an active enzyme to a deactivated one and vice versa. Besides, for a change, presenting the dissociation kinetics alone should also provide fresh physical insights.

Kopelman (1988) indicates that surface diffusion-controlled reactions that occur on clusters (or islands) are expected to exhibit anomalous and fractal-like kinetics. These fractal kinetics exhibit anomalous reaction orders and time-dependent (e.g., binding or dissociation) coefficients. As discussed in previous chapters, fractals are disordered systems, with the disorder described by nonintegral dimensions (Pfeifer and Obert, 1989). Kopelman further indicates that as long as surface irregularities show dilatational symmetry scale invariance, such irregularities can be characterized by a single number, the fractal dimension. The fractal dimension is a global property and is insensitive to structural or morphological details (Pajkossy and Nyikos, 1989). Markel *et al.* (1991) indicate that fractals are scale, self-similar mathematical objects that possess nontrivial geometrical properties. Furthermore, these investigators indicate that rough surfaces, disordered layers on surfaces, and porous objects all possess fractal structure.

A consequence of the fractal nature is a power-law dependence of a correlation function (in our case analyte-receptor complex on the surface) on a coordinate (e.g., time). This fractal nature or power-law dependence is exhibited during both the association (or binding) and/or the dissociation phases. This fractal power-law dependence has been shown for the binding of antigen–antibody (Sadana and Madagula, 1993; Sadana and Beelaram, 1995; Sadana, 1999), and for analyte-receptor (Ramakrishnan and Sadana, 2000), and for analyte-receptorless (protein) systems (Sadana and Sutaria, 1997).

Very recently, this dependence has been shown to be true for the dissociation phase too (Sadana and Ramakrishnan, 2000). In other words, the degree of roughness or heterogeneity on the surface affects both the association or binding of the analyte to the receptor on the surface and the dissociation of the analyte-receptor complex on the surface. The influence of the degree of heterogeneity on the surface may affect these two phases differently. Also, since this is a temporal reaction and presumably the degree of heterogeneity may be changing with (reaction) time, there may be two (or more) different values of the degree of heterogeneity for the association and the dissociation phases.

Fractal aggregate scaling relationships have been determined for both diffusion-limited processes and diffusion-limited scaling aggregation processes in spatial dimensions 2, 3, 4, and 5 (Sorenson and Roberts, 1997). These authors noted that the prefactor (in our case, the binding or the dissociation rate coefficient) displays uniform trends with the fractal dimension, D_f. Fractal dimension values for the kinetics of antigen–antibody binding (Sadana, 1997; Milum and Sadana, 1997), analyte-receptor binding (Sadana and Sutaria, 1997), and analyte-receptor binding and dissociation (Loomans *et al.*, 1997; Ramakrishnan and Sadana, 2000; Sadana and Ramakrishnan, 2000a) are available. We would like to further extend these ideas to two cases: binding and dissociation phase(s), and the dissociation phase alone. We will delineate the role of surface roughness on the speed of response, specificity, stability, sensitivity, and the regenerability or reusability of fiber-optic and other biosensors. We will obtain values for the fractal dimensions and the rate coefficient for the association (binding) as well as the dissociation phase(s), as well as the dissociation phase alone. A comparison of these values for the different biosensors analyzed and for the different reaction parameters should significantly assist in enhancing the relevant biosensor performance parameters. The noninteger orders of dependence obtained for the binding and dissociation rate coefficient(s) on their respective fractal dimension(s) further reinforce the fractal nature of these analyte-receptor binding–dissociation and dissociation systems.

7.2. THEORY

Milum and Sadana (1997) offer an analysis of the binding kinetics of the antigen in solution to antibody immobilized on the biosensor surface. The influence of lateral interactions on the surface and variable rate coefficients is also available (Sadana and Madagula, 1993). Here we present a method of estimating fractal dimensions and rate coefficients for both the association and the dissociation phases for analyte-receptor systems used in fiber-optic and other biosensors.

7.2.1. Variable Binding Rate Coefficient

Kopelman (1988) has indicated that classical reaction kinetics is sometimes unsatisfactory when the reactants are spatially constrained on the microscopic level by walls, phase boundaries, or force fields. Such heterogeneous reactions (e.g., bioenzymatic reactions) that occur at interfaces of different phases exhibit fractal orders for elementary reactions and rate coefficients with temporal memories. In such reactions, the rate coefficient is given by

$$k_1 = k' t^{-b}. \tag{7.1}$$

In general, k_1 depends on time, whereas $k' = k_1$ $(t = 1)$ does not. Kopelman indicates that in three dimensions (homogeneous space), $b = 0$. This is in agreement with the results obtained in classical kinetics. Also, with vigorous stirring, the system is made homogeneous and, again, $b = 0$. However, for diffusion-limited reactions occurring in fractal spaces, $b > 0$; this yields a time-dependent rate coefficient.

The random fluctuations in a two-state process in ligand-binding kinetics was analyzed by Di Cera (1991). The stochastic approach can be used to explain the variable binding rate coefficient. These ideas may also be extended to the dissociation rate coefficient. The simplest way to model these fluctuations is to assume that the binding (or the dissociation) rate coefficient is the sum of its deterministic value (invariant) and the fluctuation, $z(t)$. This $z(t)$ is a random function with a zero mean. The decreasing and increasing binding rate coefficients can be assumed to exhibit an exponential form (Cuypers *et al.*, 1987). A similar statement can be made for the dissociation rate coefficient.

Sadana and Madagula (1993) analyzed the influence of a decreasing and an increasing binding rate coefficient on the antigen concentration when the antibody is immobilized on the surface. These investigators noted that for an increasing binding rate coefficient, after a brief time interval, as time increases, the concentration of the antigen near the surface decreases, as expected for the cases when lateral interactions are present or absent. The diffusion-limited binding kinetics of antigen (or antibody or substrate) in solution to antibody (or antigen or enzyme) immobilized on a biosensor surface has been analyzed within a fractal framework (Sadana, 1997; Milum and Sadana, 1997). Furthermore, experimental data presented for the binding of human immunodeficiency virus (HIV) (antigen) to the antibody anti-HIV immobilized on a surface show a characteristic ordered "disorder" (Anderson, 1993). This indicates the possibility of a fractal-like surface. It is obvious that such a biosensor system (wherein either the antigen or the antibody is attached to the surface) along with its different complexities—including heterogeneities on the surface and in solution, diffusion-coupled reactions, and time-varying adsorption (or binding), and even dissociation rate coefficients—may be characterized as a fractal system. Sadana (1995) analyzed the diffusion of reactants toward fractal surfaces, and Havlin (1989) has briefly reviewed and discussed these results. Here we extend the ideas to dissociation reactions as well (that is, the dissociation of the analyte-receptor complex on the surface).

7.2.2. SINGLE-FRACTAL ANALYSIS

Binding Rate Coefficient

Havlin (1989) indicates that the diffusion of a particle (analyte [Ag]) from a homogeneous solution to a solid surface (e.g., receptor [Ab]-coated surface) on which it reacts to form a product (analyte-receptor complex, [Ag·Ab]) is given by

$$(\text{Analyte} \cdot \text{Receptor}) \sim \begin{cases} t^{(3-D_{f,\text{bind}})/2} = t^p & (t < t_c) \\ t^{1/2} & (t > t_c) \end{cases}. \tag{7.2a}$$

Here, $D_{f,\text{bind}}$ is the fractal dimension of the surface during the binding step. Equation (7.2a) indicates that the concentration of the product $\text{Ab} \cdot \text{Ag}(t)$ in a reaction $\text{Ab} + \text{Ag} \rightarrow \text{Ab} \cdot \text{Ag}$ on a solid fractal surface scales at short and intermediate time frames as $[\text{Ab} \cdot \text{Ag}] \sim t^p$, with the coefficient $p = (3 - D_{f,\text{bind}})/2$ at short time frames and $p = \frac{1}{2}$ at intermediate time frames. This equation is associated with the short-term diffusional properties of a random walk on a fractal surface. Note that in a perfectly stirred kinetics on a regular (nonfractal) structure (or surface), k_1 is a constant; that is, it is independent of time. In other words, the limit of regular structures (or surfaces) and the absence of diffusion-limited kinetics leads to k_{bind} being independent of time. In all other situations, one would expect a scaling behavior given by $k_{\text{bind}} \sim k' t^{-b}$ with $-b = p < 0$. Also, the appearance of the coefficient p different from zero is the consequence of two different phenomena: the heterogeneity (fractality) of the surface and the imperfect mixing (diffusion-limited) condition.

Havlin indicates that the crossover value may be determined by $r_c^2 \sim t_c$. Above the characteristic length, r_c, the self-similarity is lost. Above t_c, the surface may be considered homogeneous since the self-similarity property disappears and regular diffusion is now present. For the present analysis, we choose t_c arbitrarily and assume that the value of t_c is not reached. One may consider our analysis as an intermediate "heuristic" approach in that in the future one may also be able to develop an autonomous (and not time-dependent) model of diffusion-controlled kinetics.

Dissociation Rate Coefficient

We propose that a mechanism similar to the binding rate coefficient mechanism is involved (except in reverse) for the dissociation step. In this case, the dissociation takes place from a fractal surface. The diffusion of the dissociated particle (receptor [Ab] or analyte [Ag]) from the solid surface (e.g., analyte [Ag]-receptor [Ab] complex coated surface) into solution may be given, as a first approximation, by

$$(\text{Analyte} \cdot \text{Receptor}) \sim -k' t^{(3-D_{f,diss})/2} \quad (t > t_{diss}). \tag{7.2b}$$

Here $D_{f,diss}$ is the fractal dimension of the surface for the desorption or dissociation step. t_{diss} represents the start of the dissociation step and corresponds to the highest concentration of the analyte-receptor complex on the surface. Henceforth, its concentration only decreases. $D_{f,bind}$ may or may not be equal to $D_{f,diss}$. Equation (7.2b) indicates that during the dissociation step, the concentration of the product $Ab \cdot Ag(t)$ in the reaction $Ag \cdot Ab \rightarrow Ab + Ag$ on a solid fractal surface scales at short and intermediate time frames as $[Ag \cdot Ab] \sim -t^p$ with $p = (3 - D_{f,diss})/2$ at short time frames and $p = \frac{1}{2}$ at intermediate time frames. In essence, the assumptions that are applicable in the association (or binding) step are applicable for the dissociation step. Once again, this equation is associated with the short-term diffusional properties of a random walk on a fractal surface. Note that in a perfectly stirred kinetics on a regular (nonfractal) structure (or surface), k_{diss} is a constant; that is, it is independent of time. In other words, the limit of regular structures (or surfaces) and the absence of diffusion-limited kinetics leads to k_{diss} being independent of time. In all other situations, one would expect a scaling behavior given by $k_{diss} \sim -k' t^{-b}$ with $-b = p < 0$. Once again, the appearance of the coefficient p different from zero is the consequence of two different phenomena: the heterogeneity (fractality) of the surface and the imperfect mixing (diffusion-limited) condition. Besides providing physical insights into the analyte-receptor system, the ratio $K = k_{diss}/k_{bind}$ is of practical importance since it may be used to help determine (and possibly enhance) the regenerability, reusability, stability, and other biosensor performance parameters.

7.2.3. Dual-Fractal Analysis

Binding Rate Coefficient

We extend the single-fractal analysis just presented to include two fractal dimensions. At present, the time ($t = t_1$) at which the first fractal dimension changes to the second fractal dimension is arbitrary and empirical. For the most part, it is dictated by the data analyzed and the experience gained by handling a single-fractal analysis. A smoother curve is obtained in the transition region if care is taken to select the correct number of points for the two regions. In this case, the analyte-receptor complex (Ag • Ab) is given by

$$(\text{Analyte} \cdot \text{Receptor}) \sim \begin{cases} t^{(3-D_{f_1,\text{bind}})/2} = t^{p_1} & (t < t_1) \\ t^{(3-D_{f_2,\text{bind}})/2} = t^{p_2} & (t_1 < t < t_2 = t_c). \\ t^{1/2}(t > t_c) & \end{cases} \tag{7.2c}$$

Dissociation Rate Coefficient

Once again, we propose that a mechanism similar to that for the binding rate coefficient(s) is involved (except in reverse) for the dissociation step. And, again, the dissociation takes place from a fractal surface. The diffusion of the dissociated particle (receptor [Ab] or analyte [Ag]) from the solid surface (e.g., analyte [Ag]-receptor [Ab] complex coated surface) into solution may be given, as a first approximation, by

$$(\text{Analyte} \cdot \text{Receptor}) \sim \begin{cases} -t^{(3-D_{f_1,\text{diss}})/2} & (t_{\text{diss}} < t < t_{d_1}) \\ -t^{(3-D_{f_2,\text{diss}})/2} & (t_{d_1} < t < t_{d_2}) \end{cases}. \tag{7.2d}$$

Note that different combinations of the binding and dissociation steps are possible as far as the fractal analysis is concerned. Each of these steps, or phases, can be represented by either a single- or a dual-fractal analysis. For example, the binding (association) step may be adequately described by a single-fractal analysis. However, it is quite possible that the dissociation step may need a dual-fractal analysis to be adequately described. When the binding (association) step requires a dual-fractal analysis, the dissociation step may be adequately described by either a single- or a dual-fractal analysis. In effect, four possible combinations are possible: single-fractal (association)–single-fractal (dissociation); single-fractal (association)–dual-fractal (dissociation); dual-fractal (association)–single-fractal (dissociation); dual-fractal (association)–dual-fractal (dissociation). Presumably, it is only by the analysis of a large number of association–dissociation analyte-receptor data from a wide variety of systems that this point may be further clarified.

7.3. RESULTS

In this discussion, a fractal analysis will be applied to the data obtained for analyte-receptor binding and dissociation data for different biosensor systems. This is one possible explanation for analyzing the diffusion-limited binding and dissociation kinetics assumed to be present in all of the systems analyzed. The parameters thus obtained would provide a useful comparison of different situations. Alternate expressions involving saturation, first-order reaction, and no diffusion limitations are possible but seem to be deficient in describing the heterogeneity that inherently exists on the surface. Both the analyte-receptor binding and the dissociation reactions are complex reactions, and fractal analysis via the fractal dimension (either $D_{f,bind}$ or $D_{f,diss}$) and the rate coefficient for binding (k_{bind}) or dissociation (k_{diss}) provide a useful lumped-parameter analysis of the diffusion-limited situation.

Also, we present no independent proof or physical evidence for the existence of fractals in the analysis of these analyte-receptor binding–dissociation systems except by indicating that fractal analysis has been applied in other areas and is a convenient means to make more quantitative the degree of heterogeneity that exists on the surface. Thus, in all fairness this is only one possible way by which to analyze this analyte-receptor binding–dissociation data. One might justifiably argue that appropriate modeling may be achieved by using a Langmuir or other approach. However, a major drawback of the Langmuir approach is that it does not allow for the heterogeneity that exists on the surface.

The Langmuir approach was originally developed for gases (Thomson and Webb, 1968). Consider a gas at pressure, p, in equilibrium with a surface. The rate of adsorption is proportional to the gas pressure and to the fraction of the uncovered surface. Adsorption will only occur when a gas molecule strikes a bare site. Researchers in the past have successfully modeled the adsorption behavior of analytes in solution to solid surfaces using the Langmuir model even though it does not conform to theory. Rudzinski *et al.* (1983) indicate that other appropriate "liquid" counterparts of the empirical isotherm equations have been developed. These include counterparts of the Freundlich (Dabrowski and Jaroniec, 1979), Dubinin–Radushkevich (Oscik *et al.*, 1976), and Toth (Jaroniec and Derylo, 1981) empirical equations. These studies, with their known constraints, have provided some restricted physical insights into the adsorption of adsorbates on different surfaces. The Langmuir approach may be utilized to model the data presented if one assumes the presence of discrete classes of sites (for example, double-exponential analysis as compared to single-exponential analysis). Lee and Lee (1995) indicate that the fractal approach has been applied to surface science, for example, adsorption and reaction processes. These authors emphasize that the fractal approach provides a convenient means to represent the different structures and morphology at the reaction surface. They also suggest using the fractal approach to develop optimal structures and as a predictive approach.

Loomans *et al.* (1997) utilized the SPR biosensor to monitor peptide–surface and peptide–antibody interactions. These authors mention that affinity is of interest in immunoassay studies. However, they noted that their estimated values of affinity for the binding of mouse anti-hCG monoclonal antibody OT-3A in solution to Ata-3A immobilized on a surface were overestimated. They presumed that this was due to rebinding and wanted to determine the cause of the decrease in the rate coefficient for dissociation, k_{diss}. They checked this by performing studies with increasing hCG concentrations and noted that the increasing hCG concentrations correlated well with accelerated dissociation. Figures 7.1a–7.1h show the curves obtained using Eq. (7.2b) for the dissociation of the OT-3A from the SPR biosensor surface. In all cases, a single-fractal analysis adequately describes the dissociation kinetics.

Table 7.1 shows the values of the dissociation rate coefficient, k_d, and the fractal dimension for dissociation, $D_{f,d}$. (Please note that for dissociation we use either d or diss in the subscript. We use b or bind, or no subscript to refer to the binding phase. We do this to accommodate the number of columns in the tables and facilitate all the entries in the tables). The values presented in the table were obtained from a regression analysis using Sigmaplot (1993) to model the experimental data using Eq. (7.2b), wherein (analyte · receptor) $= -k_{diss}t^p$ for the dissociation step. The dissociation rate coefficient values presented in Table 7.1 are within 95% confidence limits. For example, for the binding of OT-3A in the presence of 0.2 μM hCG in solution to Ata-3A immobilized on the SPR biosensor surface, the reported k_d value is 0.321 ± 0.015. The 95% confidence limit indicates that 95% of the k_d values will lie between 0.306 and 0.336. This indicates that the values are precise and significant. The curves presented in the figures are theoretical curves.

Note that there is no nonselective adsorption of the OT-3A. Although our analysis does not include this nonselective adsorption, we do recognize that, in some cases, this may be a significant component of the adsorbed material and that this rate of association, which is of a temporal nature, would depend on surface availability. If we were to accommodate the nonselective adsorption into the model, there would be an increase in the degree of heterogeneity on the surface since by its very nature nonspecific adsorption is

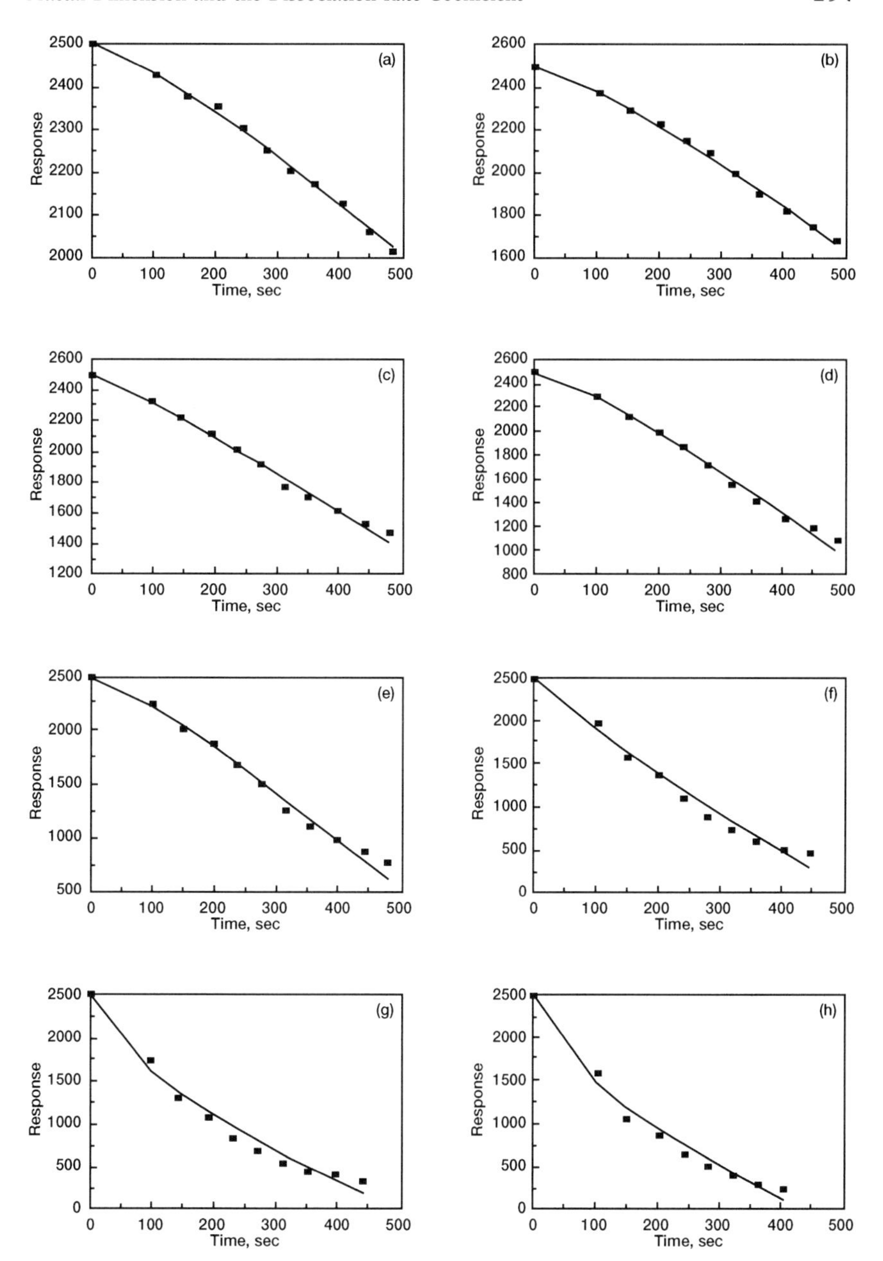

FIGURE 7.1 Binding of OT-3A in the presence of different hCG concentrations (in μM) in solution to Ata-3A immobilized on a BIACORE biosensor surface and dissociation of the analyte from the surface (Loomans *et al.*, 1997). (a) 0.2; (b) 0.4; (c) 0.6; (d) 0.8; (e) 1.0; (f) 2.0; (g) 3.0; (h) 4.0.

TABLE 7.1 Fractal Dimensions and Dissociation Rate Coefficients Using a Single-Fractal Analysis for the Dissociation (after Its Binding) of OT-3A in Solution from Ata-3A Immobilized on a BIACORE Biosensor Surface (Loomans *et al.*, 1997)

hCG concentration (μM)	Dissociation rate coefficient, k_d	Fractal dimension for dissociation, $D_{f,d}$
0.2	0.321 ± 0.015	0.640 ± 0.062
0.4	0.446 ± 0.015	0.566 ± 0.043
0.6	0.749 ± 0.03	0.640 ± 0.05
0.8	0.696 ± 0.04	0.514 ± 0.07
1.0	0.969 ± 0.09	0.548 ± 0.09
2.0	9.564 ± 0.93	1.22 ± 0.12
3.0	39.1 ± 3.5	1.65 ± 0.11
4.0	58.9 ± 5.5	1.76 ± 0.13

Different concentrations of hCG were utilized during the dissociation phase

more heterogeneous than specific adsorption. This would lead to higher fractal dimension values since the fractal dimension is a direct measure of the degree of heterogeneity that exists on the surface.

Table 7.1 indicates that k_d increases as the hCG concentration in solution increases, as expected by Loomans *et al.* (1997). Figure 7.2a shows an increase in k_d with an increase in hCG concentration in solution. Clearly, k_d varies with hCG concentration in solution in a nonlinear fashion. In the hCG concentration range of 0.2 to 4.0 μM, k_d is given by

$$k_d = (2.56 + 2.88)[\text{hCG}, \mu\text{M}]^{1.92 \pm 0.28}. \quad (7.3a)$$

The fit is not reasonable, especially at the higher hCG concentrations (above 2μM). A better fit was sought using a four-parameter model. The data was fit in two phases: 0 to 2 μM, and 2 to 4 μM hCG concentration. Using this approach, k_d is given by

$$k_d = (0.916 \pm 0.127)[\text{hCG}]^{0.671 \pm 0.102} + (1.64 \pm 0.53)[\text{hCG}]^{2.68 \pm 0.57}. \quad (7.3b)$$

More data points are required to establish this equation more firmly. Nevertheless, Eq. (7.3b) is of value since it provides an indication of the change in k_d as the hCG concentration in solution changes. The fractional exponent dependence indicates the fractal nature of the system. The dissociation rate coefficient is quite sensitive to the hCG concentration at the higher hCG concentration in solution, as indicated by the higher-than second-order value of the exponent. These results are consistent with the results obtained by Loomans *et al.* (1997), who performed a nonfractal analysis. A fractal analysis incorporates the heterogeneity that inherently

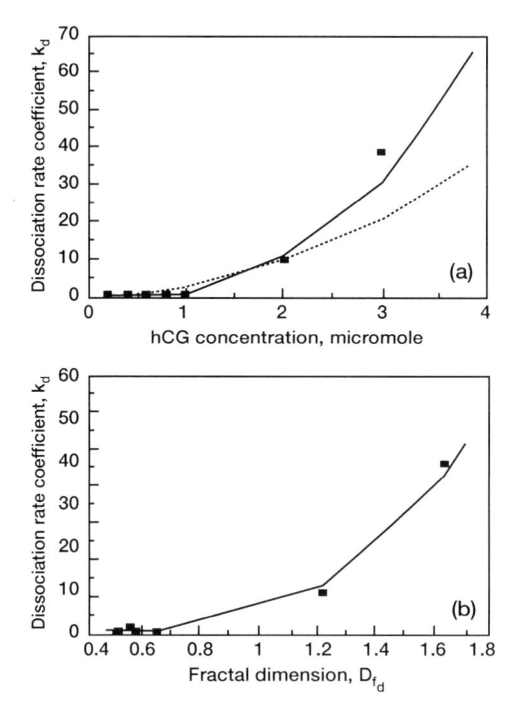

FIGURE 7.2 (a) Influence of the hCG concentration in solution on the dissociation rate

exists on the biosensor surface, which is an additional advantage of the analysis. This is reflected in the fractal dimension value: A higher value indicates a higher degree of heterogeneity on the surface.

Table 7.1 and Figure 7.2b indicate that k_d increases as the fractal dimension for dissociation, $D_{f,d}$, increases. Clearly, k_d varies with $D_{f,d}$ in a nonlinear fashion. In the hCG concentration range of 0.2 to 4.0 μmol, k_d is given by

$$k_d = (5.11 \pm 4.06)[D_{f,d}]^{3.90 \pm 043}. \tag{7.3c}$$

Although the fit is very reasonable, more data points are required to more firmly establish this equation, especially at the higher fractal dimension values. Eq. (7.3c) is of value since it provides an indication of the change in k_{diss} (or k_d) as the degree of heterogeneity on the SPR biosensor surface changes. The high exponent dependence indicates that the dissociation rate coefficient is sensitive to the degree of heterogeneity that exists on the SPR biosensor surface. The value of an expression that relates the dissociation rate coefficient to a fractal dimension is that it provides one with an avenue by which to control the dissociation rate coefficient on the surface by changing the degree of heterogeneity that exists on the surface.

Apparently, the utilization of higher hCG concentrations in solution leads to higher degrees of heterogeneity on the SPR biosensor surface, which eventually leads to higher k_d values. However, this is just one explanation of the results, and other, perhaps more suitable, explanations are also possible. Finally, since no binding rate coefficients are presented in this analysis, affinity ($K = k_d/k_{bind}$) values are not given.

Nilsson *et al.* (1995) utilized the SPR biosensor to monitor DNA manipulations in real time. These authors immobilized DNA fragments on the biosensor surface using the streptavidin–biotin system and monitored DNA hybridization kinetics, DNA strand separation, and enzymatic modifications. Figure 7.3a shows the curves obtained using Eq. (7.2c) for the binding of T7 DNA polymerase in solution to a complementary DNA immobilized on the SPR biosensor surface as well as the dissociation of the analyte from the same surface and its eventual diffusion in solution. A dual-fractal analysis is required to adequately describe the binding kinetics [Eq. (7.2c)], and a single-fractal analysis [Eq. (7.2b)] is sufficient to describe the dissociation kinetics.

Table 7.2 shows the values of the binding rate coefficients, k_{bind}, $k_{1,bind}$, $k_{2,bind}$, the dissociation rate coefficient, k_{diss}, the fractal dimensions for binding, $D_{f,bind}$, $D_{f_1,bind}$, and $D_{f_2,bind}$, and the fractal dimension for dissociation, $D_{f,diss}$. Since the dual-fractal analysis is required to adequately describe the binding phase, it will be analyzed further. The affinity, K, is equal to the ratio of the dissociation rate coefficient to the binding rate coefficient. Thus, $K_1 = k_{diss}/k_1$ has a value of 7.54 and $K_2 = k_{diss}/k_2$ has a value of 0.78 for the T7 DNA polymerase reaction. There is a decrease in the affinity value by a factor of 9.67 on going from the first phase to the second phase in the binding reaction. This is due to the increase in the binding rate coefficient value in the second phase compared to the first phase. The dissociation rate coefficient value remains the same.

In general, typical antigen–antibody affinities are in the nanomolar to picomolar range. In this case, the affinities values reported are quite a few orders of magnitude higher than normally reported. Presumably, the increase in the higher affinity value may be due to a combination of factors. One such factor is that very little or no conformational restriction of the receptor on the surface minimizes the strength of the analyte-receptor reaction (Altschuh *et al.*, 1992). This increases the dissociation rate coefficient. Furthermore, if the binding of the analyte to the receptor involves a conformational adaptation via the induced-fit mechanism, the lower structural flexibility of the analyte may

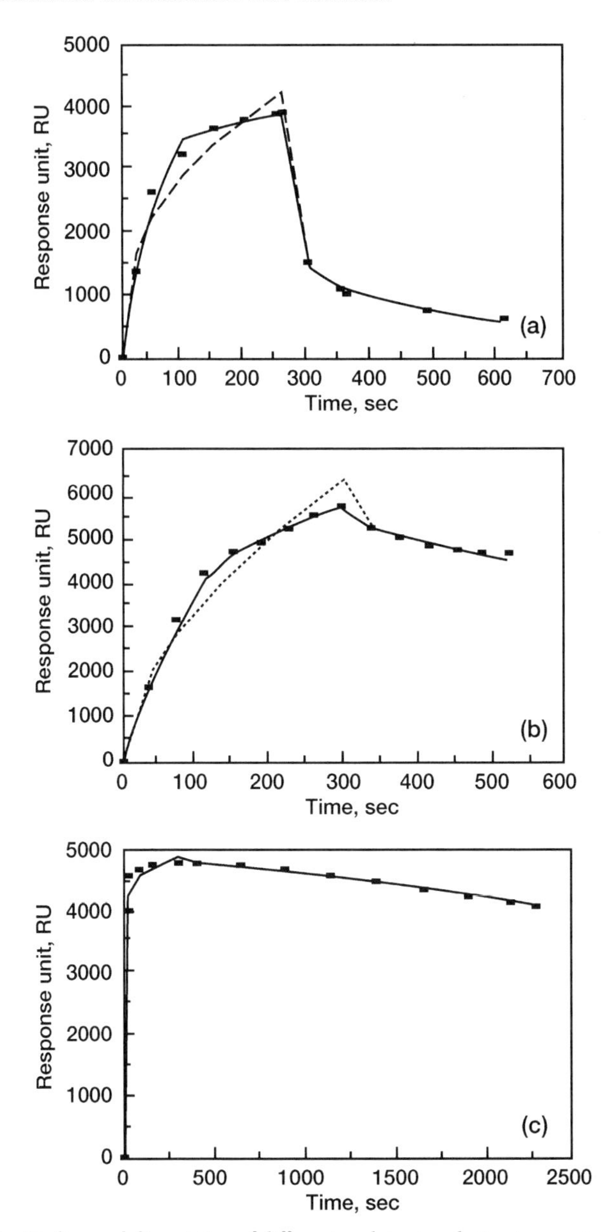

FIGURE 7.3 Binding and dissociation of different analytes in solution to receptors immobilized on a surface plasmon resonance biosensor (Nilsson *et al.*, 1995) (a) T7 DNA polymerase/complementary DNA fragment; (b) DNA polymerase I (Klenow fragment)/complementary fragment; (c) endonuclease X*ho*I/69 bp substrate complementary DNA. (- - -, single-fractal analysis; —, dual-fractal analysis).

TABLE 7.2 Fractal Dimensions and Binding and Dissociation Rate Coefficients Using a Single- and a Dual-Fractal Analysis for the Binding and Dissociation of Different Analytes in Solution to Receptors Immobilized on a Surface Plasmon Resonance (SPR) Biosensor Surface During DNA Manipulations (Nilsson *et al.*, 1995)

Analyte in solution/ receptor on surface	k_{bind}	$k_{1,bind}$	$k_{2,bind}$	k_{diss}	$D_{f,bind}$	$D_{f_1,bind}$	$D_{f_2,bind}$	$D_{f,diss}$
T7 DNA polymerase/ complementary DNA fragment	455 ± 69.6	201.2 ± 40.8	1943 ± 4.9	1517 ± 44.9	2.19 ± 0.13	1.76 ± 0.38	2.75 ± 0.01	2.73 ± 0.03
DNA polymerase I (Klenow fragment)/ complementary DNA fragment	250 ± 31	106.2 ± 9.3	1088 ± 14.5	20.1 ± 2.27	1.86 ± 0.13	1.46 ± 0.16	2.42 ± 0.05	1.55 ± 0.13
Endonuclease *Xho*I/69-bp substrate complementary DNA	3556 ± 193	na	na	0.049 ± 0.01	2.89 ± 0.049	na	na	0.496 ± 0.14

hinder dissociation (Altschuh *et al.*, 1992; Mani *et al.*, 1994). Also, the fractal nature of the surface itself may lead to higher-than-expected affinities. Other reasons for this phenomenon are also possible. In our case, the rebinding phenomenon observed (as in other BIACORE experiments) (Nieba *et al.*, 1996; Karlsson and Stahlberg, 1995; Wohlhueter *et al.*, 1994; Morton *et al.*, 1995) may be a minimum, due to the fractal nature of the surface. This too leads to an increase in the dissociation rate coefficient.

Figure 7.3b shows the curves obtained using Eq. (7.2c) for the binding of DNA polymerase I (Klenow fragment) in solution to a complementary DNA immobilized on the SPR biosensor surface as well as the dissociation of the analyte from the same surface and its eventual diffusion in solution. Once again, a dual-fractal analysis is required to adequately describe the binding kinetics [Eq. (7.2c)], and a single-fractal analysis [Eq. (7.2b)] is sufficient to describe the dissociation kinetics.

Table 7.2 shows the values of the binding rate coefficients, k_{bind}, $k_{1,bind}$, $k_{2,bind}$, the dissociation rate coefficient, k_{diss}, the fractal dimension for binding, $D_{f,bind}$, $D_{f_1,bind}$, and $D_{f_2,bind}$, and the fractal dimension for the dissociation, $D_{f,diss}$. In this case, $K_1 = 0.189$ and $K_2 = 0.018$. Once again, the affinity value decreases as one goes from the first phase to the second phase. Note that as one goes from the Klenow fragment case to the T7 DNA polymerase case, the $D_{f_1,bind}$ value increases by 20.5%—from 1.46 to 1.76, and the $k_{1,bind}$ value increases by a factor of 1.89—from 106.2 to 201.2. Similarly, the $D_{f_2,bind}$ value increases by 13.6%—from 2.42 to 2.75, and the $k_{2,bind}$ value increases by a factor of 1.78—from 1088 to 1943. Also, the fractal dimension for dissociation, $D_{f,diss}$, increases by a factor of 1.76—from 1.55 to 2.73, and the dissociation rate coefficient, k_{diss}, shows an increase by factor of 75.5. Thus, the dissociation rate coefficient is very sensitive to the degree of heterogeneity on the surface, at least in these two cases.

Figure 7.3c shows the curves obtained using Eq. (7.2a) for the binding of endonuclease *Xho*I in solution to 69-bp substrate complementary DNA immobilized on the biosensor surface, as well as the dissociation of the analyte from the same surface and its eventual diffusion in solution. In this case, the binding phase as well as the dissociation phase [Eq. (7.2b)] may be adequately described by a single-fractal analysis. The affinity, $K(= k_{diss}/k_{bind})$, is equal to 0.000014. This is an extremely low value, especially when compared to the values in the two previous cases.

Houshmand *et al.* (1999) analyzed the binding and dissociation of 80 nM large T-antigen in solution to the monoclonal antibody mAbLT1 immobilized on an SPR biosensor surface in the absence and in the presence of a competitor peptide, NH_3CPNSLTPADPTMDY-COOH. After a given time interval, the injection of the protein ligand was interrupted, and the subsequent dissociation reaction was monitored. Figure 7.4a shows the

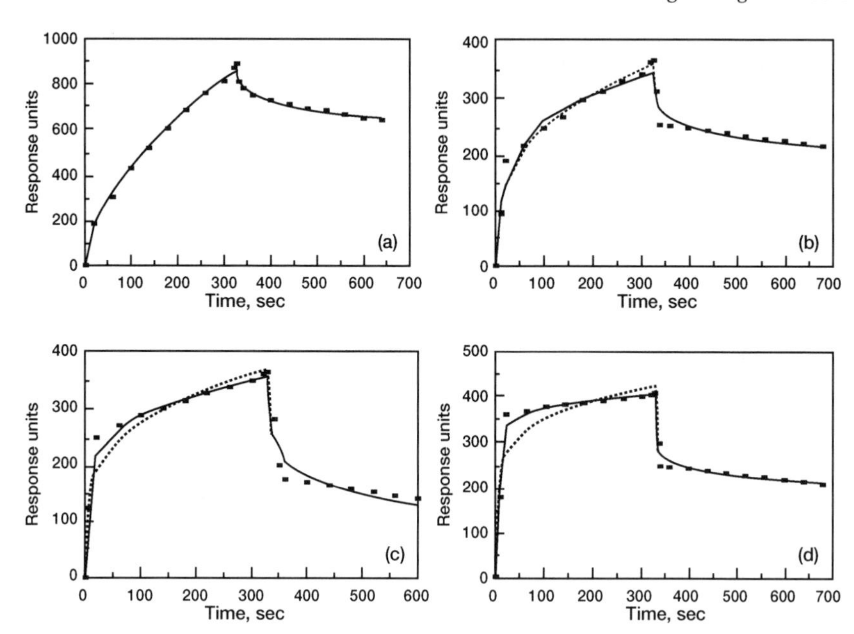

FIGURE 7.4 Binding of 80 nM large T-antigen (30 μl) in solution in the absence and in the presence of competitor peptide (NH_3-CPNSLTADPTMDY-COOH) (in μl) to mAbLT1 immobilized on a BIACORE biosensor chip surface (Houshmand *et al.*, 1999). (a) 0; (b) 50; (c) 200; (d) 800.

curves obtained using Eq. (7.2a) for the binding of the 80 nM large T-antigen in solution to the mAbLT1 immobilized on the SPR surface in the absence of the competitor peptide as well as the dissociation of the analyte from the same surface [using Eq. (7.2b)] and its eventual diffusion into solution. A single-fractal analysis is sufficient to adequately describe both the binding [Eq. (7.2a)] and the dissociation [Eq. (7.2b)] kinetics.

Table 7.3 shows the values of the binding rate coefficient, k_{bind}, the dissociation rate coefficient, k_{diss}, the fractal dimension for binding, $D_{f,bind}$, and the fractal dimension for dissociation, $D_{f,diss}$. When competitor peptide is not used the affinity, $K = 1.95$. Also, the estimated value for the fractal dimension for dissociation, $D_{f,diss}$, is larger (equal to 2.53) than the fractal dimension for binding, $D_{f,bind}$ (equal to 1.87). Figures 7.4b–7.4d show the curves obtained using Eq. (7.2c) for the binding of the large T-antigen in the presence of 50–800 μM peptide to the mAbLT1 immobilized on the SPR surface as well as the dissociation [using Eq. (7.2d)] of the analyte from the surface and its eventual diffusion into solution. When the competitor peptide (50–800 μM) is used, a dual-fractal analysis is required to adequately describe

TABLE 7.3 Fractal Dimensions and Binding and Dissociation Rate Coefficients Using a Single- and a Dual-Fractal Analysis for the Binding of 80 nM Large Antigen (30 μl) in Solution in the Absence and in the Presence of Competitor Peptide (NH_3-CPNSLTPADPTMDY-COOH) to mAbLT1 Immobilized on a BIACORE Biosensor Chip Surface (Using Surface Plasmon Resonance) (Houshmand *et al.*, 1999)

Antigen + peptide (μl) in solution/mAbLT1 on surface	k_{bind}	$k_{1,bind}$	$k_{2,bind}$	k_{diss}	$D_{f,bind}$	$D_{f_1,bind}$	$D_{f_2,bind}$	$D_{f,diss}$
0.0	32.1 ± 1.2	na	na	62.7 ± 3.4	1.87 ± 0.03	na	na	2.53 ± 0.02
50	55.9 ± 6.7	49.2 ± 12.7	84.5 ± 4.7	55.1 ± 8.6	2.36 ± 0.06	2.27 ± 0.25	2.51 ± 0.04	2.64 ± 0.09
200	94.8 ± 13.3	72.7 ± 20.4	132 ± 2.75	78.1 ± 14.5	2.53 ± 0.07	2.36 ± 0.27	2.66 ± 0.02	2.60 ± 0.09
800	172 ± 27.5	120 ± 36.4	280 ± 1.8	103 ± 9.1	2.68 ± 0.08	2.45 ± 0.29	2.87 ± 0.01	2.77 ± 0.04

the binding kinetics, and the dissociation kinetics is adequately described by a single-fractal analysis. Table 7.3 shows the values of the binding rate coefficients, k_{bind}, $k_{1,\mathrm{bind}}$, $k_{2,\mathrm{bind}}$, the dissociation rate coefficient, k_{diss}, the fractal dimension for binding, $D_{\mathrm{f,bind}}$, $D_{\mathrm{f_1,bind}}$, and $D_{\mathrm{f_2,bind}}$, and the fractal dimension for the dissociation, $D_{\mathrm{f,diss}}$.

Table 7.3 and Fig. 7.5a indicate that k_{diss} increases as the peptide concentration increases and varies with the peptide concentration in a nonlinear fashion. In the competitor peptide concentration range of 50 to 800 μM, k_{diss} is given by

$$k_{\mathrm{diss}} = (23.0 \pm 0.66)[\text{peptide concentration, } \mu\mathrm{M}]^{0.226 \pm 0.015}. \qquad (7.4a)$$

Although the fit is reasonable for the three data points presented, more data points are required to establish this equation more firmly. Nevertheless, Eq. (7.4a) is of value since it provides an indication of the change in k_{diss} as the competitor peptide concentration in solution changes. The dissociation rate

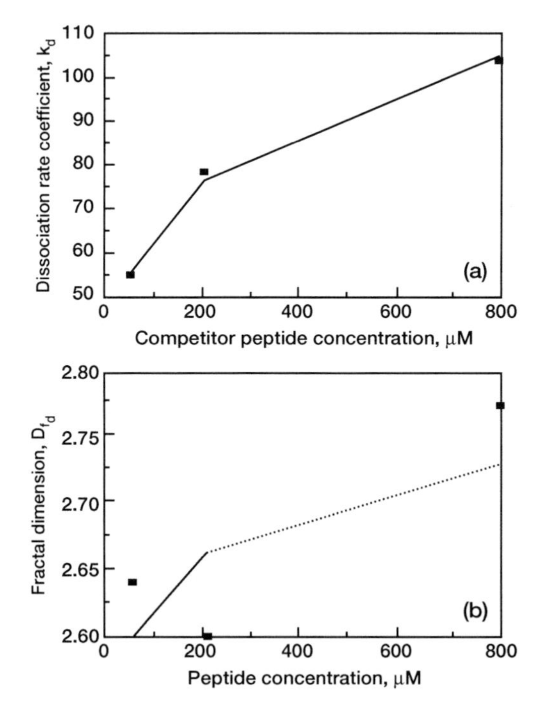

FIGURE 7.5 Influence of competitor peptide concentration on (a) the dissociation rate coefficient, k_{d}; (b) the fractal dimension, $D_{\mathrm{f_d}}$.

coefficient is only mildly sensitive to the competitor peptide concentration in solution, as noted by the low value of the exponent.

Table 7.3 and Fig. 7.5b indicate that $D_{f,diss}$ increases as the peptide concentration in solution increases. There is much scatter in the data, and $D_{f,diss}$ varies with the peptide concentration in a nonlinear fashion. In the competitor peptide concentration range of 50–800 μM, $D_{f,diss}$ is given by

$$D_{f,diss} = (2.43 \pm 0.08)[\text{peptide concentration, } \mu\text{M}]^{0.017 \pm 0.016}. \qquad (7.4b)$$

There is considerable scatter in the data. More data points are required to establish this equation more firmly. However, the very low exponent dependence of $D_{f,diss}$ on the peptide concentration indicates a negligible dependence of the dissociation rate coefficient on the peptide concentration. One might justifiably argue that if the biosensor surface is truly fractal in nature, then one would expect a self-similarity that spans the entire concentration range and hence no changes in the fractal dimension. However, our analysis does indicate a concentration dependence even though it is weak (reaction order of 0.017). For example, an increase in the peptide concentration by a factor of 16—from 50 to 800 μl in solution—increases the fractal dimension for dissociation, $D_{f,diss}$, by 6.5%—from 2.60 to 2.77. No explanation is offered at present for this slight increase.

Table 7.3 and Fig. 7.6a indicate that the binding rate coefficient, $k_{1,bind}$, increases as the fractal dimension, D_{f_1}, increases. In the competitor peptide range of 50 to 800 μM, $k_{1,bind}$ is given by

$$k_{1,bind} = (0.0033 \pm 0.0002) D_{f_1}^{11.67 \pm 0.97}. \qquad (7.5a)$$

This predictive equation fits the values presented in Table 7.3 reasonably well. The very high exponent dependence indicates that the binding rate coefficient is very sensitive to the degree of heterogeneity that exists on the surface. More data points are required to more firmly establish this equation. The fractional exponent dependence exhibited further reinforces the fractal nature of the system.

Table 7.3 and Fig. 7.6b indicate that the binding rate coefficient, $k_{2,bind}$, increases as the fractal dimension, D_{f_2}, increases. In the competitor peptide range of 50 to 800 μM, $k_{2,bind}$ is given by

$$k_{2,bind} = (0.021 \pm 0.001) D_{f_2}^{8.99 \pm 0.62}. \qquad (7.5b)$$

This predictive equation fits the values presented in Table 7.3 reasonably well. The very high exponent dependence indicates that the binding rate coefficient is very sensitive to the degree of heterogeneity that exists on the surface. More data points are required to more firmly establish this equation. The fractional

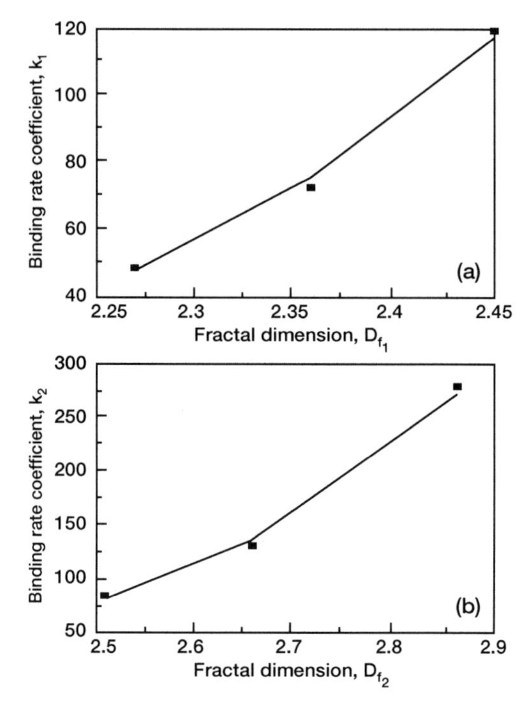

FIGURE 7.6 Influence of (a) the fractal dimension, D_{f_1}, on the binding rate coefficient, k_1; (b) the fractal dimension, D_{f_2}, on the binding rate coefficient, k_2.

exponent dependence exhibited further reinforces the fractal nature of the system. Note that $k_{1,\text{bind}}$ is more sensitive than $k_{2,\text{bind}}$ to the nature of surface, as noted by the exponent values on their respective fractal dimensions.

Satoh and Matsumoto (1999) utilized the surface plasmon resonance (SPR) biosensor to analyze the binding of lectin molecules in solution to carbohydrate molecules immobilized on a SPR surface using hydrazide groups. The hydrazide groups were attached to the surface of the CM5 sensor chip by reaction between the activated carboxy groups of the sensor surface and the hydrazide groups of adipic acid dihydrazide (ADHZ). The surface was treated with 0.1 M lactose in 10 mM hepes-buffered saline (pH 7.4). Figure 7.7a shows the curves obtained using Eq. (7.2a) for the binding of 125 nM *Sophora japonica* agglutinin (SJA) in solution to a lactose-immobilized surface as well as the dissociation of the SJA–lactose complex from the same surface and its eventual diffusion into solution. A single-fractal analysis is sufficient to adequately describe both the binding and the dissociation kinetics.

Table 7.4 shows the values of the binding rate coefficient, k_{bind}, the dissociation rate coefficient, k_{diss}, the fractal dimension for binding, $D_{f,\text{bind}}$,

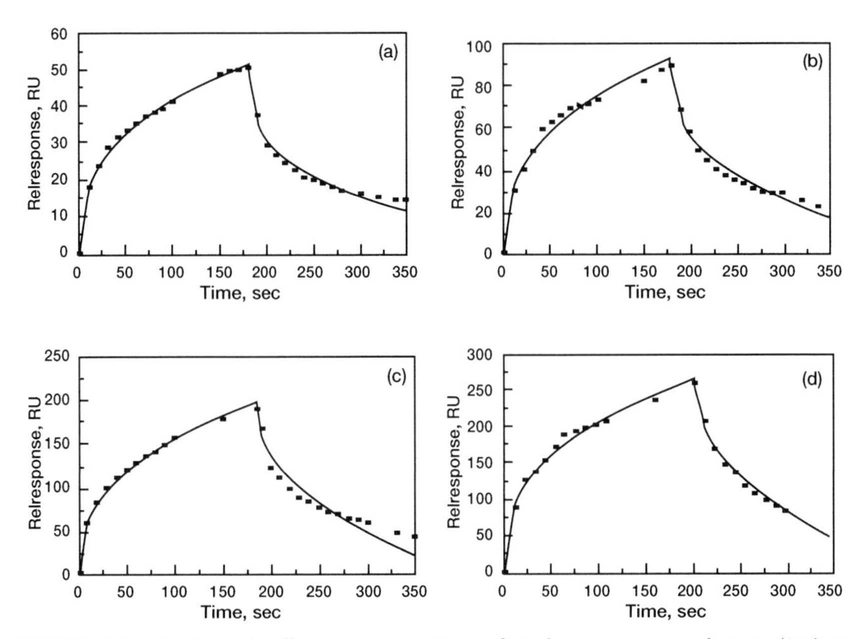

FIGURE 7.7. Binding of different concentrations of *Sophora japonica* agglutinin (SJA) in solution to a lactose-immobilized surface and dissociation of the SJA–lactose complex from the surface (Satoh and Matsumoto, 1999). (a) 125 nM; (b) 250 nM; (c) 500 nM; (d) 1000 nM.

and the fractal dimension for dissociation, $D_{f,diss}$. The values presented in the table were obtained from a regression analysis using Sigmaplot (1993) to model the experimental data using Eq. (7.2a), wherein (analyte $\cdot$ receptor) $= k_{bind}t^p$ for the binding step and (analyte $\cdot$ receptor) $= -k_{diss}t^p$ for the dissociation step. The binding and dissociation rate coefficient values presented in Table 7.4 are within 95% confidence limits. For example, for the binding of 125 nmol SJA in solution to the lactose-immobilized SPR surface,

TABLE 7.4 Influence of *Sophora japonica* Agglutinin (SJA) Concentration on Binding and Dissociation Rate Coefficients and Fractal Dimensions for Its Binding to and Dissociation from a Lactose-Immobilized BIACORE Surface (Satoh and Matsumoto, 1999)

SJA concentration, nmol	k_{bind} gm $(cm)^{-2}(sec)^{-p}$	k_{diss} gm $(cm)^{-2}(sec)^{-p}$	K	$D_{f,bind}$	$D_{f,diss}$
125	8.4 ± 0.2	7.2 ± 0.6	0.85	2.3 ± 0.0	2.3 ± 0.0
250	14 ± 0.9	9.9 ± 0.8	0.70	2.3 ± 0.0	2.2 ± 0.0
500	24 ± 0.8	14 ± 2.0	0.59	2.2 ± 0.0	2.0 ± 0.1
1000	43 ± 2.0	20 ± 2.0	0.46	2.3 ± 0.0	2.0 ± 0.1

the reported k_{bind} value is 8.43 ± 0.22. The 95% confidence limit indicates that 95% of the k_{bind} values will lie between 8.21 and 8.65. This indicates that the values are precise and significant. The curves presented in the figures are theoretical curves.

Figure 7.7b, 7.7c, and 7.7d show the binding of 250, 500, and 1000 nM SJA in solution to a lactose-immobilized surface as well as the dissociation of the SJA agglutinin–lactose complex from the surface. In each case, a single-fractal analysis is sufficient to describe both the binding and the dissociation kinetics. The values of the binding and dissociation rate coefficients are given in Table 7.4. Note that as the SJA concentration in solution increases from 125 to 1000 nM both k_{bind} (or k_{ads}) and k_{diss} (or k_{des}) exhibit increases.

Since there is no nonselective adsorption of the SJA, our analysis does not include this nonselective adsorption. We do recognize that, in some cases, this may be a significant component of the adsorbed material and that this rate of association, which is of a temporal nature, would depend on surface availability. If we were to accommodate the nonselective adsorption into the model, there would be an increase in the degree of heterogeneity on the surface since by its very nature nonspecific adsorption is more heterogeneous than specific adsorption. This would lead to higher fractal dimension values since the fractal dimension is a direct measure of the degree of heterogeneity that exists on the surface.

Table 7.4 and Fig. 7.8a indicate that k_{bind} increases as the SJA concentration in solution increases. Clearly, k_{bind} varies with SJA concentration in solution in a nonlinear fashion. Table 7.4 and Fig. 7.8b indicate that k_{diss} increases as the SJA concentration in solution increases. Again, k_{diss} varies with SJA concentration in solution in a nonlinear fashion.

The ratio $K = k_{\text{diss}}/k_{\text{bind}}$ is important since it provides a measure of affinity of the receptor for the analyte, in this case the lactose-immobilized surface for SJA in solution. Table 7.4 indicates that K decreases as the SJA concentration in solution increases from 125 to 1000 nmol in solution. Figure 7.8c shows the decrease in K as the SJA concentration in solution increases. At the lower SJA concentrations, the K value is higher. Thus, if the affinity is of concern and one has the flexibility of selecting the analyte concentration to be analyzed, then one should utilize lower concentrations of SJA. This is true, at least, in the 125–1000 nmol SJA concentration range.

Table 7.4 and Fig. 7.8d indicate that $D_{\text{f,diss}}$ decreases as the SJA concentration in solution increases. $D_{\text{f,diss}}$ is rather insensitive to the SJA concentration. Table 7.4 and Fig. 7.8e indicate that an increase in the fractal dimension for dissociation, $D_{\text{f,diss}}$, leads to a decrease in the binding rate coefficient, k_{diss}. At the outset, it may not appear to be appropriate to relate the binding rate coefficient to the fractal dimension. However, the topology of the surface, in our case the biosensor surface, does influence the binding rate

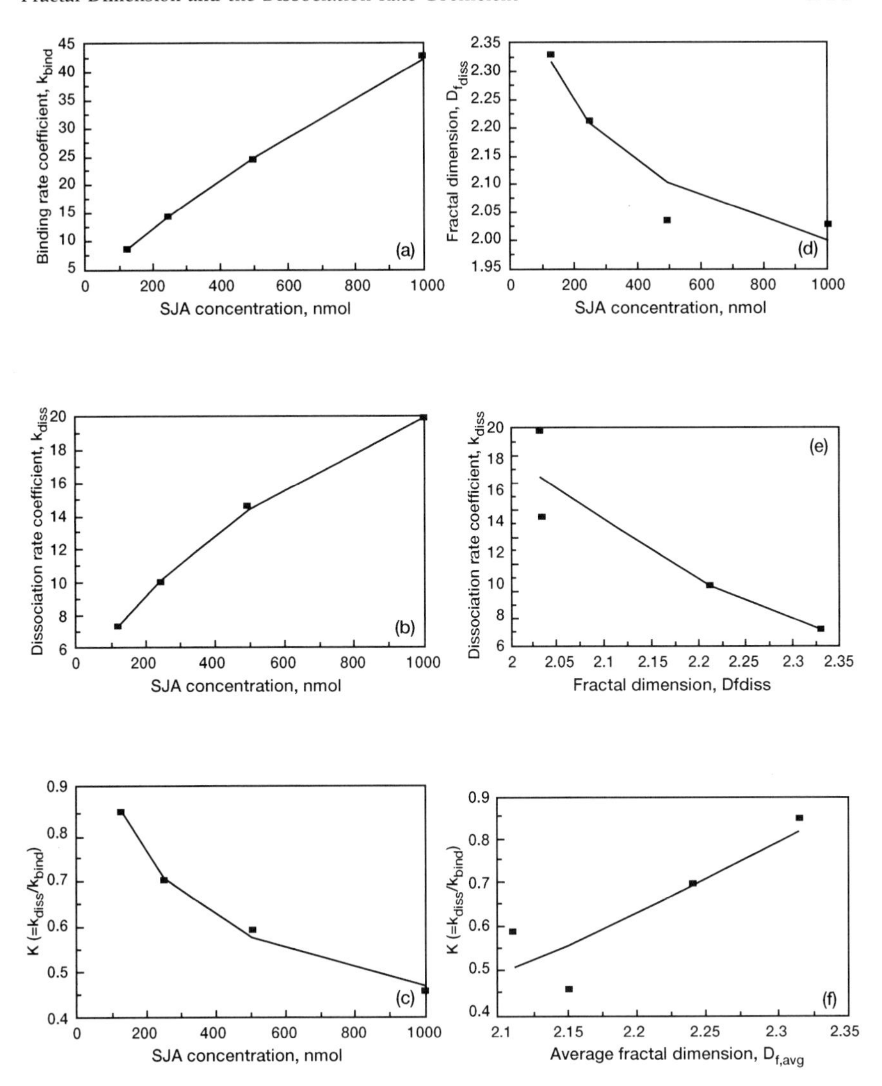

FIGURE 7.8 Influence of the *Sophora japonica* agglutinin in solution on (a) the binding rate coefficient, k_{bind}; (b) the dissociation rate coefficient, k_{diss}; (c) $K(=k_{diss}/k_{bind})$; (d) the fractal

coefficient. It would be of interest to analyze the influence of the heterogeneity of the surface on the constant, $K(=k_{bind}/k_{diss})$. Since both the association and dissociation phases are involved, and in lieu of any other indication, we will use $D_{f,avg}$, where $D_{f,avg}$ is equal to $(D_{f,bind}+D_{f,diss})/2$. Figure 7.8f shows that K increases as $D_{f,avg}$ increases.

Corr *et al.* (1994) utilized the SPR biosensor to analyze the binding of T lymphocytes bearing $\alpha\beta$ T cell receptors (TCRs) to a major histocompatability complex (MHC)-encoded class I or class II peptide. The authors employed a purified soluble analog of the murine MHC class I molecule H-2L^d (SH-2L^d) and a synthetic octamer peptide p2CL. The TCRs were covalently linked to the dextran-modified matrix of an SPR biosensor surface. Figure 7.9a shows the curves obtained using Eq. (7.2*a*) for the binding of 8 μmol SH-2L^d–p2CL complex (plus a constant concentration [10 μg/ml, 0.17 μM] of active SH-2L^d) in solution to 2C TCR immobilized on the SPR biosensor surface. A single-fractal analysis is sufficient to adequately describe the binding kinetics. The entire association curve is used to obtain the fractal dimension and the binding rate coefficient. Similarly, the entire dissociation curve is used to obtain the fractal dimension for dissociation and the dissociation rate coefficient. The dissociation curve may also be described by a single-fractal analysis.

Table 7.5 shows the values of the binding rate coefficient, k_{bind}, the dissociation rate coefficient, k_{diss}, the fractal dimension for binding, $D_{f,bind}$, and the fractal dimension for dissociation, $D_{f,diss}$. Note in this case that the values of the binding and dissociation rate coefficients are close to each other (187.4 and 187.2, respectively). Also note that the ratio $K(=k_{diss}/k_{bind})$ is close to a value of 1. The values of the fractal dimension for binding and dissociation are close to each other as well (2.57 and 2.53, respectively). This would indicate that the degree of heterogeneity on the SPR biosensor surface exhibits a negligible change, in this case, on going from one phase (binding) to the other (dissociation). Thus, in this case, the binding and dissociation rate coefficients correlate well with the degree of heterogeneity on the SPR biosensor surface.

Figure 7.9b shows the curves obtained using Eq. (7.2b) for the binding of 80 μmol SH-2L^d–p2CL complex [plus a constant concentration (10 μg/ml, 0.17 μM) of active SH-2L^d] in solution to 2C TCR immobilized on the SPR biosensor surface. In this case, a dual-fractal analysis [Eq. (7.2b)] is required to adequately describe the binding, or association, kinetics. The dissociation kinetics is, however, adequately described by a single-fractal analysis. The values of the association and dissociation rate coefficients and the fractal dimensions for the association and dissociation phases for both a single- and a dual-fractal analysis are given in Table 7.5.

Note that increasing the analyte concentration in solution by an order of magnitude, from 8 to 80 μM p2CL, leads to a change in the mechanism, in the sense that at the lower concentration a single-fractal analysis is sufficient but at the higher concentration a dual-fractal analysis is required to adequately describe the binding kinetics. For the dual-fractal analysis presented, note that an increase in the fractal dimension value by 29.5%—from $D_{f_1,bind} = 2.2$

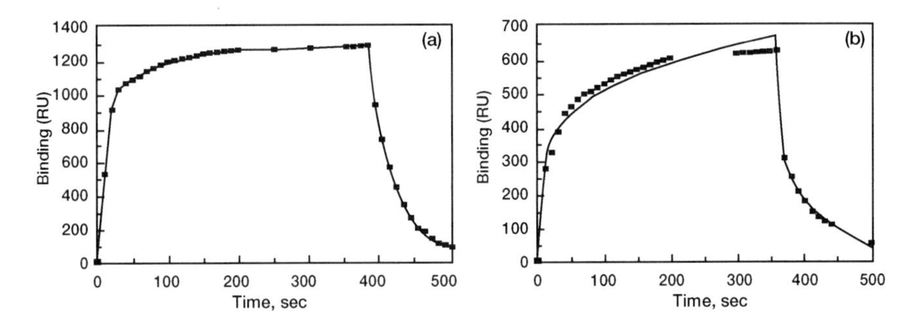

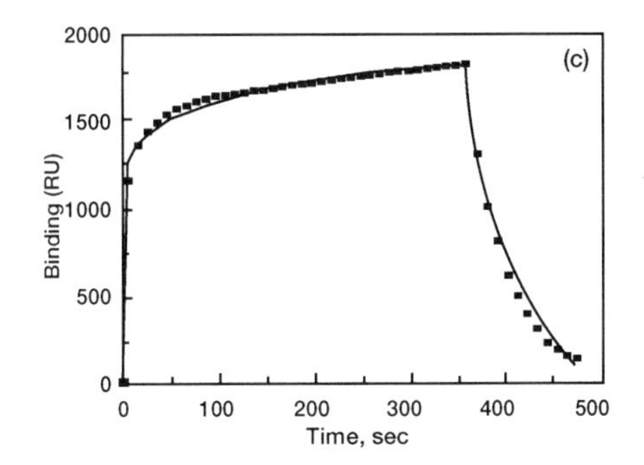

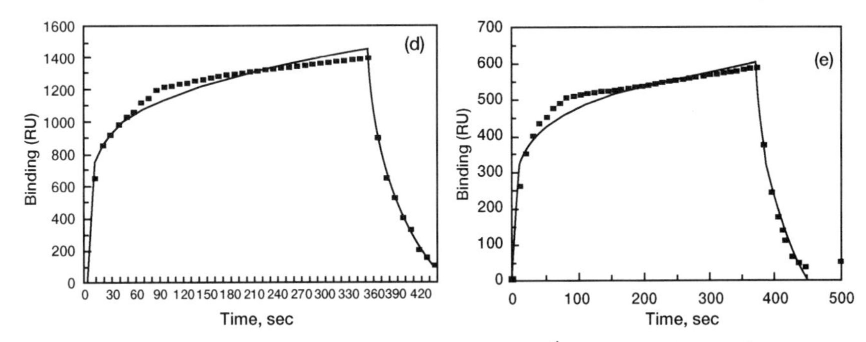

FIGURE 7.9 Binding of (a) 8 μmol and (b) 80 μmol SH-2L^d–p2CL complex in solution to 2C TCR immobilized on a SPR biosensor surface. Binding of (c) 1 μM , (d) 0.1 μM , and (e) 0.03 μM SH-2L^d in solution to 2C TCR immobilized on a SPR biosensor surface (Corr *et al.*, 1994).

to $D_{f_2,\mathrm{bind}} = 2.85$—leads to an increase in the binding (or association) rate coefficient by a factor of 3.52—from $k_{1,\mathrm{bind}} = 234.4$ to $k_{2,\mathrm{bind}} = 824.4$. In this case, the binding or adsorption rate coefficient is quite sensitive to the degree

TABLE 7.5. Fractal Dimensions and Binding and Dissociation Rate Coefficients for the Binding and Dissociation of SH2L^{d}-p2CL Complexes to Immobilized 2C-TCR on a SPR Surface (Corr *et al.*, 1994)

Analyte in solution	k_{bind}	$k_{1,bind}$	$k_{2,bind}$	k_{diss}	$D_{f,bind}$	$D_{f_1,bind}$	$D_{f_2,bind}$	$D_{f,diss}$
8 μM p2CL (+10 μg/ml (SH-2L^{d})	187 ± 11	na	na	187 ± 3	2.57 ± 0	na	na	2.53 ± 0.0
80 μM p2CL (+10 μg/ml (SH-2L^{d})	na	234 ± 31	824 ± 15	129 ± 10	na	2.2 ± 0.2	2.8 ± 0.0	2.03 ± 0.1
1 μM SH-2L^{d} (+250 μM p2CL)	1014 ± 21	na	na	219.3 ± 18	2.8 ± 0.0	na	na	2.11 ± 0.1
0.1 μM SH-2L^{d} (+250 μM p2CL)	499.4 ± 21	na	na	183 ± 7	2.6 ± 0	na	na	2.1 ± 0
0.03 μM SH-2L^{d} (+250 μM p2CL)	213 ± 11	na	na	81 ± 6	2.6 ± 0	na	na	2.1 ± 0

The units on the binding and the dissociation rate coefficients are: $gm(cm)^{-2}(sec)^{-p}$

of heterogeneity on the SPR biosensor surface. An increase in the degree of heterogeneity on the surface does lead to an increase in the binding (adsorption) or association rate coefficient. Note that the curve for single-fractal analysis is not shown in Fig.7.9b, and the corresponding binding rate coefficient and fractal dimension value are not given in Table 7.5.

It is of interest to compare the values of the rate coefficient and fractal dimension obtained for the dissociation phase on going from 8 to 80 μM p2CL in the analyte concentration in solution. For the dissociation phase only, an increase in the $D_{f,diss}$ value from 2.03 to 2.53 (on going from 80 to 8 μM p2CL in solution) leads to an increase in the k_{diss} value from 129 to 187. Thus, an increase in the degree of heterogeneity in the surface, or the fractal dimension on the surface, leads to an increase in the dissociation rate coefficient. Thus, apparently during both the association and dissociation phases, an increase in the degree of heterogeneity on the SPR biosensor surface leads to an increase in both the binding and dissociation rate coefficients.

Figure 7.9c shows the curves obtained using Eq. (7.2a) for the binding of 1 μM SH-2L^{d} (plus a constant concentration of 250 μM p2CL) in solution to 2C TCR immobilized on the SPR biosensor surface. In this case, a single-fractal analysis is sufficient to adequately describe the binding (or association) as well as the dissociation kinetics. The values of the binding and dissociation rate coefficients and the fractal dimensions for the association and dissociation phases are given in Table 7.5. In this case, the fractal dimension for the binding (adsorption) or association phase ($D_{f,bind} = 2.80$) is 32.7% higher than the fractal dimension for the dissociation phase ($D_{f,diss} = 2.11$). This leads to an increase in the binding (or association or adsorption) rate coefficient ($k_{bind} = 1014$) by a factor of 4.62 when compared to the dissociation (or desorption) rate coefficient ($k_{diss} = 219.3$). The ratio, $K(= k_{bind}/k_{diss})$ is equal to 0.216.

Figure 7.9d shows the curve obtained using Eq. (7.2a) for the binding of 0.1 μM SH-2L^{d} (plus a constant concentration of 250 μM p2CL) in solution to 2C TCR immobilized on the SPR biosensor surface. Once again, a single-fractal analysis is sufficient to describe the binding as well as the dissociation kinetics. The values of the association and dissociation rate coefficients and the fractal dimensions for the association and dissociation phases are given in Table 7.5. Note that a decrease in the SH-2L^{d} (analyte) concentration in solution from 1.0 to 0.1 μM leads to decreases in the rate coefficients for binding and dissociation as well as the fractal dimensions for the association and dissociation phases. For example, on going from 1.0 to 0.1 μM SH-2L^{d} in solution, the binding rate coefficient decreases from 1014 to 499.4. In this case, a decrease in the analyte concentration by an order of magnitude from 1.0 to 0.1 μM leads to an increase in the K value by a factor of 1.7—from 0.216 to 0.367.

Figure 7.9e shows the curve obtained using Eq. (7.2a) for the binding of 0.03 μM SH-2L^d (plus a constant concentration of 250 μM p2CL) in solution to 2C TCR immobilized on the SPR biosensor surface. Once again, a single-fractal analysis is sufficient to describe the binding as well as the dissociation kinetics. The values of the association and dissociation rate coefficients and the fractal dimensions for the association and dissociation phases are given in Table 7.5. In this case, the K value is 0.383. As noted above, a decrease in the analyte (SH-2L^d) concentration in solution from 1.0 to 0.03 μM leads, in general, to decreases in the rate coefficients and the fractal dimensions for the association and dissociation phases. The decreases in the fractal dimension are, however, relatively small (less than 5.7%). Also, note that a decrease in the analyte concentration in solution (this time from 1.0 μM to 0.03 μM) leads to a gradual increase in the K value. Thus, if this parameter is of importance, then one should use as low an analyte concentration as is possible.

In the SH-2L^d concentration range of 0.1 to 1.0 μM, Table 7.5 indicates that k_{bind} and k_{diss} increase as the SH-2L^d concentration in solution increases. Also note the value of $K(=k_{diss}/k_{bind})$ for the three concentrations of SH-2L^d analyzed. As one goes from 0.03 to 0.01 to 1.0 μM SH-2L^d concentration in solution, the K value decreases from 0.38 to 0.37 to 0.22. Thus, at the higher concentration there is a significant reduction in the K value. Thus, if a high K value is of interest, then one should employ lower concentrations of SH-2L^d if it is possible to do so.

In the concentration range of 0.03 to 1.0 μM SH-2L^d, Table 7.5 indicates that an increase in the fractal dimension for binding, $D_{f,bind}$, leads to an increase in the binding rate coefficient for binding, k_{bind}. In the concentration range of 0.03 to 1.0 μM SH-2L^d, Table 7.5 also indicates that an increase in the fractal dimension for dissociation, $D_{f,diss}$, leads to an increase in the dissociation rate coefficient, k_{diss}.

7.4. CONCLUSION

A confirmable fractal analysis only of the binding of antigen (or antibody) in solution to antibody (or antigen) immobilized on the biosensor surface provides a quantitative indication of the state of disorder (fractal dimension, $D_{f,bind}$) and the binding rate coefficient, k_{bind}, on the surface. Including the fractal dimensions for the dissociation step, $D_{f,diss}$, and dissociation rate coefficients, k_{diss} provides a more complete picture of the analyte-receptor reactions occurring on the surface (Sadana, 1999). One may also use the numerical values for the rate coefficients for the binding and dissociation steps to classify the analyte-receptor biosensor system as, for example, moderate binding, extremely fast dissociation; moderate binding, fast dissociation; moderate binding, moderate dissociation; moderate binding, slow dissociation; fast binding, extremely fast dissociation; fast binding, fast dissociation; fast binding, moderate dissociation; or fast binding, slow dissociation.

The fractal dimension value provides a quantitative measure of the degree of heterogeneity that exists on the surface for the analyte-receptor systems. The degree of heterogeneity for the binding and dissociation phases is, in general, different for the same reaction. This indicates that the same surface exhibits two degrees of heterogeneity, one for the binding and one for the dissociation reaction. In our discussion, we gave examples wherein either a single- or a dual-fractal analysis was required to describe the binding kinetics. The dual-fractal analysis was used only when the single-fractal analysis did not provide an adequate fit. This was determined using the regression analysis provided by Sigmaplot (1993). The dissociation step was adequately described by a single-fractal analysis for all the examples presented.

In accord with the prefactor analysis for fractal aggregates (Sorenson and Roberts, 1997), quantitative (predictive) expressions were developed for the binding rate coefficient, k_{bind}, as a function of the fractal dimension for binding, $D_{f,bind}$, for a single-fractal analysis, and for the dissociation rate coefficient, k_{diss}, as a function of the fractal dimension for dissociation, $D_{f,diss}$, for a single-fractal analysis. The $K(=k_{diss}/k_{bind})$ values presented provide an indication of the stability, reusability, and regenerability of the biosensor. Also, depending on one's final goal, a higher or a lower K value may be beneficial for a particular analyte-receptor system.

The fractal dimensions for the binding and the dissociation phases, $D_{f,bind}$ and $D_{f,diss}$, respectively, are not typical independent variables (as is, for example, the analyte concentration) that may be directly manipulated. These fractal dimensions are estimated from Eqs. (7.2a) and (7.2b), and one may consider them as derived variables. The predictive relationships developed for the binding rate coefficient as a function of the fractal dimension as well as for the dissociation rate coefficient as a function of the fractal dimension are of considerable value because these relationships directly link the binding or the dissociation rate coefficients to the degree of heterogeneity that exists on the surface and provide a means by which the binding or the dissociation rate coefficients may be manipulated by changing the degree of heterogeneity that exists on the surface. Note that a change in the degree of heterogeneity on the surface would, in general, lead to changes in both the binding and the dissociation rate coefficients. Thus, this may require a little thought and manipulation. The binding and the dissociation rate coefficients are rather sensitive to their respective fractal dimensions, or the degree of heterogeneity that exists on the biosensor surface, as may be seen by the high orders of dependence. It is suggested that the fractal surface (roughness) leads to turbulence, which enhances mixing, decreases diffusional limitations, and leads to an increase in the binding rate coefficient (Martin *et al.*, 1991).

More such studies are required to determine whether the binding and the dissociation rate coefficients are sensitive to their respective fractal dimensions. If they are, experimentalists may find it worth their effort to pay more attention to the nature of the surface as well as how it may be manipulated to control the relevant parameters and biosensor performance in desired directions. Also, in a more general sense, the treatment should also be applicable to non-biosensor applications wherein further physical insights could be obtained.

REFERENCES

Altschuh, D., Dubs, M.-C., Weiss, E., Zeder-Lutz, G., and Van Regenmortel, M. H. V., *Biochemistry* **31**, 6298 (1992).

Anderson, J., Unpublished results, NIH panel meeting, Case Western Reserve University, Cleveland, OH, 1993.

Bluestein, R. C., Diaco, R., Hutson, D. D., Miller, W. K., Neelkantan, N. V., Pankratz, T. J., Tseng, S. Y., and Vickery, E. K., *Clin. Chem.* **33**(9), 1543–1547 (1987).

Bluestein, B. I., Craig, M., Slovacek, G., Stundtner, L., Uricouli, C., Walczak, I., and Lisderer, A., (1991) in "Biosensors with Fiber-Optics," D. Wise and L. B. Wingard, Jr., eds., p. 181, Humana, New York.

Christensen, L. L. H., *Anal. Biochem.* **249**, 153–164 (1997).

Corr, M., Salnetz, A. E., Boyd, L. F., Jelonek, M. T., Khilko, S., Al-Ramadi, B. K., Kim, Y. S., Maher, S. E., Bothwell, A. L. M., Margulies, D. H., *Science* **265**, 946–949 (1994).

Cuypers, P. A., Willems, G. M., Kop, J. M., Corsel, J. W., and Hermens, W. T., in *Proteins at Interfaces: Physciochemical and Biochemical Studies* (J. L. Brash, and T. A. Horbett, eds.), American Chemical Society, Washington, DC, pp. 208–211, 1987.

Dabrowski, A. and Jaroniec, M., *J. Colloid & Interface Sci.* **73**, 475–482 (1979).

Di Cera, E., *J. Chem. Phys.* **95**, 5082–5086 (1991).

Eddowes, E., *Biosensors* **3**, 1–15 (1987/1988).

Edwards, P. R., Gill, A., Pollard-Knight, D. V., Hoare, M., Bucke, P. E., Lowe, P. A., and Leatherbarrow, R. J., *Anal. Biochem.* **231**, 210–217 (1995).

Elbicki, J. M., Morgan, D. M., and Weber, S. G., *Anal. Chem.* **56**, 978–985 (1984).

Fischer, R. J., Fivash, M., Casa-Finet, J., Bladen, S., and McNitt, K. L., *Methods* **6**, 121–133 (1994).

Giaver, I., *J. Immunol.* **116**, 766–771 (1976).

Glaser, R. W., *Anal. Biochem.* **213**, 152–158 (1993).

Havlin, S., in *The Fractal Approach to Heterogeneous Chemistry: Surfaces, Colloids, and Polymers* (D. Avnir, ed.), Wiley, New York, pp. 251–269, 1989.

Houshmand, H., Froman, G., and Magnusson, G. *Anal. Biochem.* **268**, 363–370 (1999).

Jaroniec, M. and Derylo, A., *Chem. Eng. Sci.* **36**, 1017–1019 (1981).

Jonsson, U., Fagerstam, L., Ivarsson, B., Johnsson, B., Karlsson, R., Lundh, K., Lofas, S., Persson, B., Roos, H., and Ronnberg, I., *Biotechniques* **11**, 620 (1991).

Karlsson, R. and Stahlberg, R., *Anal. Biochem.* **228**, 274 (1995).

Kopelman, R., *Science* **241**, 1620–1626 (1988).

Lee, C. K. and Lee, S. L., *Surface Sci.* **325**, 294 (1995).

Loomans, E. E. M. G., Beumer, T. A. M., Damen, K. C. S., Bakker, M. A., and Schielen, W. J. G., *J. Colloid & Interface Sci.* **192**, 238–249 (1997).
Mani, J. C., Marchi, V., and Cucurou, C., *Mol. Immunol.* **31**, 439 (1994).
Markel, V. A., Muratov, L. S., Stockman, M. I., and George, T. F., *Phys. Rev. B* **43**(10), 8183–8195 (1991).
Martin, S. J., Granstaff, V. E., and Frye, G. C., *Anal. Chem.* **65**, 2910–2922 (1991).
Matsuda, H., *J. Electroanal. Chem.* **179**, 107–117 (1967).
Milum, J. and Sadana, A., *J. Colloid & Interface Sci.* **187**, 128–138 (1997).
Morton, T. A., Myszka, D. G., and Chaiken, I. M., *Anal. Biochem.* **227**, 176–185 (1995).
Myszka, D. G., Morton, T. A., Doyle, M. L., and Chaiken, I. M., *Biophys. Chem.* **64**, 127–137 (1997).
Nieba, L., Krebber, A., and Pluckthun, A., *Anal. Biochem.* **234**, 155 (1996).
Nilsson, P., Persson, B., Uhlen, M., and Nygren, P. A., *Anal. Biochem.* **224**, 400–408 (1995).
Nygren, P. A. and Stenberg, M., *J. Immunol. Methods* **80**(1), 15–24 (1985).
Oscik, J., Dabrowski, A., Jaroniec, M., and Rudzinski, W., *J. Colloid & Interface Sci.* **56**, 403–412 (1976).
Pajkossy, T., and Nyikos, L., *Electrochim. Acta* **34**(2), 71–179 (1989).
Pfeifer, P. and Obert, M., in *The Fractal Approach to Heterogeneous Chemistry: Surfaces, Colloids, and Polymers* (D. Avnir, ed.), Wiley, New York, pp. 11–43 (1989).
Pisarchick, M. L., Gesty, D., and Thompson, N. L., *Biophys. J.* **63**, 215–233 (1992).
Place, J. F., Sutherland, R. M., and Dahne, C., *Anal. Chem.* **64**, 1356–1361 (1985).
Place, J. F., Sutherland, R. M., Riley, A., and Mangan, C., in "Biosensors with Fiberoptics," D. Wise and L. B. Wingard, Jr., eds., p. 253, Humana, New York. (1991).
Ramakrishnan, A. and Sadana, A., *J. Colloid & Interface Sci.* **229**, 628–640 (2000).
Rudzinski, W., Lattar, L., Zajac, J., Wofram, E., and Puszli, J., *J. Colloid Interface Sci.*, **96**, 339–359 (1983).
Sadana, A., *Biotech. Progr.* **11**, 50–59 (1995).
Sadana, A., *J. Colloid & Interface Sci.* **190**, 232–240 (1997).
Sadana, A., *Biosen. & Bioelectr.* **14**, 515–531 (1999).
Sadana, A. and Beelaram, A., *Biosen. & Bioelectr.* **10**, 301–316 (1995).
Sadana, A. and Chen, Z., *Biosen. & Bioelectr.* **11**(8), 769–782 (1996).
Sadana, A. and Madagula, A., *Biotech. Progr.* **9**, 259–269 (1993).
Sadana, A. and Madagula, A., *Biosen. & Bioelectr.* **9**, 45–55 (1994).
Sadana, A. and Ramakrishnan, A., *J. Colloid & Interface Sci.* **224**, 219–230 (2000a).
Sadana, A. and Sii, D., *J. Colloid & Interface Sci.* **151**, 166–177 (1992a).
Sadana, A. and Sii, D., *Biosen. & Bioelectr.* **7**, 559–568 (1992b).
Sadana, A. and Sutaria, M., *Biophys. Chem.* **65**, 29–44 (1997).
Satoh, A. and Matsumoto, I., *Anal. Biochem.* **275**, 268–270 (1999).
Sigmaplot, Scientific Graphing Software, User's manual, Jandel Scientific, San Rafael, CA, 1993.
Sjolander, S. and Urbaniczky, C., *Anal. Chem.* **63**, 2338–2345 (1991).
Sorenson, C. M. and Roberts, G. C., *J. Colloid & Interface Sci.* **186**, 447–452 (1997).
Stenberg, M. and Nygren, H. A., *Anal. Biochem.* **127**, 183–192 (1982).
Stenberg, M., Stiblert, L., and Nygren, H. A., *J. Theor. Biol.* **120**, 129–142 (1986).
Thomson, S. J. and Webb, G., *Heterogeneous Catalysis*, Wiley, New York, p. 23, 1968.
Williams, C. and Addona, T. *TIBTECH* **18**, 45–48 (2000).
Wohlhueter, R. M., Parakh, K., Udhayakumar, V., Fang, S., and Lal, A. A., *J. Immunol.* **153**, 181 (1994).

CHAPTER 8

Influence of Nonspecific Binding on the Rate and Amount of Specific Binding: A Classical Analysis

8.1. INTRODUCTION

Rapid and numerous advances in health care are a major driving force for the development of biosensors. In health care it is imperative to obtain precise and accurate quantitative diagnostics, and nonspecific binding (NSB) plays a deleterious role in such analysis. Thus, an elimination of, or at least a reduction in, NSB is critical not only in the health care field, but also in areas such as agriculture, horticulture, veterinary medicine, physics, chemistry, oceanography, aviation, and environmental control. Areas for the application of biosensors are continuously increasing, and thus it is imperative to obtain a better understanding of NSB in a general sense, and then further fine-tune it for a particular application. Mathewson and Finley (1992) emphasize that the exponential growth in the interest in biosensor development is due to the biosensor's potential for the convenient detection of an almost limitless number of analytes in a wide variety of surroundings. It is anticipated that the combination of biotechnology with microelectronics will result in a variety of inexpensive, disposable biosensors (Dambrot, 1993; Wise and Wingard, 1991). Sensors should be reliable, simple, rapid in their measurement, and able to detect low levels of analytes, often in a mixture of similar substances. Understandably, NSB becomes more and more critical with a decrease in (or the dilution of) the analyte of interest.

Biosensors, as the name indicates, use biologically derived molecules as sensing elements. The main feature of the biosensor is the spatial unity of the biomolecules with a signal transducer (Lowe, 1985). Scheller *et al.* (1991) emphasize the importance of providing a better understanding of the mode of operation of biosensors to improve their sensitivity, stability, specificity, and speed of response. NSB plays a critical role in such analysis but, unfortunately, very little is known about NSB. Thus, its influence on biosensor performance parameters is difficult to estimate using a scientific basis.

The solid-phase immunoassay technique provides a convenient means for the separation of reactants (for example, antigen) in solution. Such a separation is possible because of the high specificity of the analyte for the immobilized antibody. As indicated in earlier chapters, external diffusional limitations play a role in the analysis of such assays. We need to emphasize that in most biosensor analysis, nonspecific binding has not been seriously considered. However, according to Place *et al.* (1985) there is a balance inherent in the practical utility of biosensor systems. These authors estimate that although acceptable response times of minutes or less should be obtainable at μM concentration levels, inconveniently lengthy response times will be found at nM or lower concentration levels. The presence of high degrees of nonspecific binding will necessarily increase these response times. Nonspecific binding of the antigen may occur directly on the biosensor surface area where no antibody is present. Also, during the immobilization procedure on the biosensor surface, nonspecific binding of the antibody may occur. The latter case can lead to heterogeneities on the biosensor surface. Thus there is a need for analyzing nonspecific binding and for providing some factor that relates the extent of nonspecific binding to the total binding (specific and nonspecific) (Scheller *et al.*, 1991; Sadana and Sii, 1992).

We will now analyze some examples wherein nonspecific binding is presumed present. Walczak *et al.* (1992) have developed an evanescent biosensor to analyze the human enzyme creatine kinase (CK; EC 2.7.3.2) isoenzyme MB form (CK MB) with a molecular weight of 84,000. Custom β-phycoeythin CK-MB antibody conjugates were immobilized on fused silica fiber-optic sensors. There is considerable interest in the development of a biosensor for the detection of this cardiac isoenzyme creatine kinase since it would permit early detection of myocardial infarctions. Walczak *et al.* indicate that it is unlikely that nonspecific adsorption of the antibody conjugate accounts for a large fraction of the signal observed. These authors estimate that, for their case, nonspecific adsorption mechanisms at best can account for 25% of the accumulated signal for the lowest CK-MB concentrations used. As expected, the relative amount of this nonspecific adsorption mechanism when compared to the specific adsorption mechanism is proportionately smaller when higher CK-MB isoenzyme samples are analyzed.

Betts *et al.* (1991) utilized dansylated F(ab′) antibody fragments in the fabrication of a selective, sensitive, and regenerable biosensor. These authors indicate that one of the negative points is the nonspecific adsorption of the immobilized F(ab′) onto the quartz substrate. The authors planned to use different substrata to help alleviate this problem.

Byfield and Abuknesha (1994) indicate that antibodies or immunoglobulins (Ig) consist of four polypeptide subunits. There are two identical small, or light, chains (L chains) and two identical large, or heavy, chains (H chains) held together by noncovalent forces. See Fig. 8.1. Also, covalent interchain disulfide bonds are involved. These authors indicate that the Fab fragments contain the key portions of the antibody molecule that contain the binding sites for the antigen. The Fab fragment itself contains a light chain and a heavy

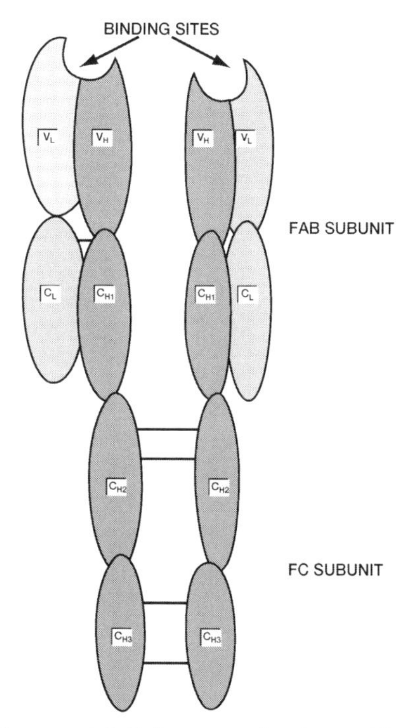

FIGURE 8.1 Antibody (immunoglobulin) structure showing two identical light chains (VL-CL) and two identical heavy chains ($VH\text{-}CH_1\text{-}CH_2\text{-}CH_3$) (Byfield and Abuknesha, 1994).

chain. The advantage of using Fab fragments in biosensor applications is very clear: it minimizes the nonspecific binding. This is because the entire Ig molecule contains Fc sections to which nonantigen components may stick. This is different from the binding of the actual antigen to anything (such as directly to the biosensor surface), from which no signal is obtained. Extending the definition of nonspecific binding to include distortion of the measured signal, Byfield and Abuknesha indicate that a single binding site on a Fab as compared to two binding sites on an Ig molecule will minimize or prevent the formation of antigen-bridged complexes. These complexes distort the quantitative aspects of conventional immunoassays, which could lead to disastrous results, especially in the health care field.

Since not much is currently known about the science of nonspecific binding and how it influences the amount of specific binding, various attempts have been made to minimize NSB. It is certain that only under rare or unusual circumstances will NSB actually be useful in immunoassay analysis. Thus, it is useful to examine and analyze at least some examples where the influence of NSB has been minimized. Owaku *et al.* (1993) indicate that employing the Langmuir–Blodgett (LB) film technique produces a highly ordered and alligned antibody layer on a biosensor surface, which should help minimize this type of nonspecific binding.

Ahluwalia *et al.* (1991) have utilized the LB film technique to deposit oriented antibodies on a surface; the antibodies form reproducible expanded films at a gas–water interface. These films are then transferred to a solid surface to produce high-density surface coatings of the material of interest. These authors indicate that for an efficient surface one requires high-surface density of active molecules, the absence of nonspecific binding, and stability and durability.

Figure 8.2 shows the procedure for the immobilization of the antibody using hydrophilic adsorption, hydrophobic adsorption, and amino silanization. The monoclonal antiprolactin (IgG_1) forms a complex with the conjugated antibody (IgG_2-HRP). Here HRP is a horse radish peroxidase. The antigen is human prolactin in solution with the salts and the preservatives. Ahluwalia *et al.* (1991) indicate that the nonspecific binding here is due to the adsorption of salts or preservatives. Out of the three immobilization procedures utilized, the APTS (5% aminopropyltriethoxysilane) method yielded the smallest nonspecific binding (0.1) for the highest surface density of antibody ($\rho = 320\,ng/cm^2$). In comparison, the adsorption on a hydrophilic surface gave ρ and NSB values of $41\,ng/cm^2$ and 0.67, respectively. Similarly, ρ and NSB values of $217\,ng/cm^2$ and 0.21, respectively, were obtained for adsorption on a hydrophobic surface. It appears that NSB is inversely proportional to the surface densities of the antibody, at least for this case.

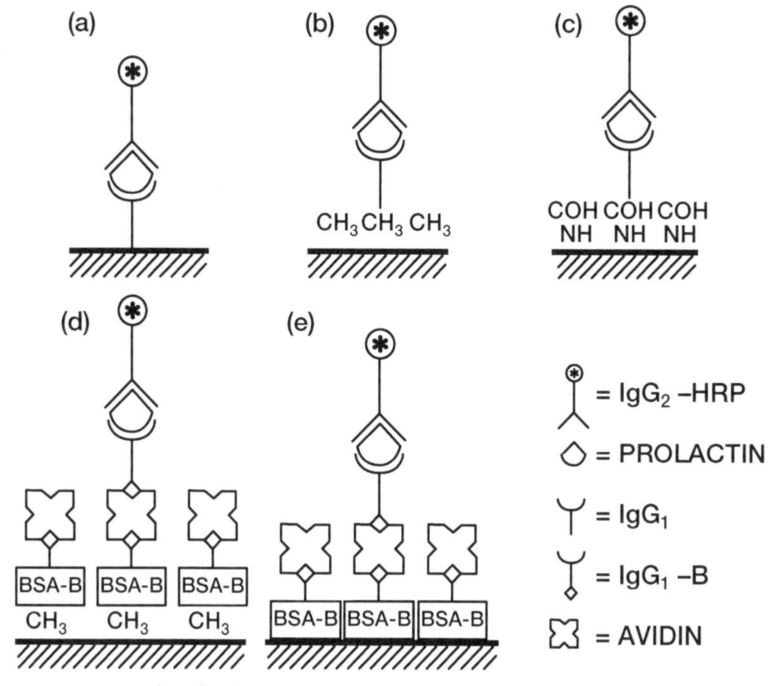

FIGURE 8.2 Procedure for the immobilization of the antibody using hydrophillic adsorption, hydrophobic adsorption, and amino silanization (Ahluwalia *et al.*, 1991).

Other methods by which nonspecific binding may be minimized exist. For example, one may coat the surface with a hydrophilic material that exhibits a low interfacial energy (Scheller *et al.*, 1991). The ratio of nonspecific to the total binding should be carefully examined, and if it is negligible (say, less than 5–10%, an arbitrary number) it should be explicitly stated (Sadana and Sii, 1992). Furthermore, the spatial distribution of antibody-coated colloidal gold particles over an antigen-coated surface by electron microscopy has been analyzed by Nygren (1988). This author noted that the nonspecific as well as the specific binding for this case exhibits a spatial distribution. Thus, nonspecific binding should be considered in the development of an appropriate model.

Okano *et al.* (1992) have developed a heterogeneous immunoassay for attomole-level detection. These authors indicate that the sensitivity depends on both the reaction efficiency and the nonspecific adsorption of the microparticles on the microplate surface. More specifically, they indicate that there are two ways to increase the sensitivity of the assay. In the first method, the reaction equilibrium needs to be shifted between the microparticle-conjugated antibody and antigen toward the binding state. The other method is to minimize the nonspecific adsorption. Often, these are at cross-purposes: One may increase the number of surface antibodies, which would enhance the reactivity of the microparticle-conjugated antibody, but overloading the antibodies on the surface increases the nonspecific adsorption. Okano *et al.* also indicate that homogeneity of the antibody on the detection surface plays an important role in the sensitivity. One must take care to ensure that the capture antibody is distributed uniformly on the microplate surface. A certain amount of capture antibody is essential, otherwise this uniformity is not maintained. The authors added bovine serum albumin, which increased the total amount of protein and helped the uniformity of the capture antibody, thus assisting in minimizing the nonspecific adsorption. In essence, the bovine serum acts as a protein matrix, which helps promote the uniform immobilization. For example, the authors indicate that adding 25 μg of bovine serum albumin to 1 μg of capture antibody could lower the nonspecific adsorption of microparticles to one-fourth of that without serum albumin.

Byfield and Abuknesha (1994) indicate that different types of proteins, ions, or small organic molecules can interact with the surface in a nonspecific manner. This, as expected, causes a major reduction in the signal-to-noise ratio for antibody–antigen measurements (as compared to when only pure antibody or pure antigen is utilized). These authors indicate that nonspecific adsorption is minimized in the hydrogel matrix developed by Pharmacia based on carboxy-methylated dextran. This matrix is linked to the gold surface of a surface plasmon resonance biosensor chip by a linker layer of a hydroxyalkyl thiol. Nonspecific adsorption is reduced by the hydrophilic nature of the dextran matrix. Biological binding reactions are promoted due to the dominance of the electrostatic forces. This prevents nonspecific interactions, which are hydrophobic in nature. As an additional benefit, the physical barrier provided by the dextran matrix prevents other (undesirable) components from reaching the reaction interface, and the specific antibody–antigen reaction is optimized. See Fig. 8.3.

Nonspecific binding may occur directly on the biosensor surface area where no antibody is present. Also, during the immobilization procedure on the biosensor surface, nonspecific binding of the antibody may occur. Often, this immobilization procedure can lead to heterogeneities on the biosensor surface. These must be taken into account in the analysis of specific binding by an appropriate procedure. Ideally, of course, it is better to minimize or, if possible, eliminate nonspecific adsorption. Collison *et al.* (1994) have emphasized that although many biomolecule immobilization strategies have been developed—including entrapment, chemical cross-linking, electropolymerization, and covalent attachment—they often lack the reproducibility and reliability required for the fabrication of biosensors on a commercial scale.

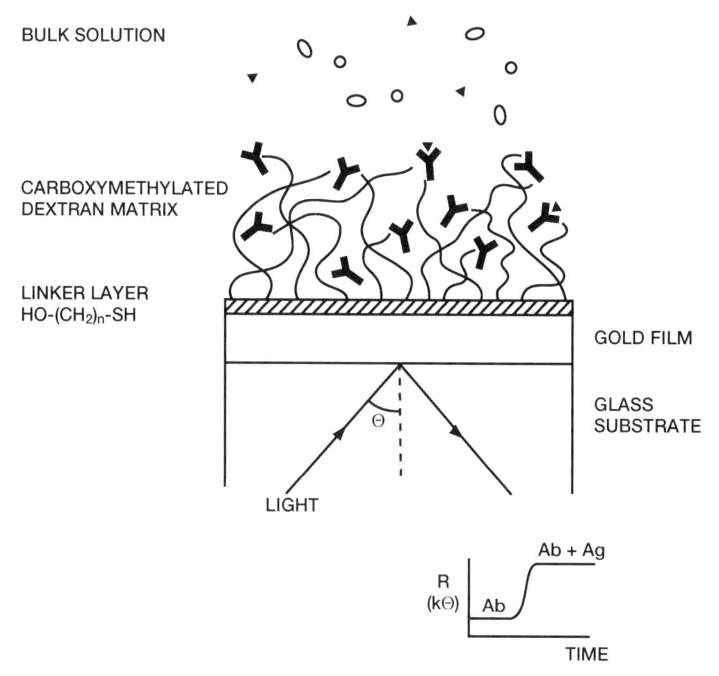

FIGURE 8.3 Pharmacia BIACORE schematic representation. Inset shows typical response versus time curve (Byfield and Abuknesha, 1994).

Understandably, the mode of attachment or immobilization will significantly affect both nonspecific and specific binding of the antigen to the antibody or vice versa.

Disley *et al.* (1994) indicate the need to develop compatible interfaces to which the antibodies (or antigens) can be attached in a controlled and oriented manner, thereby assisting in either eliminating or minimizing nonspecific adsorption. The method suggested by these authors resulted in a reduction in the amount of nonspecific binding while simultaneously observing an increase in the specific binding of the polythiolated IgG interface for its target antigen. These authors suggest that the decrease in nonspecific effects was due to more mobile and hydrophilic surfaces, due to entropically driven protein monolayer repulsive interactions. More such studies are required to clearly delineate the method and the reason(s) why the nonspecific binding decreased.

Sadik and Wallace (1994) have emphasized that a significant cause of frustration is the fact that no surface is truly inert. This is derogatory as far as the development of stable and reversible biosensors is concerned. The authors indicate that "unruly" chemical processes occurring at biosensor surfaces will affect biosensor performance to different degrees. In essence, NSB occurs at surface sites where no antigen is adsorbed. Thus, the careful preparation of surfaces where the surface coverage is high would be helpful. In addition, free adsorption sites should be neutralized by inert molecules. Arwin and Lundstrom (1985) emphasize the importance of microorientation to minimize nonspecific adsorption wherein either the antigen or the antibody is immobilized on the surface. Here, due to the small number of determinants on the antibody molecules, it is important to orient the receptors on the surface toward the analyte in solution.

Immobilization of reagents for optical sensors is achieved in different ways. For example, immobilization can be achieved by adsorption on polymeric supports such as PTFE (Wyatt *et al.*, 1987). Blair *et al.* (1993) indicate that PTFE is an excellent surface for the immobilization of reagents since the reagent phase adsorbed on the PTFE is easily accessible to the analyte as compared to the adsorption of the reagent on the resin beads. Besides, the PTFE surface is inert, which eliminates or minimizes the nonspecific adsorption. Of course, it also makes the adsorption of the reagent more difficult.

Recently, there has been a significant effort to analyze and utilize surface-oriented affinity methods such as surface plasmon resonance (SPR). This technique should significantly assist in revealing the nature of the nonspecific and specific binding of the analyte in solution to the biosensor surface. Such a qualitative description that yields the details of the interactions of the molecules on the biosensor surface is of considerable utility. Pritchard *et al.* (1994) have emphasized that the major challenge faced during the design of a multianalyte immunosensor is the patterning of specific antibodies at discrete transducer sites and retaining the immunological activity. In addition, it is essential to minimize nonspecific effects.

Ekins and Chu (1994) have also discussed the development of multianalyte assays and their importance in microanalytical technology. Further development in multianalyte assays is required to increase their sensitivity and their reproducibility characteristics. These authors emphasize a miniaturized multianalyte system. Miniaturization would permit one to measure different analytes in very small samples. The extension of such multianalyte assay systems into conventional binding assay type techniques would permit their application into areas from which they were previously excluded due to cost and other complexities. Ekins and Chu are working the "microspot principle", which is basically that highly sensitive assays can be performed using much smaller amounts of antibody (confined to a "microspot") than previously thought necessary. Each microspot would then be used for different analytes. The authors do indicate that nonspecific binding will be a problem. Ways to minimize the effects of nonspecific binding include the selection of better supports, better antibodies or antibody fragments, and better instrumentation techniques. A time resolution of the fluorescent signal may be utilized to eliminate solid-support-generated backgrounds. Finally, Connolly (1994) indicates that the application of micro- and nano-fabrication techniques to biosensors should produce more oriented or specified arrays of antibody or antigen structures immoblized on the biosensor surface, besides minimizing nonspecific binding.

Marose *et al.* (1999) analyzed the use of optical systems for bioprocess monitoring. They indicate that although these systems exhibit potential for useful applications, there are some problems. These problems include the temperature dependence and that the optical sensors are subject to fouling. Of direct concern to us at present is that such systems need to be recalibrated often, which implies nonspecific binding. Since the possibility of continuous optical bioprocess monitoring is coming closer to reality, the issue of NSB needs to be addressed carefully. Very possibly, different methods of NSB minimization will be needed for different bioprocess applications.

Regarding another type of NSB, Williams and Addona (2000) have analyzed the successful integration of biospecific interaction analysis based on surface plasmon resonance combined with mass spectrometry. According to these authors, this technique combines the benefits of sensitive affinity capture and characterization of binding events with the ability to characterize interacting molecules. The amount of protein recoverable is in femtomoles. As expected, whenever extremely small amounts of analytes are involved, then NSB becomes increasingly important to maintain accuracy and precision. These authors indicate that due to NSB and dilution there may be a significant loss in the analyte due to the tubing and the recovery. In later versions of the instrument, NSB has been minimized by eliminating the recovery cup and minimizing the use of tubing. These improvements in minimizing the NSB and dilution effects (by using very small regenerant (for reverse flow elution)) permits a more accurate identification and characterization of the analyte molecules bound to the biosensor surface.

Ohlson *et al.* (2000) have analyzed the weak biological interactions that occur throughout biological systems either alone or in concert. They have designed a system that provides continuous monitoring using an on-line immunosensing device. One of the problems they encountered was nonspecific binding. This can be minimized by the selection of an appropriate reference system. Otherwise, nonspecific interactions may distort the interpretation of data.

Kortt *et al.* (1997) have indicated that nonspecific binding may be a cause for error in BIACORE binding measurements. These authors analyzed the interaction of monovalent forms of NC41 (an antiviral neuraminidase antibody) and 11-1G10 (an anti-idiotype antibody). The authors used this as a model system to demonstrate problems that may arise due to nonspecific amine coupling. Their results indicate that nonspecific immobilization (improper orientation) by one or more lysine residues close to or within the CDR2 region of the 11-1G10V_H domain is presumably responsible for the decrease in interaction strength with NC41. This resulted in a reduction in the measured binding affinity. The authors emphasized that one should utilize site-specific immobilization strategies when accurate kinetic measurements are necessary.

Finally, Kubitschko *et al.* (1997) have attempted to enhance the sensitivity of optical immunosensors using nanoparticles. They indicate that due to low sensitivity many substances are not analyzable in serum. Nonspecific binding, along with the low molecular weight of the analytes, prevents an accurate quantitative determination of these analytes. They indicate that nonspecific binding makes a significant contribution to the signal and falsifies the result. This is especially true for smaller analytes, which cause only a small change in the refractive index on the sensor surface. The authors emphasize that for *in vitro* diagnostics small molecules, such as thyroid-stimulating hormone (TSH), need to be analyzed in serum so nonspecific binding needs to be minimized.

In this chapter, we will analyze theoretically and in some detail the influence of nonspecific binding and temporal model parameters on the specific binding of the antigen in solution to the antibody immobilized on the surface. The temporal nature of the specific binding parameters of the antigen in solution to the antibody immobilized on the surface, along with the inclusion of the nonspecific binding of the antigen in solution directly to the fiber-optic surface, is a more realistic approach to the actual situation. The analysis should provide fresh physical insights into first-, one-and-one-half-, second-, and other-order antigen–antibody reactions occurring on the fiber-optic biosensor surface under external diffusion-limited conditions.

8.2. THEORY

Figure 8.4 describes the steps involved in the binding of the antigen in solution to the antibody covalently attached to the surface (specific binding) and to the fiber-optic biosensor surface (nonspecific binding). External diffusion limitations play a significant role in the specific and the nonspecific binding of the antigen. The diffusion-limited reaction can be determined by

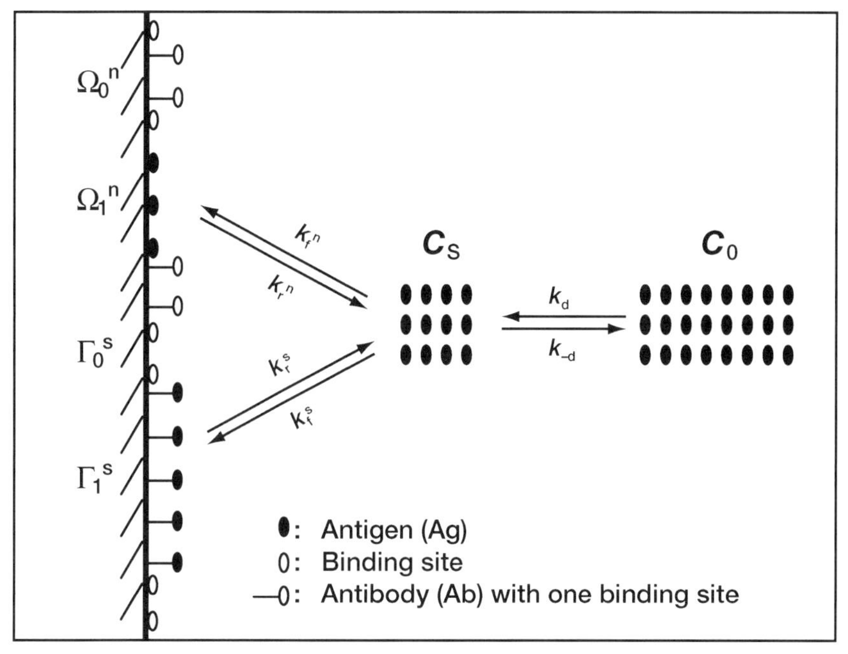

FIGURE 8.4 Elementary steps involved in the binding of the antigen in solution to the antibody covalently attached to the fiber-optic surface for the first-order reaction. Γ_0^s and Ω_0^n are the total concentration of the specific binding sites on the antibody and of the nonspecific binding sites on the fiber-optic surface, respectively.

the equation

$$\frac{\partial c}{\partial t} = D\nabla^2 c = D\frac{\partial^2 c}{\partial x^2}. \tag{8.1a}$$

Here, t is the reaction time, x is the distance away from the fiber-optic surface, and D is the diffusion coefficient. Eq. (8.1a) may be rewritten in dimensionless form as

$$\frac{\partial y}{\partial \theta} = \frac{\partial^2 y}{\partial z^2}. \tag{8.1b}$$

Here, $y = c/c_0$, $z = x/L$ (where L is a characteristic length dimension—for example, the diameter of a fiber-optic biosensor), and $\theta = t/(L^2/D)$.

The appropriate initial condition for Eq. (8.1a) in dimensionless form is

$$\begin{aligned} y(z,0) &= 1 \quad \text{for } z>0, \theta=0 \\ y(0,0) &= 0 \quad \text{for } z=0, \theta=0. \end{aligned} \tag{8.2a}$$

This initial condition is equivalent to the rapid immersion of a sensor into a solution with antigen.

A boundary condition in dimensionless form for Eq. (8.1a) is

$$y(\infty,\theta) = 1 \quad \text{for } \theta>0, \; z=\infty. \tag{8.2b}$$

This boundary condition has not been mentioned in similar previous analyses (Stenberg *et al.*, 1986; Sadana and Sii, 1991, 1992; Sadana and Madagula, 1993, 1994; Sadana and Beelaram, 1994). We now feel that this more correctly represents the actual situation.

Another boundary condition for Eq. (8.1a) is

$$D\frac{\partial c}{\partial x} = \frac{d\Gamma^{s}_{Ag}}{dt} + \frac{d\Omega^{n}_{Ag}}{dt} \quad \text{for } t>0, x=0. \tag{8.2c}$$

Eq. (8.2c) arises because of mass conservation, wherein the flow of antigens to the surface must be equal to the rate of the antigen reacting with the antibody on the surface (specific binding) as well as the binding of the antigen to the surface itself (nonspecific binding). Here, $d\Gamma^{s}_{Ag}/dt$ and $d\Omega^{n}_{Ag}/dt$ represent the rates of specific and nonspecific binding, respectively. The right-hand side is different for different reaction orders. We will present the details for obtaining this boundary condition for first- and second-order reactions. The analysis is then easily extended for one-and-one-half and general *n*th-order reactions. The analysis for first- and second-order reactions where only specific binding is present is available in the literature (Sadana and Sii, 1991, 1992).

8.2.1. First-Order Reactions

The rate of binding of a single antigen by an antibody is given by

$$\frac{d\Gamma^{s}_{1}}{dt} = k^{s}_{f} c_{s}(\Gamma^{s}_{0} - \Gamma^{s}_{1}) - k^{s}_{r}\Gamma^{s}_{1}, \tag{8.3a}$$

where Γ^{s}_{0} is the total concentration of the antibody sites on the surface, Γ^{s}_{1} is the surface concentration of antibodies bound to a single antigen at any time t, c_s is the concentration of the antigen close to the surface, k^{s}_{f} is the forward reaction rate constant, and k^{s}_{r} is the reverse reaction rate constant. In this case, even though the antibody molecule has two binding sites, for all practical purposes we believe that an antigen molecule reacts with an antibody as if it had only one binding site.

For initial binding kinetics, after some simplification, we obtain (Sadana and Sii, 1991, 1992)

$$\frac{d\Gamma_1^s}{dt} = k_f^s \Gamma_0^s c_s. \tag{8.3b}$$

Integration of this equation yields the relative concentration of the antigen bound by specific binding. This is proportional to the optical signal of the sensor at any time t. Therefore,

$$\frac{\Gamma_{Ag}^s}{c_0} = k_f^s \Gamma_0^s \int_0^t y dt. \tag{8.3c}$$

The rate of nonspecific binding of the antigen is given by

$$\frac{d\Omega_1^n}{dt} = k_f^n c_s (\Omega_0^n - \Omega_1^n) - k_r^n \Omega_1^n, \tag{8.4a}$$

where Ω_0^n is the total concentration of the nonspecific binding sites on the fiber-optic surface, Ω_1^n is the concentration of the antigen bound by nonspecific binding, and k_f^n and k_r^n are the forward and reverse binding rate constants for nonspecific binding, respectively.

In the initial kinetics regime, $\Omega_1^n \ll \Omega_0^n$ and $k_f^n c_s \Omega_0^n \gg k_r^n \Omega_1^n$. Then, from Eq. (8.4a) we obtain the following for nonspecific binding:

$$\frac{d\Omega_1^n}{dt} = k_f^s \Omega_0^n c_s. \tag{8.4b}$$

The simplified reaction scheme for the binding of the antigen by specific as well as nonspecific binding is given by

$$\begin{gathered} \mathrm{Ab} + \mathrm{Ag} \rightleftharpoons \mathrm{Ab} \cdot \mathrm{Ag} \\ \mathrm{NS} + \mathrm{Ag} \rightleftharpoons \mathrm{NS} \cdot \mathrm{Ag}. \end{gathered} \tag{8.5}$$

Here, Ab represents the antibody binding site, NS represents the nonspecific binding site, and Ag represents the antigen.

Substituting the rates of specific [Eq. (8.3b)] and nonspecific [Eq. (8.4b)] binding into the boundary condition [Eq. (8.2c)] yields, in dimensionless form,

$$\frac{\partial y}{\partial z} = \mathrm{Da}(1 + \alpha) y. \tag{8.6}$$

Here, $\alpha = (k_f^n \Omega_0^n)/(k_f^s \Gamma_0^s)$, Da is the Damkohler number, and α is the ratio of the maximum binding rate due to nonspecific and specific binding. Also, $Da = k_f^s \Gamma_0^s L/D$. Recall that the Damkohler number is the ratio between the maximum reaction rate and the maximum rate of external mass transport.

8.2.2. Second-Order Reactions

The modeling of the specific binding of an antigen in solution to an antibody immobilized on the surface is a two-step process. The elementary steps involved in the reaction scheme are shown in Fig. 8.5. (Sadana and Chen, 1996).

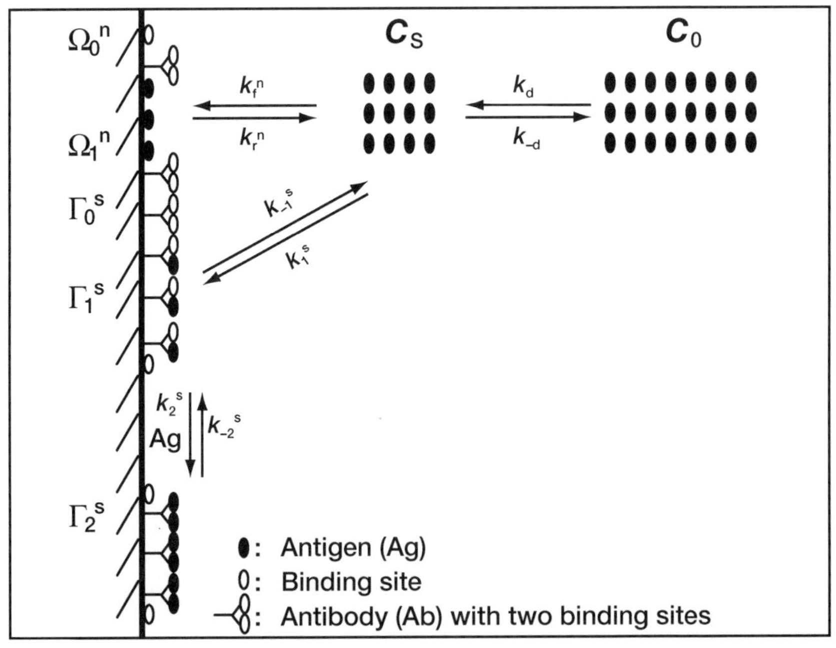

FIGURE 8.5 Elementary steps involved in the binding of the antigen in solution to the antibody covalently attached to the fiber-optic surface for the second-order reaction. Γ_0^s and Ω_0^n are the total concentration of the specific binding sites on the antibody and of the nonspecific binding sites on the fiber-optic surface, respectively. Γ_1^s and Ω_1^n are the concentration of the antibody bound to one antigen due to specific binding and of the filled nonspecific binding sites on the fiber-optic surface, respectively. Γ_2^s is the concentration of the antibody bound to two antigens due to specific binding (Sadana and Chen, 1996).

The rate of specific binding of a single antigen by an antibody is given by

$$\frac{d\Gamma_1^s}{dt} = k_1^s c_s(\Gamma_0^s - \Gamma_1^s - \Gamma_2^s) - k_{-1}^s\Gamma_1^s + k_{-2}^s\Gamma_2^s - k_2^s c_s\Gamma_1^s, \tag{8.7a}$$

where Γ_0^s is the total concentration of the antibody sites on the surface, Γ_1^s is the surface concentration of antibodies bound to a single antigen at any time t, and Γ_2^s is the surface concentration of the antibody that binds two antigens. The rate at which the antibody specifically binds two antigens is given by

$$\frac{d\Gamma_2^s}{dt} = k_2^s c_s\Gamma_1^s - k_{-2}^s\Gamma_2^s. \tag{8.7b}$$

For initial binding kinetics, after some simplification, we obtain from Eq. (8.7a) (Sadana and Sii, 1991)

$$\frac{d\Gamma_1^s}{dt} = k_f^s c_s^2\Gamma_0^s. \tag{8.7c}$$

The second-order dependence on antigen concentration is not surprising since two molecules of the antigen can bind to two binding sites on the same antibody molecule. On integrating the Eq. (8.7c), we obtain the concentration of the antigen bound due to specific binding. This bound concentration is given by

$$\frac{\Gamma_1^s}{c_0} = \frac{\Gamma_{Ag}^s}{c_0} = k_f^s\Gamma_0^s c_0\int_0^t y^2 dt. \tag{8.7d}$$

The rate of nonspecific binding of the antigen to the surface is given by

$$\frac{d\Omega_1^n}{dt} = k_f^n c_s(\Omega_0^n - \Omega_1^n) - k_r^n\Omega_1^n, \tag{8.8}$$

where Ω_0^n is the total concentration of the nonspecific binding sites on the fiber-optic surface and Ω_1^n is the concentration of the filled nonspecific binding sites on the surface. In the initial regime, $\Omega_1^n \ll \Omega_0^n$ and $k_f^n c_s\Omega_0^n \gg k_r^n\Omega_1^n$. Then, the total rate of antigen bound by specific and nonspecific binding is given by

$$\frac{d\Gamma_{Ag}^t}{dt} = k_f^s\Gamma_0^s c_s^2 + k_s^n\Omega_0^n c_s. \tag{8.9}$$

On substituting the total rate of antigen bound into Eq. (8.2c), we obtain the boundary condition for the second-order reaction at $x=0$, in dimensionless form, as

$$\frac{\partial y}{\partial z} = \mathrm{Da}(y^2 + \alpha' y). \tag{8.10}$$

The Damkohler number, Da, now is equal to $k_f^s \Gamma_0^s L c_0^{n-1}/D$, and $\alpha' = \alpha c_0^{n-1}$. Here, $n-1$ is the order of reaction.

8.2.3. Other-Order Reaction

For the present, no reaction mechanisms are proposed for the one-and-one-half-order reaction kinetics. Nevertheless, it is useful to display curves of Γ_{Ag}^s with respect to time for these reaction orders.

One-and-one-half-order reaction

For the one-and-one-half-order reaction, the rate of antigen bound specifically is given by

$$\frac{d\Gamma_{\mathrm{Ag}}^s}{dt} = k_f^s \Gamma_0^s c_s^{3/2}. \tag{8.11a}$$

On integrating Eq. (8.11a), we obtain the relative concentration of the antigen bound by specific binding:

$$\frac{\Gamma_{\mathrm{Ag}}^s}{c_0} = k_f^s \Gamma_0^s c_0^{1/2} \int_0^t y^{3/2} dt. \tag{8.11b}$$

The boundary condition in dimensionless form is

$$\frac{\partial y}{\partial z} = \mathrm{Da}(y^{3/2} + \alpha' y). \tag{8.12}$$

In this case, the Damkohler number is given by $k_f^s \Gamma_0^s L c_0^{1/2}/D$. Similar expressions can be derived for the one-half-order case by substituting 1/2 for 3/2 where the reaction order exponent is used.

The solution for the diffusion equation [Eq. (8.1a)] for the different reaction orders may be obtained by using different but appropriate boundary conditions at $x=0$, and the same initial condition at $t=0$ [Eqn. (8.2a)], and the same boundary condition at $x=\infty$. Note that since the boundary condition is nonlinear, except for the first-order reaction, and the initial condition exhibits a discontinuity, the solution to the diffusion equation is obtained by a numerical method. After obtaining the numerical solution of the diffusion equation, one can obtain the concentration of the antigen bound specifically to the antibody on the biosensor surface by numerically integrating the appropriate equations for the different reaction orders.

Different numerical techniques were considered in the solution of Eq. (8.1a). The explicit finite difference method was unsuitable due to severe restrictions placed by the stability conditions on the interval size. The Crank–Nicholson implicit finite difference method was also unsuitable since in this method very slowly decaying finite oscillations can occur in the neighborhood of discontinuities in the initial values or between the initial and boundary values. In our model, the initial condition in the neighborhood of $z = 0$ is discontinuous.

The technique found to be suitable was that in which the partial differential equation is reduced to a system of ordinary differential equations. Appropriate expressions for the different reaction orders can easily be obtained (Chen, 1994). Once the solution of the diffusion equation is obtained, the concentration of the antigen bound to the antibody due to specific binding can be obtained using the Hermite cubic quadrature. Chen (1994) used a computer subroutine called SDRIV2 for solving the initial value problem. This subroutine is particularly useful for solving a variety of initial value problems.

8.2.4. Influence of Nonspecific Binding

Forward Binding Rate Constant, k_f^s

When nonspecific binding is absent ($\alpha = 0$), an increase in the forward binding rate constant, k_f^s, leads to an increase in the specific binding of the antigen in solution to the antibody immobilized in the biosensor surface, Γ_{Ag}^s/c_0 for first-, one-and-one-half-, and second-order reactions. Since these results can be found in the literature (Sadana and Sii, 1991, 1992), they are not repeated here. As expected, an increase in the forward binding rate constant, k_f^s, leads to a decrease in the normalized concentration of the antigen near the surface, c_s/c_0, and an increase in the antigen specifically bound to the antibody on the surface, Γ_{Ag}^s/c_0.

Figures 8.6a and 8.6b show the influence of k_f^s on c_s/c_0 and on Γ_{Ag}^s/c_0 for a first-order reaction when $\alpha = 0.5$. Note that, as expected, an increase in the k_f^s value leads to a decrease in the c_s/c_0 value and an increase in the Γ_{Ag}^s/c_0 values.

When nonspecific binding is present ($\alpha > 0$), the influence of k_f^s on the binding curve for Γ_{Ag}^s/c_0 becomes complicated for one-and-one-half- and second-order reactions. Apparently, there is an optimum value of k_f^s that leads

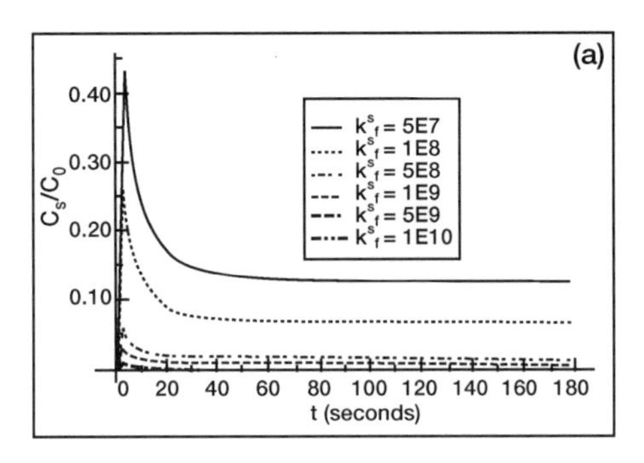

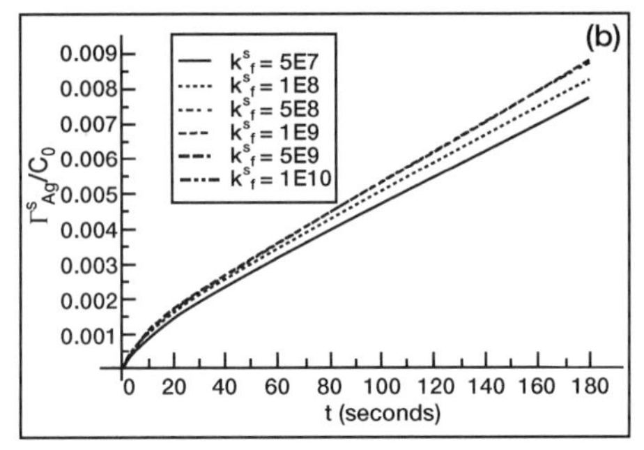

FIGURE 8.6 The influence of the forward binding rate constant, k_f^s, for a first-order reaction when $\alpha = 0.5$ on (a) the normalized antigen concentration near the fiber-optic surface, c_s/c_0, and (b) the amount of antigen specifically bound to the antibody immobilized on the biosensor surface, Γ^s_{Ag}/C_0.

to the maximum amount of antigen that can be specifically bound to an antibody immobilized on the surface. Figures 8.7a and 8.7b show that for a one-and-one-half-order reaction initially as k_f^s increases, Γ^s_{Ag}/c_0 increases. However, as k_f^s increases further, Γ^s_{Ag}/c_0 begins to decrease. Similar behavior is observed for a second-order reaction, as seen in Figs. 8.7c and 8.7d. The curves in Fig. 8.7 are for α values of 0.01 and 0.1 for both the reaction orders. Note that for the first-order reaction no such complicated behavior is observed.

Time-Dependent Forward Binding Rate Constant, k_f^s

Due to complexities and heterogeneities on the reaction surface in real-life situations, the specific binding forward rate coefficient of the antigen in solution to the antibody immobilized on the surface may exhibit a temporal

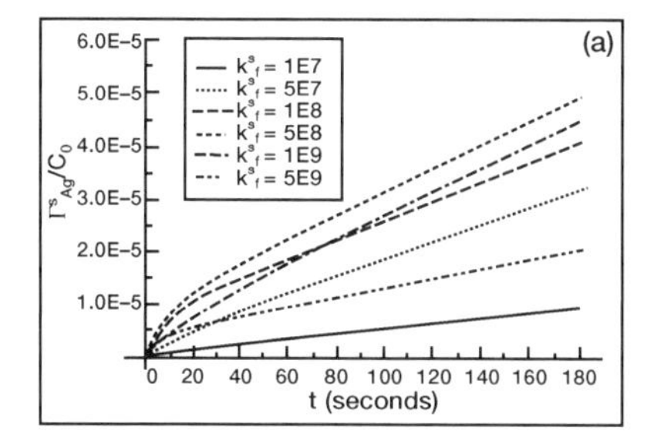

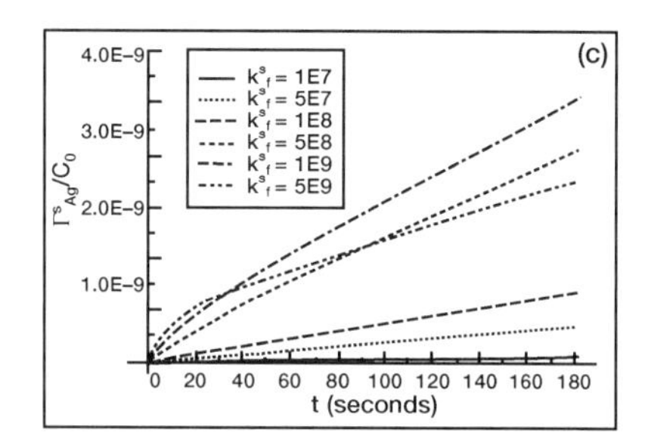

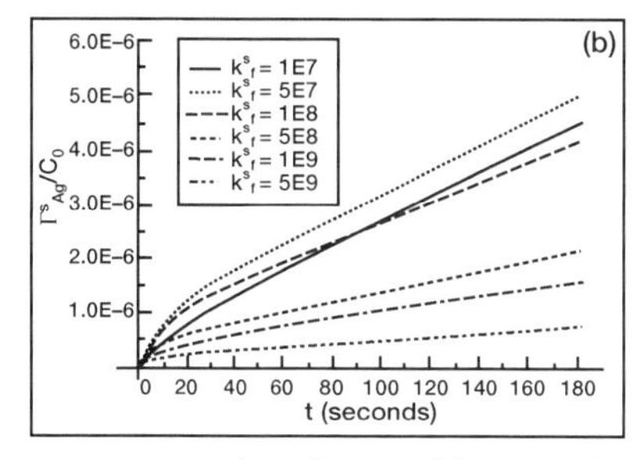

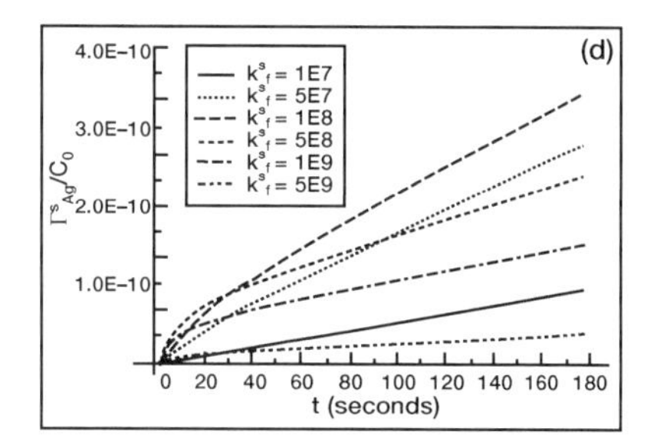

FIGURE 8.7 The influence of the forward binding rate constant, k_f^s, for different α values on the amount of antigen specifically bound to the antibody immobilized on the biosensor surface, Γ^s_{Ag}/c_0. One-and-one-half-order reaction (a) $\alpha = 0.01$; (b) $\alpha = 0.1$. Second-order reaction (c) $\alpha = 0.01$; $\alpha = 0.1$.

nature. We will examine the influence of a temporal forward specific binding rate coefficient, k_f^s, on the normalized concentration of the antigen in solution near the biosensor surface, c_s/c_0, and on the amount of antigen bound specifically to the antibody immobilized on the surface, Γ^s_{Ag}/c_0, for cases when $\alpha = 0$ and when $\alpha > 0$.

The decreasing and increasing specific binding rate coefficients are assumed to exhibit following the exponential forms (Sadana and Sii, 1991, Cuypers *et al.*, 1987):

$$k_f^s = k_{f,0}^s e^{(-\beta t)} \tag{8.13a}$$

$$k_f^s = k_{f,0}^s e^{(\beta t)}. \tag{8.13b}$$

Here, β and $k_{f,0}^s$ are constants.

Nonspecific Binding Absent

Figure 8.8 shows the influence of a decreasing forward binding rate coefficient, k_f^s, on the amount of normalized antigen concentration in solution near the biosensor surface, c_s/c_0, for first-, one-and-one-half-order, and second-order reactions when nonspecific binding is absent ($\alpha = 0$). As expected, after a brief initial period, a decrease in the k_f^s value leads to an increase in the c_s/c_0 value for a first-order reaction (Fig. 8.8a). The c_s/c_0 value is rather sensitive to the β value of the forward binding rate coefficient. For example, for a reaction time of 3 min, the c_s/c_0 value changes from about 0.25 to about 0.01 as the β value changes from 0.02 to 0 (time-invariant forward binding rate constant). Note that the changes in c_s/c_0 values are almost nonexistent for the one-and-one-half-order reaction (Fig. 8.8b) and for the second-order reaction (Fig. 8.8c). This is because very little antigen in solution is bound specifically to the antibody immobilized on the surface or nonspecifically to the biosensor surface under these conditions.

Figure 8.9 shows the influence of a decreasing k_f^s on Γ_{Ag}^s/c_0. As expected, a decrease in the binding rate coefficient leads to a decrease in the amount of the antigen bound specifically to the antibody on the surface. Note how sensitive Γ_{Ag}^s/c_0 is to the order of the reaction. The Γ_{Ag}^s/c_0 values are considerably lower (by orders of magnitude) for the one-and-one-half- and the second-order reaction when compared to the first-order reaction for the same t and β values. Note that higher β values lead to increasing tendencies toward earlier exhibition of "saturation type" behavior by the binding curves for the reaction orders analyzed. Since the curves for an increasing k_f^s exhibit trends for the c_s/c_0 and Γ_{Ag}/c_0 curves that are similar to a decreasing k_f^s, they are not presented here. (See Chen, 1994).

Nonspecific Binding Present

Figures 8.10a and 8.10b show the influence of a decreasing forward binding rate coefficient, k_f^s on the c_s/c_0 and the Γ_{Ag}^s/c_0 curves when nonspecific binding is present ($\alpha = 0.5$) for a first-order reaction. As expected, an increase in the β value leads to an increase in the c_s/c_0 value and a corresponding decrease in the Γ_{Ag}^s/c_0 values.

Figures 8.11a–8.11d show the influence of a decreasing k_f^s on the c_s/c_0 and the Γ_{Ag}^s/c_0 values when nonspecific binding is present for a one-and-one-half-order reaction. Figure 8.11a shows that for an α value of 0.01 an increase in the β value leads to an increase in the c_s/c_0 value, as expected. When $\beta = 0$, the c_s/c_0 value decreases continuously. However, for $\beta > 0$, the c_s/c_0 curve exhibits an initial decrease followed by an increase that asymptotically approaches a value of 1 for large time t. Figure 8.11b shows that an increase in

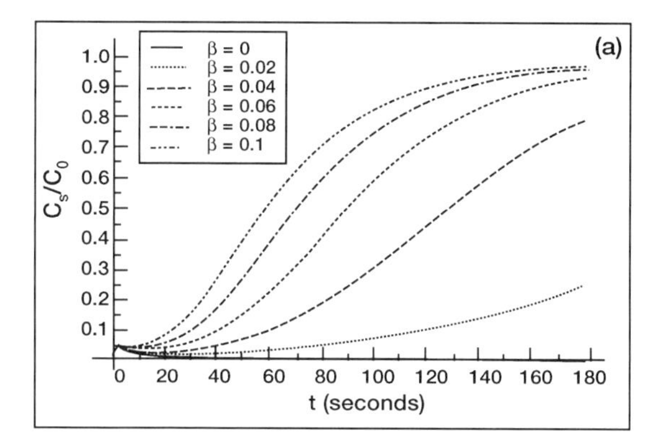

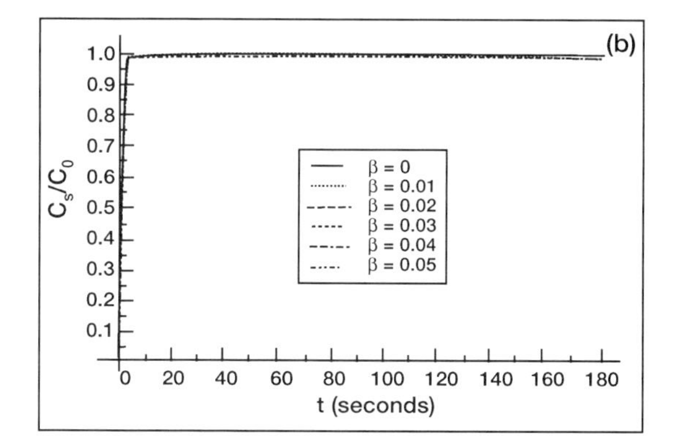

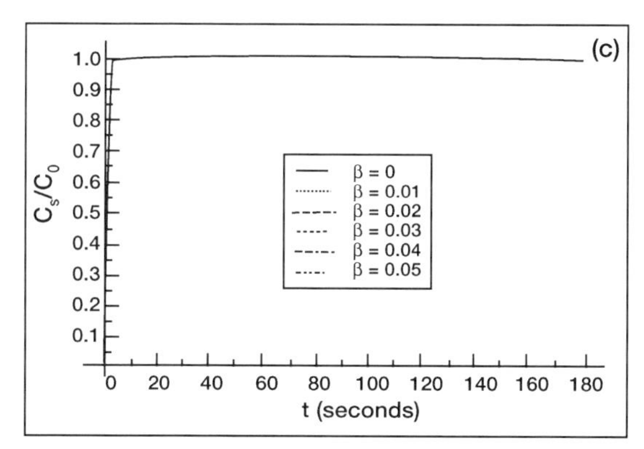

FIGURE 8.8 Influence of a decreasing forward binding rate coefficient, k_f^s, on the amount of normalized antigen concentration in solution near the surface, c_s/c_0, for different reaction orders when nonspecific binding is absent. (a) First; (b) one and one half; (c) second order.

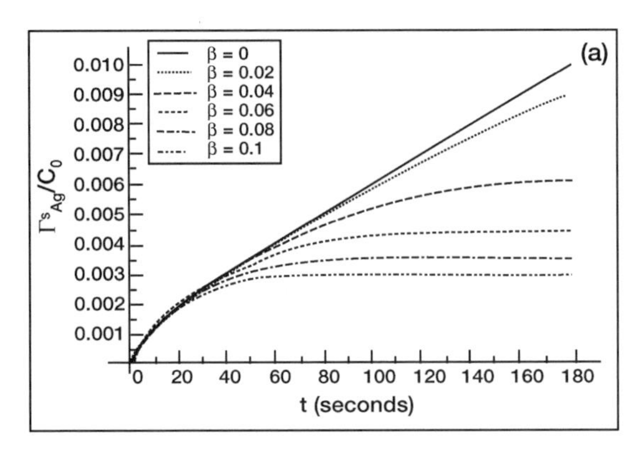

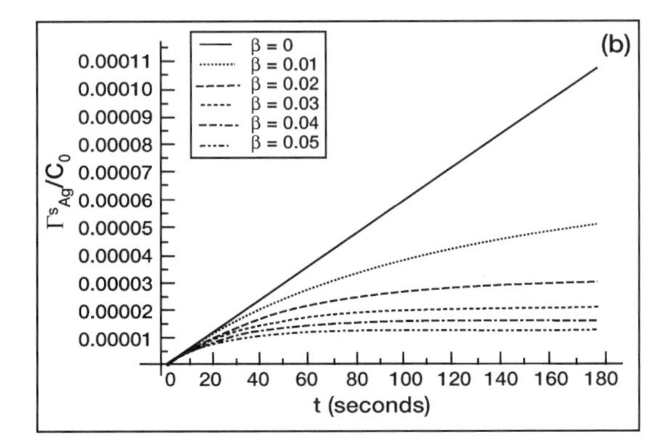

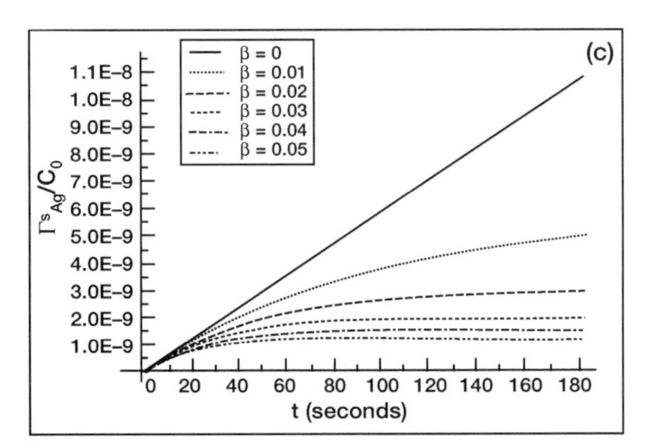

FIGURE 8.9 Influence of a decreasing forward binding rate coefficient, k_f^s, on the amount of antigen in solution specifically bound to the antibody immobilized on the biosensor surface, Γ^s_{Ag}/c_0, for different reaction orders when nonspecific binding is absent. (a) First; (b) one and one half; (c) second order.

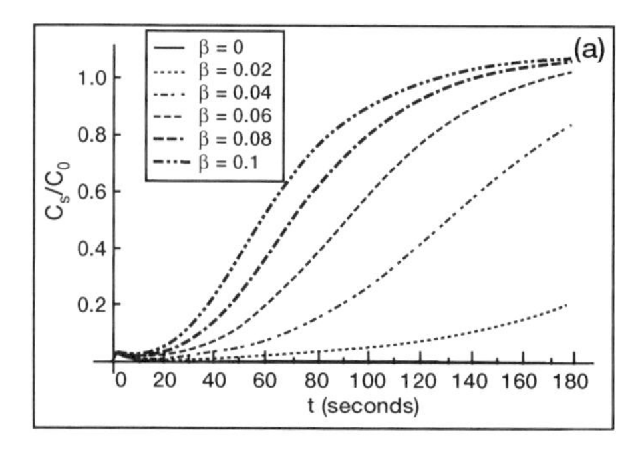

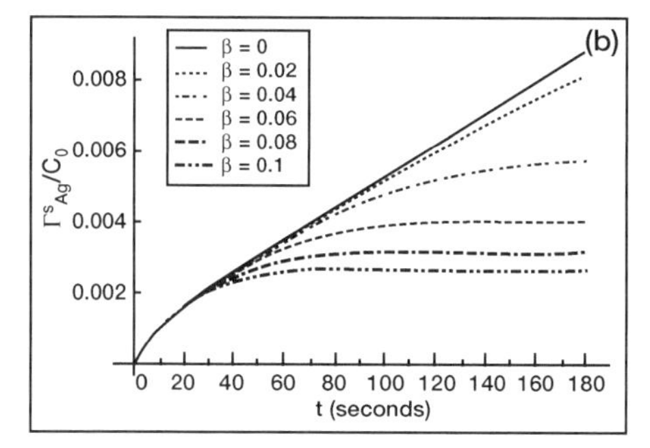

FIGURE 8.10 Influence of a decreasing forward binding rate coefficient, k_f^s, on (a) the normalized antigen concentration in solution near the surface, c_s/c_0, and (b) the amount of antigen specifically bound to the antibody immobilized on the biosensor surface, Γ^s_{Ag}/c_0, for a first-order reaction when nonspecific binding is present ($\alpha = 0.5$).

the β value also leads to a decrease in the amount of antigen in solution bound specifically to the antibody immobilized on the surface, Γ^s_{Ag}/c_0. Higher β values also lead to lower rates of specific binding as well as lower levels of "apparent saturation". Figure 8.11c shows that for an α value of 0.1 an increase in the β value once again leads to an increase in the c_s/c_0 value, as expected. Also, comparing Figs. 8.11a and 8.11c reveals that an increase in the α value means higher nonspecific binding, which leads to lower c_s/c_0 values for identical β and t values, as expected. Figure 8.11d shows the complexities involved when $\alpha = 0.1$ in the specific binding of the antigen in solution to the antibody immobilized on the biosensor surface. There is an optimum value of β that leads to the highest rate of specific binding as well as the amount of antigen bound specifically to the antibody on the surface. In the present case, for a reaction time of 3 min the highest amount of Γ^s_{Ag}/c_0 and the rate of specific binding is obtained for a β value of 0.01.

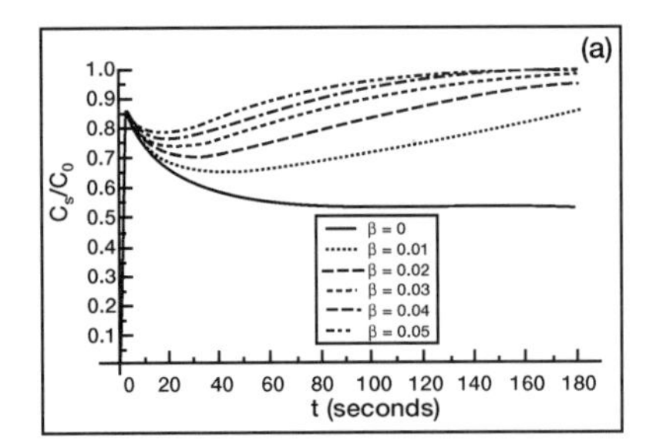

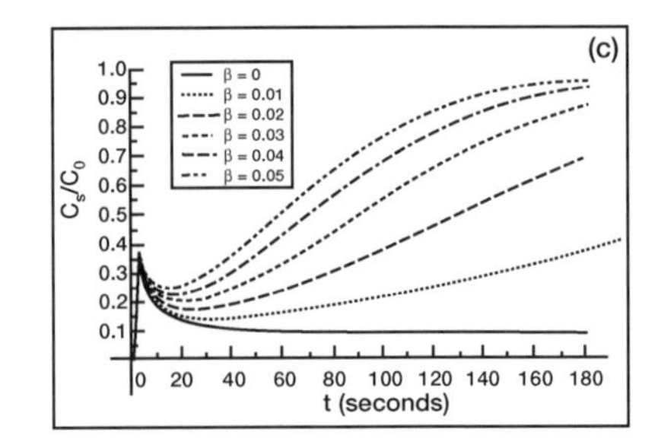

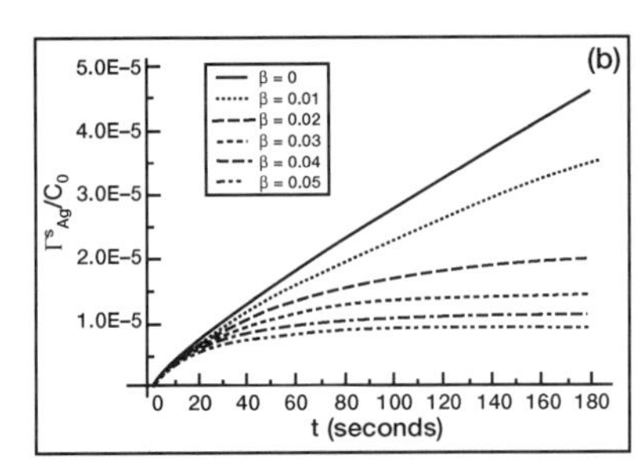

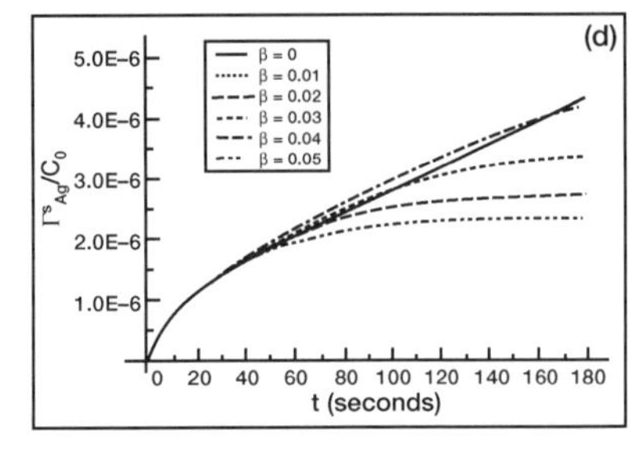

FIGURE 8.11 Influence of a decreasing forward binding rate coefficient, k_f^s, on the normalized antigen concentration in solution near the surface, c_s/c_0, when nonspecific binding is present for (a) $\alpha = 0.01$ and (c) $\alpha = 0.1$, and on the amount of antigen specifically bound to the antibody immobilized on the biosensor surface, Γ_{Ag}^s/c_0, for a one-and-one-half-order reaction for (b) $\alpha = 0.01$, and (d) $\alpha = 0.1$.

Figures 8.12a–8.12d show the influence of a decreasing k_f^s on the c_s/c_0 and the Γ_{Ag}^s/c_0 values when nonspecific binding is present for a second-order reaction. Figure 8.12a shows that for an α value of 0.01 an increase in the β value leads to an increase in the c_s/c_0 value, as expected. When $\beta = 0$, the c_s/c_0 value decreases continuously. Similar behavior was observed for the one-and-one-half-order reaction. Note that for $\beta > 0$ the c_s/c_0 curve exhibits an initial decrease followed by an increase that asymptotically approaches a value of 1 for large time t. Similar behavior was observed for the one-and-one-half-order reaction for the same α value. Similarly, Fig. 8.12b shows that an increase in the β value also leads to a decrease in the amount of antigen in solution bound specifically to the antibody immobilized on the surface. Once again (as observed for the one-and-one-half-order reaction), higher β values also lead to lower rates of specific binding as well as lower levels of apparent saturation. Figure 8.12c shows that for an α value of 0.1 an increase in the β value once again leads to an increase in the c_s/c_0 value, as expected. Figure 8.12d shows the complexities involved when $\alpha = 0.1$ in the specific binding of

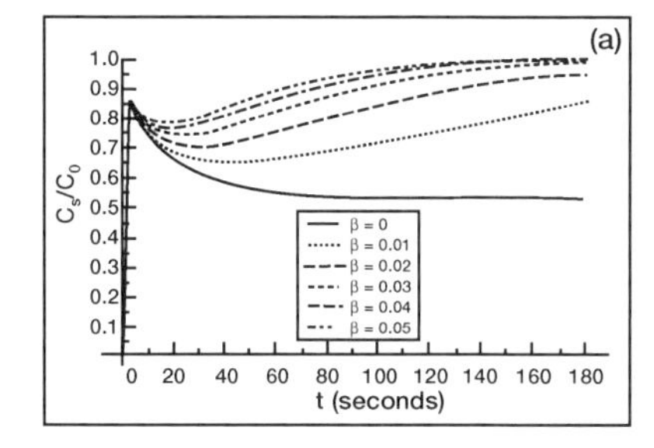

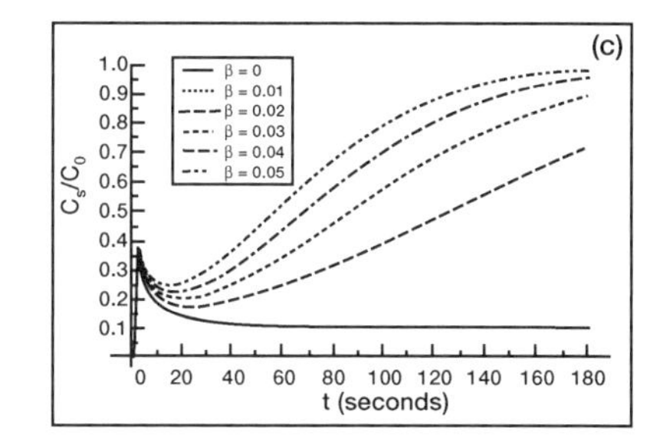

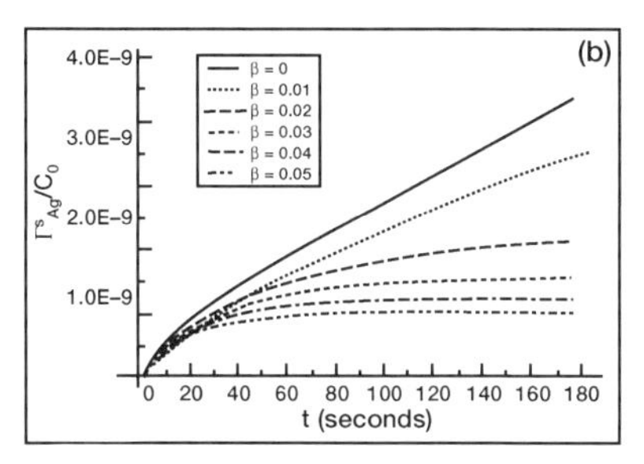

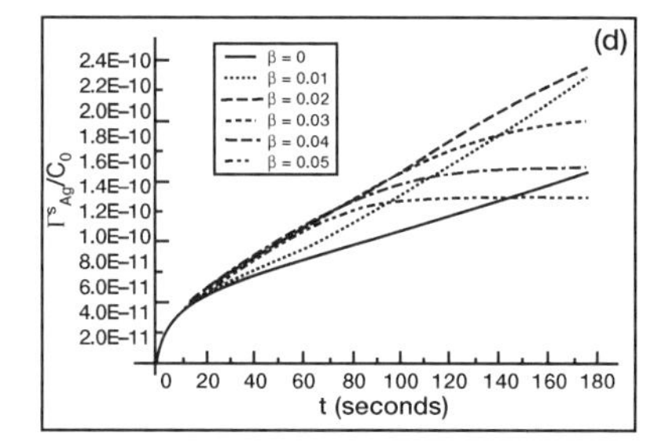

FIGURE 8.12 Influence of a decreasing forward binding rate coefficient, k_f^s on the normalized antigen concentration in solution near the surface, c_s/c_0 when nonspecific binding is present ((a) $\alpha = 0.01$) and (c) $\alpha = 0.1$), and on the amount of antigen bound specifically to the antibody immobilized on the biosensor surface, Γ_{Ag}^s/c_0 for a second-order reaction for (b) $\alpha = 0.01$, and (d) $\alpha = 0.1$.

the antigen in solution to the antibody immobilized on the surface. There is an optimum value of β that leads to the highest rate of specific binding as well as the amount of antigen bound specifically to the surface. Note that for a reaction time of 3 min the highest amount of Γ_{Ag}^s/c_0 occurs at a β value of 0.02, and the highest rate of specific binding occurs at a β value of 0.01. The β values are identical for the one-and-one-half- and second-order reactions as far as obtaining the highest Γ_{Ag}^s/c_0 for this time interval.

Figures 8.13a and 8.13b show the influence of an increasing k_f^s on the c_s/c_0 and the Γ_{Ag}^s/c_0 values when nonspecific binding is present ($\alpha = 0.5$) for a first-order reaction. As expected, an increase in the β value leads to a decrease in the c_s/c_0 value and a very slight (almost imperceptible) change in the Γ_{Ag}^s/c_0 value.

Figures 8.14a–8.14d show the influence of an increasing k_f^s on the c_s/c_0 and the Γ_{Ag}^s/c_0 values when nonspecific binding is present for a one-and-one-half-order reaction. Figure 8.14a shows that for an α value of 0.01 an increase in

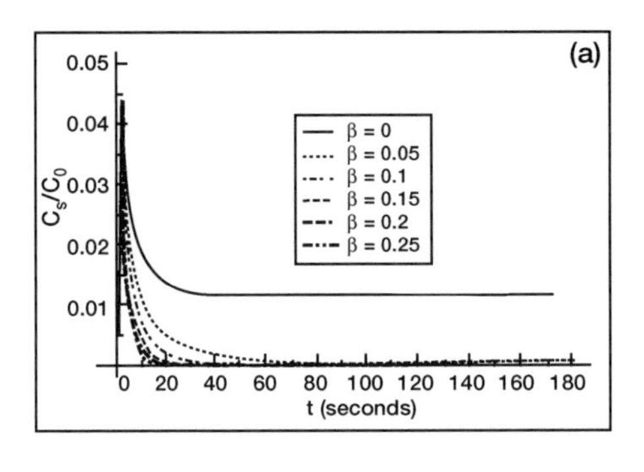

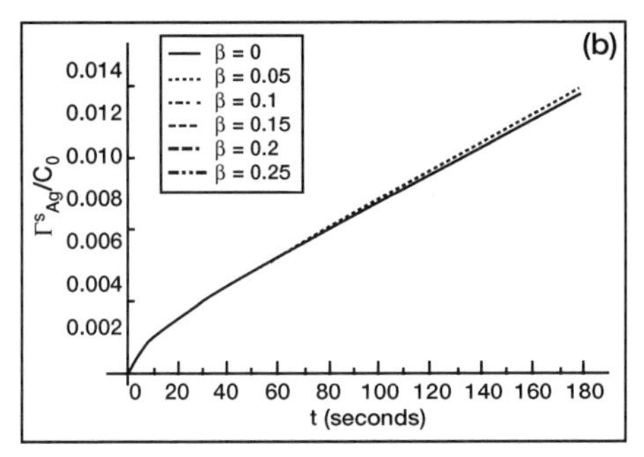

FIGURE 8.13 Influence of an increasing forward binding rate coefficient, k_f^s, on (a) the normalized antigen concentration in solution near the surface, c_s/c_0, and on (b) the amount of antigen specifically bound to the antibody immobilized on the biosensor surface, Γ^s_{Ag}/c_0, for a first-order reaction when nonspecific binding is present ($\alpha = 0.5$).

the β value leads to a decrease in the c_s/c_0 value. The c_s/c_0 curve decreases continuously for β values ranging from 0 to 0.025. Figure 8.14b shows the complexities involved when $\alpha = 0.01$ in the specific binding of the antigen in solution to the antibody immobilized on the biosensor surface. Up to about 80 sec, the highest β ($= 0.025$) specific binding curve exhibits the highest rate of binding and the highest amount of antigen in solution specifically bound to the antibody immobilized on the surface. After about 140 sec, the optimum value of β is 0.01. This β value leads to the highest rate of binding and the maximum amount of antigen specifically bound to the antibody on the biosensor surface. It seems that the optimum value of β depends on the time interval of measurement. Interestingly, a similar maximum was obtained for a decreasing value of k_f^s. Figure 8.14c shows that for an α value of 0.1 an increase in the β value leads to a continuous decrease in the c_s/c_0 value, as expected. Figure 8.14d indicates that there is a decrease in the amount of

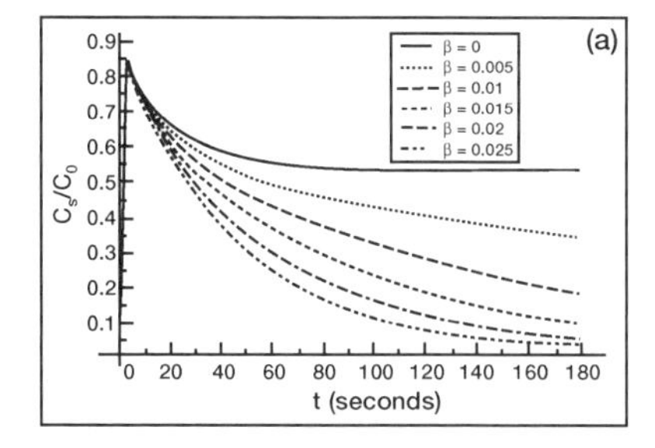

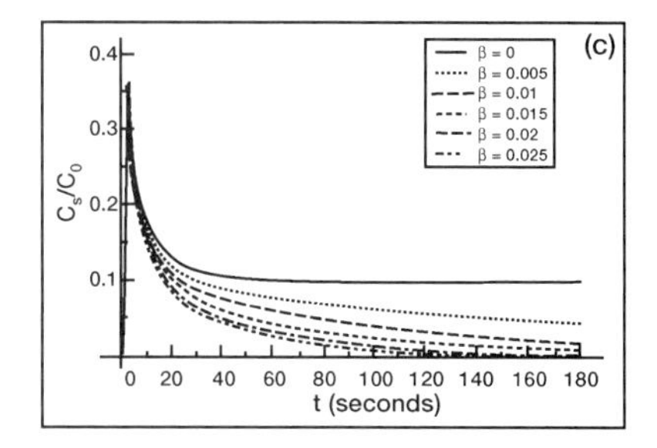

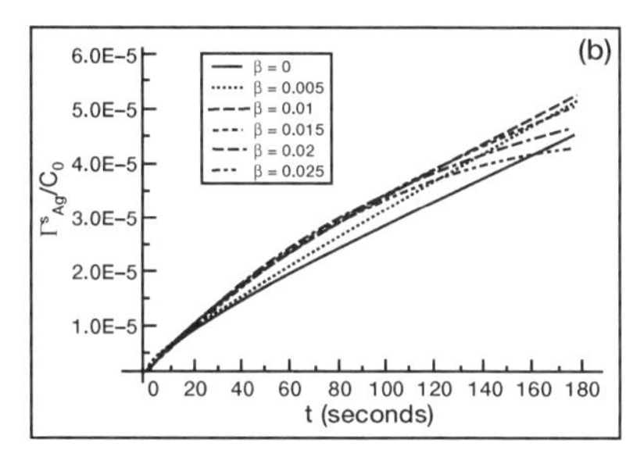

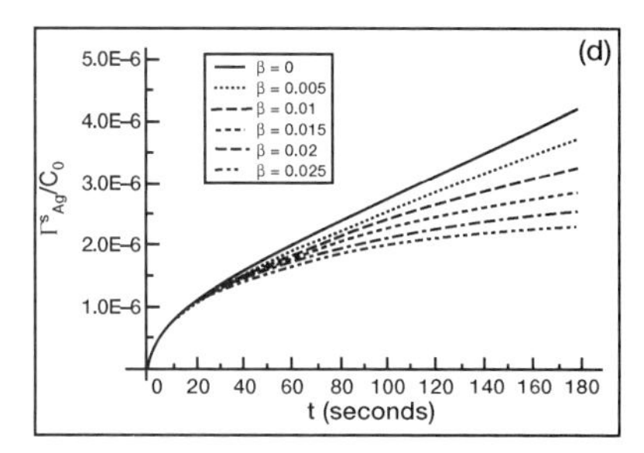

FIGURE 8.14 Influence of an increasing forward binding rate coefficient, k_f^s, on the normalized antigen concentration in solution near the surface, c_s/c_0, when nonspecific binding is present for (a) $\alpha = 0.01$ and (c) $\alpha = 0.1$, and on the amount of antigen specifically bound to the antibody immobilized on the biosensor surface, Γ_{Ag}^s/c_0, for a one-and-one-half-order reaction for (b) $\alpha = 0.01$, and (d) $\alpha = 0.1$.

antigen specifically bound to the surface as β increases. In this case, an increase in β leads to an increase in the Damkohler number, which leads to a decrease in the Γ_{Ag}^s/c_0 value. It would appear that these conditions would favor nonspecific binding rather than the required specific binding. Note that a decrease in the amount of antigen in solution specifically bound to the antibody immobilized on the surface was also observed for an increasing forward binding rate coefficient when nonspecific binding was absent (Sadana and Sii, 1991).

Figures 8.15a–8.15d show the influence of an increasing k_f^s on the c_s/c_0 and the Γ_{Ag}^s/c_0 values when nonspecific binding is present for a second-order reaction. Figure 8.15a shows that for an α value of 0.01 an increase in the β value leads to a decrease in the c_s/c_0 value. The c_s/c_0 decreases continuously for β values ranging from 0 to 0.025. Figure 8.15b shows the complexities involved when $\alpha = 0.01$ in the specific binding of the antigen in solution to the antibody immobilized on the biosensor surface. Up to about 50 sec, the

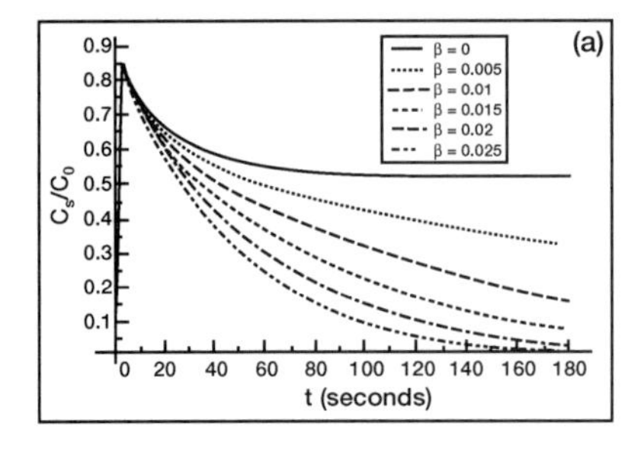

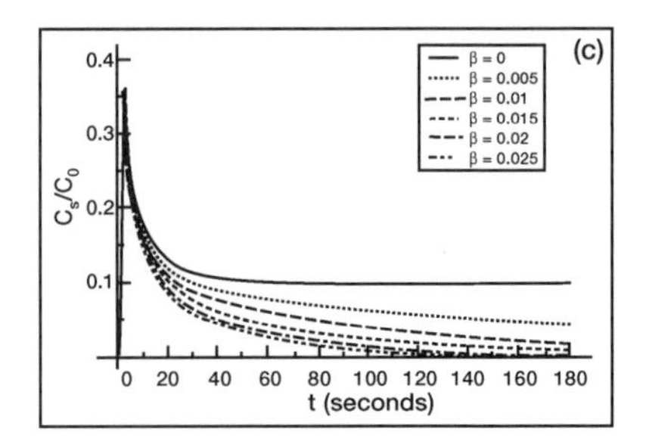

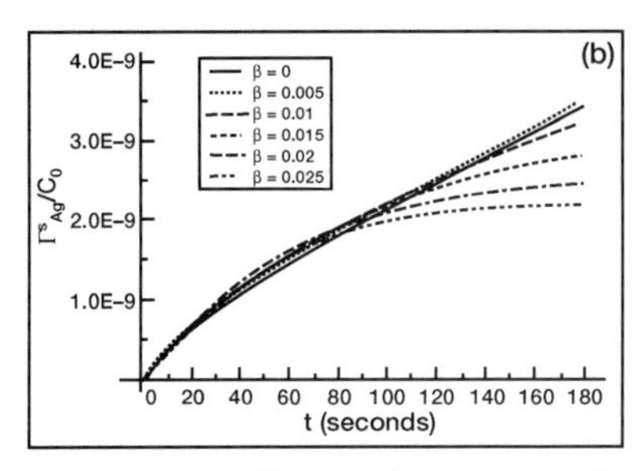

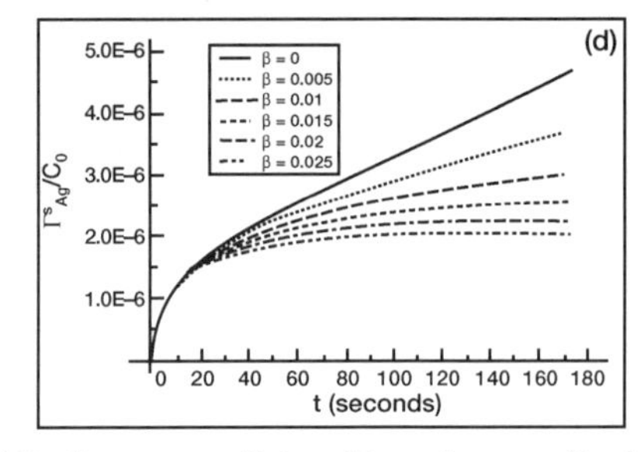

FIGURE 8.15 Influence of an increasing forward binding rate coefficient, k_f^s, on the normalized antigen concentration in solution near the surface, c_s/c_0, when nonspecific binding is present for (a) $\alpha = 0.01$ and (c) $\alpha = 0.1$, and on the amount of antigen specifically bound to the antibody on the biosensor surface, Γ_{Ag}^s/c_0, for a second-order reaction for (b) $\alpha = 0.01$, and (d) $\alpha = 0.1$.

highest β ($= 0.025$) specific binding curve exhibits the highest rate of binding and the highest amount of antigen in solution specifically bound to the antibody immobilized on the surface. After about 120 sec, the optimum value of β is 0.005. This β value leads to the highest rate of specific binding and the maximum amount of antigen in solution specifically bound to the antibody immobilized on the surface. It seems, once again, that the optimum value of β depends on the time interval of measurement. Figure 8.15c shows that for an α value of 0.1 an increase in k_f^s leads to a decrease in the c_s/c_0 value. Figure 8.15d exhibits interesting behavior in that an increase in the binding rate coefficient leads to a decrease in the Γ_{Ag}^s/c_0 value. Apparently, in this case an increase in the β value in the forward binding rate coefficient leads to an increase in the Damkohler number (increase in the external mass transfer limitations), which leads to a decrease in the Γ_{Ag}^s/c_0 values. Similar behavior was reported by Sadana and Sii (1991), as previously indicated for an increasing forward binding rate coefficient when nonspecific binding was absent ($\alpha = 0$).

8.3. CONCLUSIONS

The presence of nonspecific binding does affect the external diffusion-limited specific binding of antigen in solution to antibody covalently or noncovalently immobilized on a surface for first-, one-and-one-half-, and for second-order reactions. An increase in α (the ratio of nonspecific binding to specific binding) and an increase in the forward binding rate coefficient leads to a decrease in the normalized antigen concentration near the surface, c_s/c_0, and the amount of antigen bound specifically to the antibody immobilized on the surface, Γ^s_{Ag}/c_0, for a first-order reaction. The presence of nonspecific binding complicates the influence of the forward binding rate constant, k^s_f, on the Γ^s_{Ag}/c_0 amount for the one-and-one-half- and second-order reactions. Apparently, there is an optimum value of k^s_f that leads to the maximum amount of antigen that can be specifically bound to an antibody immobilized on the surface. Also, the Γ^s_{Ag}/c_0 value is rather sensitive to the order of reaction and decreases considerably (by orders of magnitude) as one goes from first-, to one-and-one-half-, to second-order reactions.

Temporal forward binding rate coefficients, k^s_f, more correctly represent the complexities and heterogeneities involved during the binding of the antigen in solution to the antibody immobilized on the biosensor surface. When nonspecific binding is absent ($\alpha = 0$), a decrease in the binding rate coefficient leads to a decrease in the amount of antigen bound specifically to the antibody on the surface, as expected. Note, once again, how sensitive the Γ^s_{Ag}/c_0 value is to the order of reaction. The Γ^s_{Ag}/c_0 values are considerably lower (by orders of magnitude) for the one-and-one-half- and for the second-order reaction when compared to the first-order reaction for the same t and β values.

When nonspecific binding is present ($\alpha = 0.01$) an increase in the β value for the decreasing forward binding rate coefficient leads to an increase in the c_s/c_0 value and a decrease in the Γ^s_{Ag}/c_0 values, as expected, for the one-and-one-half- and second-order reactions. Once again, note how sensitive the amount of antigen in solution specifically bound to the antibody on the surface is to the order of reaction. The amount specifically bound decreases by orders of magnitude as one goes from the one-and-one-half- to the second-order reaction. Of course, everything else being equal, an increase in the α value leads to decreasing c_s/c_0 and Γ^s_{Ag}/c_0 values, as expected. When $\alpha = 0.1$, the one-and-one-half- and second-order reactions exhibit complexities when a decreasing (temporal) forward binding rate coefficient is involved. Apparently, there exists an optimum value of β that leads to the highest rate of specific binding as well as the amount of antigen in solution specifically bound to the antibody immobilized on the biosensor surface. For a reaction time of 3 min and for the case analyzed, the optimum value of β was 0.01 for the one-and-one-half-order reaction. Interestingly, for a second-order reaction with this same reaction time and the same α value (0.1), once again the optimum value of β is 0.01 for the maximum amount of antigen specifically bound to the antibody on the surface. This optimum value of β changes when α is changed. Note that no such complexities are exhibited for the first-order reaction.

For $\alpha = 0.5$, an increase in the forward binding rate coefficient, k_f^s, leads to a decrease in the c_s/c_0 value and an insignificant increase in the Γ_{Ag}^s/c_0 values for a first-order reaction. Apparently, under these conditions most, if not all, of the antigen in solution is nonspecifically bound to the biosensor surface at this high value of α. Note that when nonspecific binding is present an increase in the forward binding rate coefficient leads to a decrease in the c_s/c_0 and Γ_{Ag}^s/c_0 values. Apparently, in this case, the increasing forward binding rate coefficients lead to increasing Damkohler numbers, which in turn lead to lower amounts of antigen in solution specifically bound to the antibody immobilized on the surface.

The inclusion of nonspecific binding in the analysis of specific binding of the antigen in solution to the antibody immobilized on the surface provides a more realistic picture of what is happening on the surface, besides leading to temporal binding rate coefficients. Furthermore, quantitative estimates of the amount of antigen specifically bound are available for different degrees of specific binding (α). In many cases, this higher α value considerably decreases the rate and amount of specifically bound antigen especially for the higher reaction orders. The estimates and trends provided by the modeling should help in controlling these reactions to advantage, besides helping to improve the stability, sensitivity, reproducibility, and reaction time of biosensors. Nonspecific binding should, in the ideal case, be minimized. Nevertheless, it will be present to some degree in most (if not all) cases, and the modeling presented helps estimate the derogatory influence it has on biosensor performance. Having suitable experimental data available would considerably assist in developing better models and more realistic estimates of the influence of nonspecific binding on biosensor performance. Such studies, which should include both experimental and theoretical approaches, are strongly recommended.

REFERENCES

Ahluwalia, A., De Rossi, D., Ristori, C., Schirone, A., and Serra, G., *Biosen. & Bioelectr.* 7, 207–214 (1991).

Angel, S. M., *Spectroscopy* 2(4), 38–46 (1987).

Arwin, H. and Lundstrom, I., *Anal. Biochem.* 145, 106–112 (1985).

Betts, T. A., Catena, G. C., Huang, J., Litwiler, K. S., Zhang, J., Zagrobelny, J., and Bright, F. V., *Anal. Chim. Acta* **246**, 55–63 (1991).
Blair, T. L., Cynowski, T., and Bachas, L. G., *Anal. Chem.* **65**, 945–947 (1993).
Byfield, M. P. and Abuknesha, R. A., *Biosen. & Bioelectr.* **9**, 373–400 (1994).
Chen, Z., "Influence of Nonspecific Binding on Antigen–Antibody Binding Kinetics: Biosensor Applications," Master's thesis, University of Mississippi, MS, 1994.
Collison, M. E., Tarlov, M. J., and Plant, A. L., "Immobilization of Enzymes at Biosensor Surfaces Using Spontaneous Self-Assembly Phenomena," oral presentation, Biosensors 94, New Orleans, LA, June 1–3, 1994.
Connolly, P., *TIBTECH* **12**(4), 123–127 (1994).
Cuypers, P. A., Willems, G. M., Henker, H. C., and Hermens, W. T., in *Blood in Contact with Natural and Artificial Surfaces* (E. F. Leonard, V. T. Turitto, and L. Vroman, eds.), *Annals NY Acad. Sci.* **516**, pp. 244–252, 1987.
Dambrot, S. M., *Bio/Technology* **10**, 120–120 (1993).
Disley, D. M., Cullen, D. C., and Lowe, C. R., "Interfacial Protein Studies at Silver and Chemically Modified Silver Surfaces for Immunoassay Applications," oral presentation, Biosensors 94, New Orleans, LA, June 1–3, 1994.
Ekins, R. P. and Chu, F., *TIBTECH* **12**(3), 89–94 (1994).
Giaver, I., *J. Immunol.* **116**, 766–771 (1976).
Kopelman, R., *Science* **241**, 1620–1626 (1988).
Kortt, A. A., Oddie, G. W., Iliades, P., Gruen, L. C., and Hudson, P.J., *Anal. Biochem.* **253**, 103–111 (1997).
Kubitschko, S., Spinke, J., Bruckner, T., Pohl, S., and Oranth, N., *Anal. Biochem.* **253**, 112–122 (1997).
Lowe, C., *Biosensors* **1**, 3–16 (1985).
Marose, S., Lindermann, C., Ulber, R., and Scheper, T., *TIBTECH* **17**, 30–34 (1999).
Mathewson, P. R. and Finley, J. W., eds., *Biosensor Design and Application* ACS Symposium Series 511, American Chemical Society, Washington, DC, pp. ix–xi, 1992.
Nygren, H., *J. Immunol. Meth.* **114**, 107–114 (1988).
Nygren, H. and Stenberg, M., *J. Colloid & Interface. Sci.* **107**, 560–566 (1985).
Ohlson, S., Jungar, C., Strandh, M., and Mandenius, C. F., *TIBTECH* **18**, 49–52 (2000).
Okano, K., Takahashi, S., Yasuda, K., Tokinaga, D., Imai, K., and Koga, M., *Anal. Biochem.* **202**, 120–125 (1992).
Owaku, K., Goto, M., Ikariyama, Y., and Aizawa, M., *Sensors & Actuators* **13–14**, 723–724 (1993).
Place, J. F., Sutherland, R. M., and Dahne, C., *Biosensors* **1**, 321–353 (1985).
Pritchard, D. J., Morgan, H., and Cooper, J. M., "Multianalyte Biosensors Employing a Photoactive Ligand for Immobilization," poster presentation, Biosensors 94, New Orleans, LA, June 1–3, 1994.
Sadana, A. and Beelaram, A., *Biotech. Progr.* **10**, 291–298 (1994).
Sadana, A. and Chen, Z., *Biosen. & Bioelectron.* **11**(8), 769–782 (1996).
Sadana, A. and Madagula, A., *Biotech. Progr.* **9**, 259–266 (1993).
Sadana, A. and Madagula, A., *Biosen. & Bioelectr.* **9**, 45–55 (1994).
Sadana, A. and Sii, D., *J. Colloid & Interface Sci.* **151**(1), 166–177 (1991).
Sadana, A. and Sii, D., *Biosen. & Bioelectr.* **7**, 559–568 (1992).
Sadik, O. A. and Wallace, G. G., "Electrodynamic Surfaces: Key to a New Biosensing Technology," oral presentation, Biosensors 94, New Orleans, LA, June 1–3, 1994.
Scheller, F. W., Hintsche, R., Pfeiffer, D., Schubert, F., Reidel, K., and Kindervater, R., *Sensors & Actuators B* **4**, 197–206 (1991).
Stenberg, M., Elwing, H, and Nygren, M., *J. Theor. Biol.* **98**, 307–315 (1982).
Stenberg, M. and Nygren, H., *Anal. Biochem.* **127**, 183–192 (1982).

Stenberg, M., Stiblert, L., and Nygren, H., *J. Theor. Biol.* **120**, 129–142 (1986).
Walczak, I. M., Love, W. F., Cook, T. A., and Slovacek, R. E., *Biosen. & Bioelectr.* 7, 39–48 (1992).
Wise, D. L. and Wingard Jr., L. B., eds., *Biosensors with Fiberoptics* pp. vii–viii, Humana Press, New York, NY, 1991.
Williams, C. and Addona, T. A., *TIBTECH*, **18**, 45–48 (2000).
Wyatt, W. A., Bright, F. V., and Hieftje, G. M., *Anal. Chem.* **59**, 2272–2276 (1987).

CHAPTER 9

Influence of Nonspecific Binding on the Rate and Amount of Specific Binding: A Fractal Analysis

9.1. INTRODUCTION

The preceding chapter analyzed the influence of nonspecific binding (NSB) on the rate and amount of specific binding using a classical analysis. Since very little is known about nonspecific binding and how it influences specific binding, it is worthwhile to also use a fractal analysis to analyze this influence. The binding constants for antigen–antibody reactions at interfaces and for protein adsorption systems (which exhibit behavior similar to antibody–antigen systems) are often of a temporal nature. This temporal nature of the binding constants and the inherent heterogeneities present in these systems may be described using fractals (Sadana and Madagula, 1994; Sadana and Beelaram, 1994, 1995). As indicated in earlier chapters, Kopelman (1988) states that surface diffusion-controlled reactions that occur on clusters (or islands) are expected to exhibit anomalous or fractal-like kinetics. These fractal kinetics exhibit anomalous reaction orders and time-dependent rate (e.g., binding) coefficients. The analysis of time-dependent NSB should assist in providing a more realistic approach to the reaction (including specific binding) occurring on the reaction or biosensor surface. The fractal analysis approach helps describe (and is but one possible way) the heterogeneities that exist on the biosensor surface. NSB would lead to increasing heterogeneites on the biosensor surface. Furthermore, in multicomponent environments the nonspecific binding of other molecules plays a significant role that leads to further heterogeneities on the surface.

Havlin (1989) indicates that from an experimental point of view, diffusion toward fractal surfaces has been studied more extensively than diffusion on fractal surfaces, due to the number of applications (such as catalytic reactions). However, some studies on diffusion toward fractal surfaces are available. For example, Giona (1992) analyzed first-order reaction-diffusion kinetics in complex media. This author emphasizes that the analysis of the temporal nature of the diffusion-limited reaction on the surface could play an important role in understanding both the reaction kinetics and the reactions at the interface/surface. Fractal kinetics also have been reported in other biochemical reactions. Li *et al.* (1990) emphasize that the nonintegral dimensions of the Hill coefficient used to describe the allosteric effects of proteins is a direct consequence of the fractal property of proteins. The protein is a biological macromolecule composed of amino acid residues whose branches are fractals. The substrate molecules randomly walk on the enzyme surface until they hit or react on an active site. Thus, for a better understanding of reactions at interfaces, a fractal analysis can be used to examine, for example, the influence of nonspecific binding on the diffusion-limited antigen–antibody binding on biosensor surfaces. Such an analysis should assist considerably in providing novel physical insights into the reactions occurring at the biosensor surface, besides helping to improve the sensitivity, stability, and speed of response of biosensors.

In this chapter, as in the previous chapter, we will analyze theoretically and in some detail the influence of nonspecific binding and temporal model parameters on the specific binding of the antigen in solution to the antibody immobilized on the surface. However, in this chapter we will use a fractal analysis rather than the classical analysis used in the previous chapter. The temporal nature of the specific binding parameters of the antigen in solution to the antibody immobilized on the surface, along with the inclusion of the nonspecific binding of the antigen in solution directly to the fiber-optic surface, is a more realistic approach to the actual situation. Thus our theoretical model will include both specific and nonspecific binding. Note that a part of nonspecific binding that arises due to both the nonspecific binding of the antibody itself and the nonspecific binding of the other molecules to the sensing surface in multicomponent environments is included in the temporal (specific) binding rate coefficient. Both factors lead to heterogeneities on the surface. The fractal analysis should provide insights into first-, one-and-one-half-, second-, and other-order antigen–antibody (and, in general, analyte-receptor) reactions occurring on the fiber-optic biosensor surface under external diffusion-limited conditions.

9.2. THEORY

The theory is almost the same as in Chapter 8. Equations (8.1)–(8.12) apply here, as do Figs. 8.4–8.6. These are therefore not repeated here. Since we will be examining a fractal analysis, we will now provide the additional fractal material required to provide the basis for the analysis. Kopelman (1988) indicated that classical reaction kinetics is sometimes unsatisfactory when the reactants are spatially constrained on the microscopic level by either walls, phase boundaries, or force fields. Such heterogeneous reactions (e.g., bioenzymatic reactions) that occur at interfaces of different phases exhibit fractal orders for elementary reactions and rate coefficients with temporal memories. In such reactions, the rate coefficient exhibits the form

$$k_1 = k^{'}t^{-b} \quad 0 \leq b \leq 1 \quad (t \geq 1). \tag{9.1}$$

In general, k_1 depends on time, whereas $k^{'} = k_1(t = 1)$ does not. Kopelman indicates that in three dimensions (homogeneous space) $b = 0$. This is in agreement with the results in classical kinetics. Also, with vigorous stirring, the system is made homogeneous and, again, $b = 0$. However, for diffusion-limited reactions occurring in fractal spaces, $b > 0$; this yields a time-dependent rate coefficient.

Di Cera (1991) has analyzed the random fluctuations in a two-state process in ligand-binding kinetics. The stochastic approach can be used as a means to explain the variable adsorption rate coefficient. The simplest way to model these fluctuations is to assume that the adsorption rate coefficient, $k_1(t)$, is the sum of its deterministic value (invariant) and the fluctuation, $z(t)$. This $z(t)$ is a random function with a zero mean. The decreasing and increasing adsorption rate coefficients can be assumed to exhibit the following exponential forms (Cuypers *et al.*, 1987):

$$\begin{aligned} k_f^s &= k_{f,0}^s \exp(-\beta t) \\ k_f^s &= k_{f,0}^s \exp(\beta t). \end{aligned} \tag{9.2}$$

Furthermore, experimental data presented by Anderson (1993) for the binding of HIV virus (antigen) to anti-HIV (antibody) immobilized on a surface displays a characteristic ordered disorder. This indicates the possibility of a fractal-like surface. It is obvious that the biosensor system (wherein either the antigen or the antibody is attached to the surface) along with its different complexities—including heterogeneities on the surface and in solution, diffusion-coupled reactions, time-varying adsorption rate coefficients, etc.—can be characterized by a fractal system.

Sadana and Madagula (1994) performed a theoretical analysis using fractals for the time-dependent binding of antigen in solution to antibody immobilized on a fiber-optic biosensor surface. The authors noted that an increase in the fractal parameter utilized in their studies decreased both the rate of antigen and the amount of antigen bound. They recommended obtaining or estimating a fractal parameter (or perhaps a range of such) to help characterize the antibody–antigen interactions for fiber-optic biosensor systems. Note that in this case the diffusion is in the Euclidean space surrounding the fractal surface (Giona, 1992). Sadana and Madagula did not consider the effect of nonspecific binding on the specific binding of the antigen in solution to the antibody immobilized on the fiber-optic surface. However, it is beneficial to analyze the influence of nonspecific binding on both the rate of specific binding and the amount of antigen bound to the antibody immobilized on the biosensor surface.

Equation 9.1 is associated with the short-term diffusional properties of a random walk on a fractal surface. Also, in a perfectly stirred kinetics on a regular (nonfractal) structure (or surface), k_1 is a constant; that is, it is independent of time. In other words, and as indicated in earlier chapters, the limit of regular structures (or surfaces) and the absence of diffusion-limited kinetics leads to k_1 being independent of time. In all other situations, one would expect a scaling behavior given by $k_1 \sim k' t^{-b}$ with $-b < 0$. Also, the appearance of the coefficient $-b$ different from $b = 0$ is the consequence of two different phenomena—the heterogeneity (fractality) of the surface and the imperfect mixing (diffusion-limited condition).

It is of practical interest to evaluate or estimate the parameter b in Eq. (9.1) for a real surface. In a brief analysis of diffusion or reactants toward fractal surfaces, Havlin (1989) indicates that the diffusion of a particle (antibody or antigen, as the case may be) from a homogeneous solution to a solid surface (the biosensor surface) where it reacts to form a product (antibody–antigen complex) is given by (De Gennes, 1982; Pfeifer *et al.*, 1984; Nyikos and Pajkossy, 1986)

$$(\mathrm{Ab \cdot Ag}) \sim \begin{cases} t^{(3-D_f)/2} = t^p & t < t_c \\ t^{1/2} & t > t_c. \end{cases} \tag{9.3}$$

Here, $p = -b$ and D_f is the fractal dimension of the surface. Havlin (1989) states that the crossover value may be determined by $r_c \sim t_c^2$. Above the characteristic length, r_c, the self-similarity of the surface is lost and the surface may be considered homogeneous. Note that the product (Ab · Ag) in a reaction (Ab + Ag→Ab · Ag) on a solid fractal surface corresponds to $p = (3 - D_f)/2$ at short time frames and $p = \frac{1}{2}$ at intermediate time frames. The values of the parameters k, p, and D_f may be obtained for antigen–

antibody kinetics data where nonspecific binding is either present or absent. This may be done by a regression analysis using, for example, Sigmaplot (1993), along with Eq. (9.3), where $(\mathrm{Ab} \cdot \mathrm{Ag}) = kt^p$ (Sadana and Beelaram, 1994, 1995). The fractal dimension may be obtained from the parameter p. Higher values of the fractal dimension would indicate higher degrees of disorder or heterogeneity on the surface. It is reasonable to anticipate that increasing levels of nonspecific binding would lead to higher levels of heterogeneity on the sensing surface.

Note that antigen–antibody binding is unlike reactions in which the reactant reacts with the active site on the surface and the product is released. In this sense, the catalytic surface exhibits an unchanging fractal surface to the reactant in the absence of fouling and other complications. In the case of antigen–antibody binding, the biosensor surface exhibits a changing fractal surface to the antigen or antibody (analyte) in solution. This occurs because as each binding reaction takes place, fewer sites are available on the biosensor surface to which the analyte may bind. This is in accord with Le Brecque's (1992) comment that the active sites on a surface may themselves form a fractal surface. Furthermore, the inclusion of nonspecific binding sites on the surface would increase the fractal dimension of the surface.

In general, log–log plots of the distribution of molecules, $M(r)$, as a function of the radial distance, (r), from a given molecule are required to demonstrate fractal-like behavior (Nygren, 1993). This plot should be close to a straight line. The slope of the log $M(r)$ versus $\log(r)$ plot determines the fractal dimension.

9.3. RESULTS

Figures 9.1a and 9.1b show the influence of nonspecific binding (NSB) on the normalized antigen concentration near the fiber-optic surface, c_s/c_0, for the first-order reaction when NSB is absent and when it is present. The c_s/c_0 term represents the antigen depletion in the zone close to the surface, which due to external diffusion limitations. Figure 9.1a indicates that when NSB is absent, an increase in the fractal parameter b leads to an increase in the c_s/c_0 value. In other words, an increase in the heterogeneity of the reaction surface on the fiber-optic biosensor leads to decreasing amounts of antigen bound to the antibody immobilized on the surface. As expected, the presence of nonspecific binding ($\alpha = 0.5$) (Fig. 9.1b) leads to a decrease in the c_s/c_0 value compared to when this nonspecific binding is absent. For example, a 3 min-reaction time and a fractal parameter b value of 0.8 decreases the c_s/c_0 by about 22.5%—from 0.4 to 0.31—when nonspecific binding is absent and present, respectively. This difference represents the amount of antigen in solution

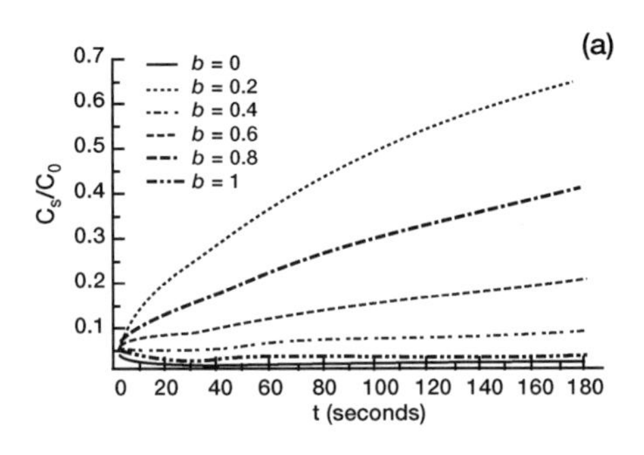

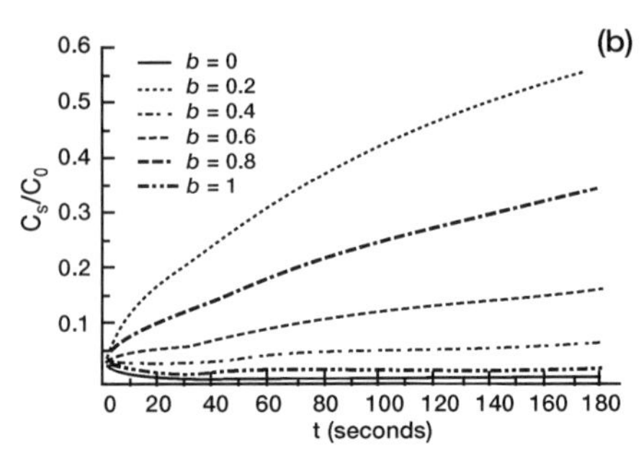

FIGURE 9.1 Influence of the fractal parameter b on the normalized antigen concentration near the fiber-optic surface, c_s/c_0 for a first-order reaction when nonspecific binding is (a) absent ($\alpha = 0$) and when it is (b) present ($\alpha = 0.5$).

nonspecifically bound to the biosensor surface. Note that this difference increases as the fractal parameter b decreases.

Figures 9.2a and 9.2b show the influence of NSB on the amount of antigen in solution specifically bound to the antibody immobilized on the surface, Γ^s_{Ag}/c_0, for a first-order reaction. As expected, and as indicated above, an increase in the fractal parameter b leads to a decrease in the amount of the antigen specifically bound to the antibody immobilized on the surface. Also, as expected, the presence of NSB leads to a decrease in the amount of antigen specifically bound to the antibody on the surface. For example, for a 3 min. reaction time and a fractal parameter b value of 0.6, the presence of nonspecific binding ($\alpha = 0.5$) decreases the Γ^s_{Ag}/c_0 value by about 30%—from 0.112 to 0.0076.

Figures 9.3a and 9.3b show the influence of NSB on the normalized antigen concentration near the fiber-optic surface, c_s/c_0, for the one-and-one-half-order reaction when NSB is absent and when it is present. When nonspecific binding is absent, the fractal parameter b value does not affect the c_s/c_0 value

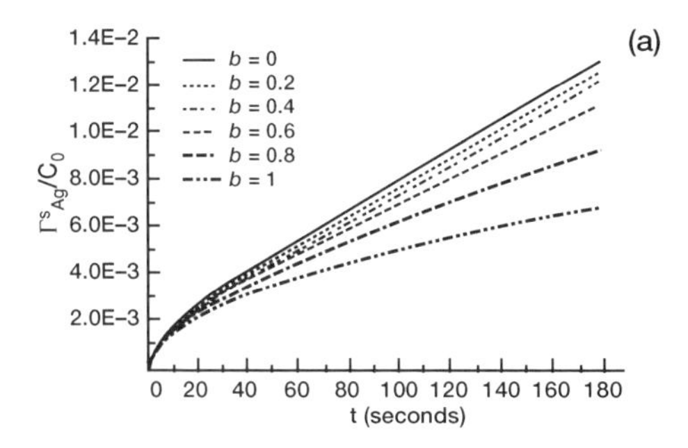

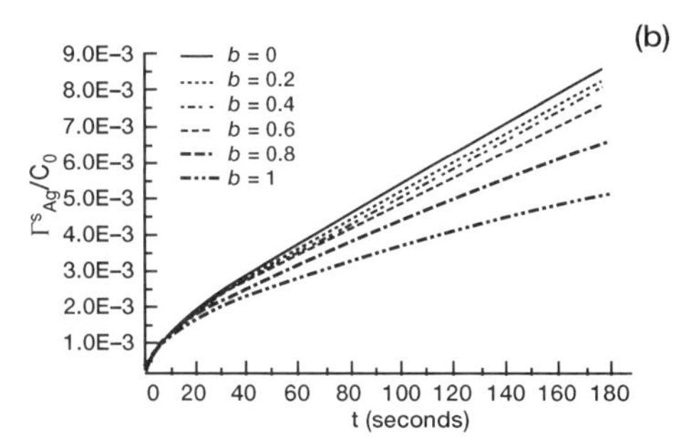

FIGURE 9.2 Influence of the fractal parameter b on the amount of antigen specifically bound to the antibody immobilized on the biosensor surface, Γ^s_{Ag}/c_0, for a first-order reaction when nonspecific binding is (a) absent ($\alpha = 0$) and when it is (b) present ($\alpha = 0.5$).

(Fig. 9.3a). However, when NSB is present (Figs. 9.3b and 9.3c), an increase in the fractal parameter b value does lead to an increase in the c_s/c_0, as expected. An increase in the heterogeneity on the reaction surface leads to lower amounts of antigen specifically bound to the biosensor surface, which leads to higher amounts of antigen in solution. Note that a value of $b = 0$ (homogeneous surface) leads to lower values of c_s/c_0 as the α value is increased from 0 to 0.01 to 0.1. The differences in the c_s/c_0 values in this case are due to the amount of antigen nonspecifically bound to the biosensor surface. For example, the c_s/c_0 values for a reaction time of 3 min. are 0.98, 0.55, and 0.12 for $\alpha = 0$, 0.01, and 0.1, respectively. Also, an increase in the α value increases the sensitivity of the c_s/c_0 curves for the different b values.

Figures 9.4a and 9.4b show the influence of NSB on the amount of antigen in solution specifically bound to the antibody immobilized on the surface, Γ^s_{Ag}, for a one-and-one-half-order reaction. Figure 9.4a shows that when NSB is absent and for a very low value of NSB (Fig. 9.4b) an increase in the fractal parameter b value leads to a decrease in the amount of antigen specifically

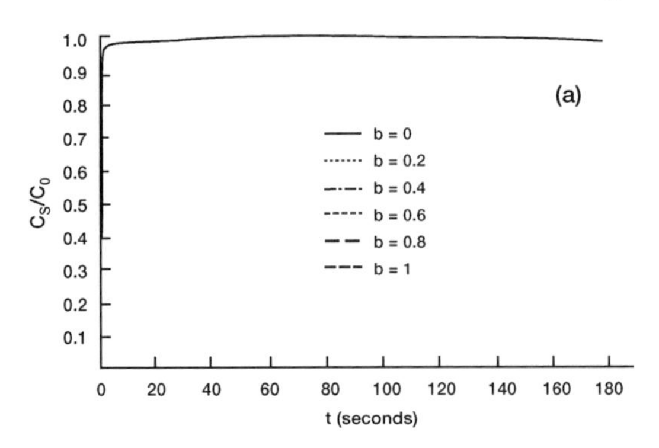

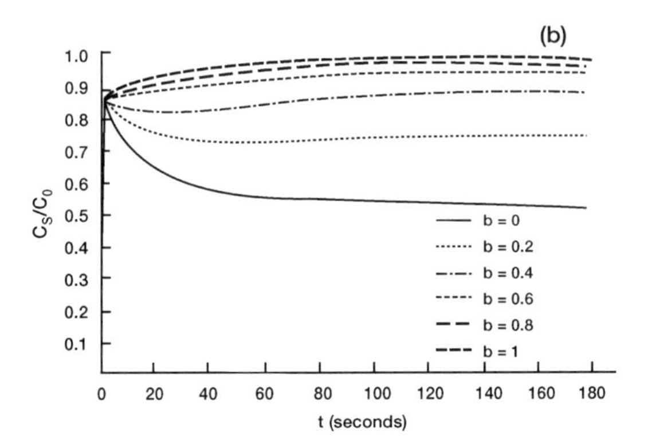

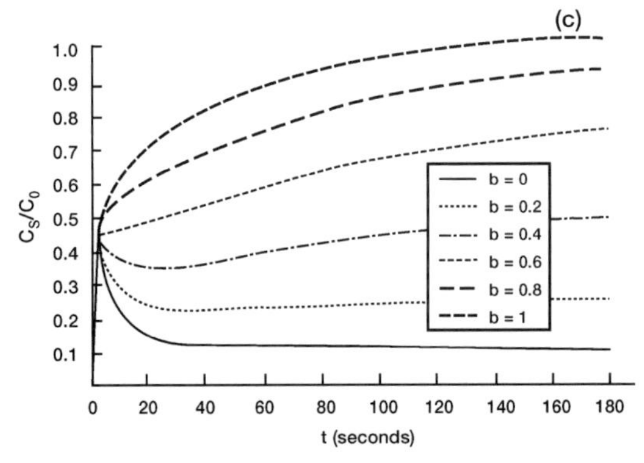

FIGURE 9.3 Influence of the fractal parameter b on the normalized antigen concentration near the fiber-optic surface, c_s/c_0, for a one-and-one-half-order reaction when nonspecific binding is (a) absent ($\alpha = 0$) and when it is present: (b) $\alpha = 0.01$; (c) $\alpha = 0.1$.

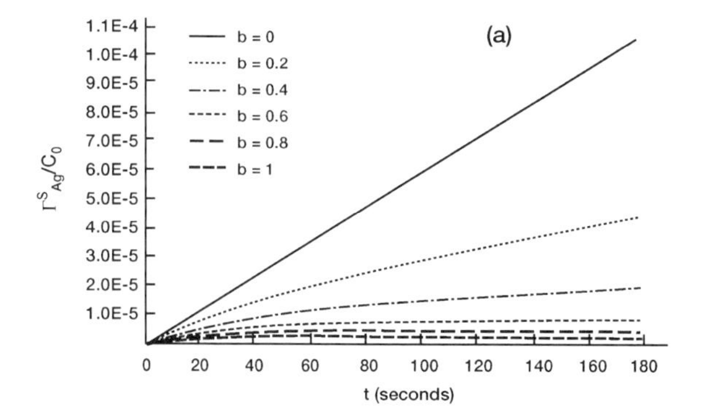

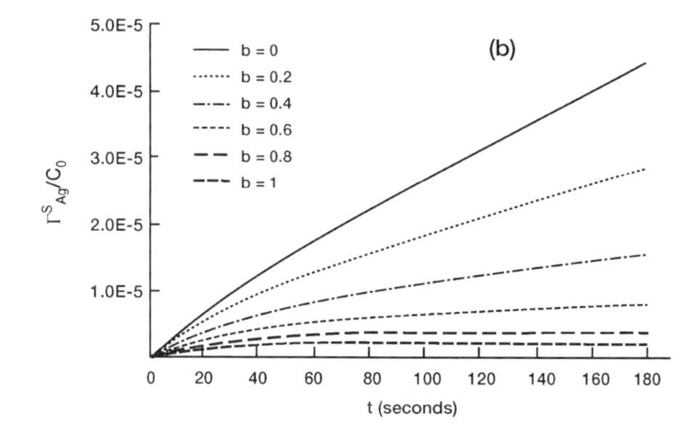

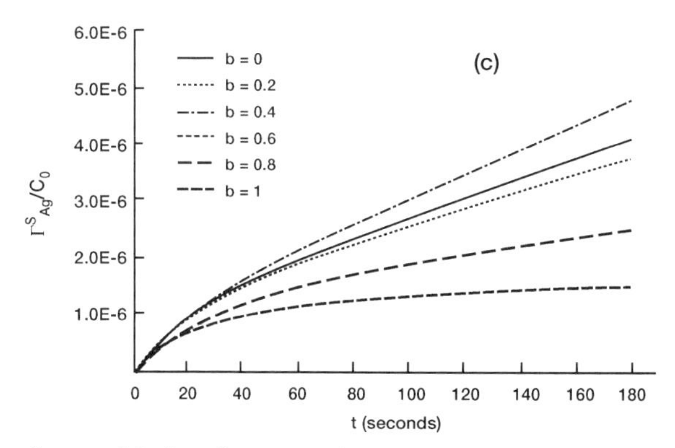

FIGURE 9.4 Influence of the fractal parameter b on the amount of antigen specifically bound to the antibody immobilized on the biosensor surface, Γ^s_{Ag}, for a one-and-one-half-order reaction when nonspecific binding is (a) absent ($\alpha = 0$) and when it is present: (b) $\alpha = 0.01$; (c) $\alpha = 0.1$.

bound to the antibody immobilized on the surface, Γ^s_{Ag}/c_0, as expected. For $\alpha = 0.1$, the maximum rate and amount of antigen specifically bound to the antibody immobilized on the surface is obtained for fractal parameter b values of 0.2 and 0.4 (Fig. 9.4c). Both of these b values yield about the same curve for Γ^s_{Ag}/c_0. Note that there is an optimum value or range of fractal parameter values that yield the maximum amount and rate of antigen specifically bound to the antibody immobilized on the surface. Apparently, some heterogeneity is helpful in obtaining the optimum amount and rate of antigen specifically bound to the antibody immobilized on the surface for the one-and-one-half-order reaction. Interestingly, practicing "biosensorists" have known all along that for a one-to-one antibody–antigen reaction, a densely packed and oriented antibody surface will be less susceptible to nonspecific binding than an irregular one with, say, holes (or gaps) or randomly oriented antibodies. Furthermore, a certain amount of heterogeneity will be required for higher-order reactions due to steric factors. This effect has been observed for avidin–biotin recognition systems.

Figures 9.5a and 9.5b show the influence of NSB on the normalized antigen concentration near the fiber-optic surface, c_s/c_0, for the second-order reaction when NSB is present and when it is absent. When NSB is absent, the fractal parameter b value does not affect the c_s/c_0 value (Fig. 9.5a). This behavior was also noted for the one-and-one-half-order reaction. Also, as exhibited by the one-and-one-half-order reaction, when NSB is present (Fig. 9.5b), an increase in the fractal parameter b value leads to a decrease in the c_s/c_0 value. For $\alpha = 0$ and $\alpha = 0.01$, the c_s/c_0 values are almost the same as for the one-and-one-half-order reaction. For $\alpha = 0.1$ (Fig. 9.5c), the second-order curve for c_s/c_0 is slightly higher than that for the one-and-one-half-order reaction for b values ranging from 0 to 1.0. The difference in the two curves is higher for the higher fractal parameter values.

Figures 9.6a and 9.6b show the influence of NSB on the amount of antigen in solution specifically bound to the antibody immobilized on the surface, Γ^s_{Ag}/c_0, for a second-order reaction. When nonspecific binding is absent (Fig. 9.6a) and for a very low value of nonspecific binding ($\alpha = 0.01$) (Fig. 9.6b), an increase in the fractal parameter b value leads to, as noted for the one-and-one-half-order case, a decrease in the amount of antigen specifically bound to the antibody immobilized on the surface, Γ^s_{Ag}/c_0. For $\alpha = 0.1$ and a 3 min reaction time, the maximum rate and amount of antigen specifically bound to the antibody immobilized on the surface is obtained for a fractal parameter b value of 0.4 (Fig. 9.6c). Note that similar behavior was observed for the one-and-one-half-order reaction, except that there the optimum fractal parameter b value was between 0.2 and 0.4. Once again, when slightly higher amounts of nonspecific binding are present, some heterogeneity does lead to higher rates and amounts of antigen specifically bound to the immobilized antibody on the

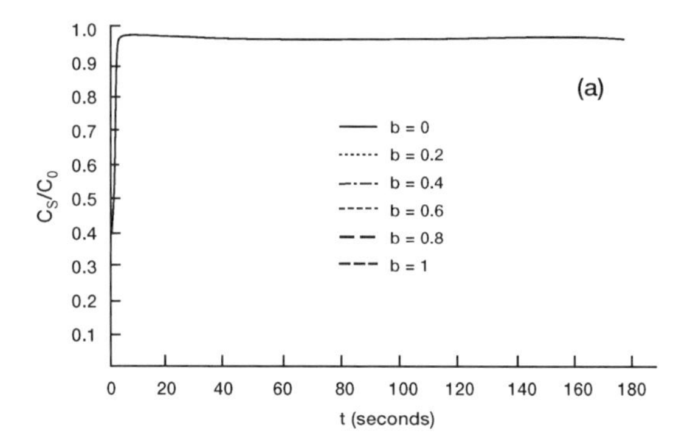

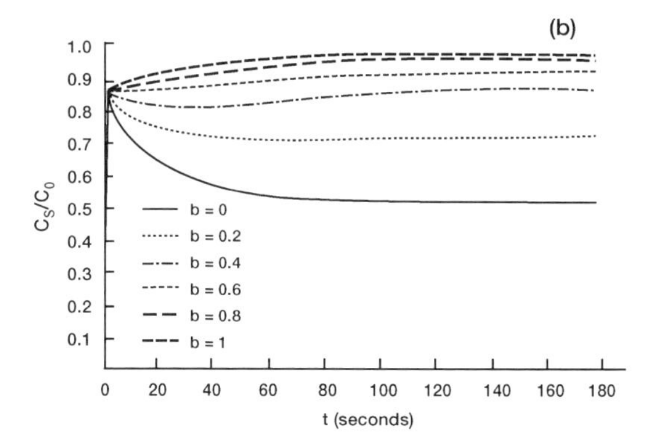

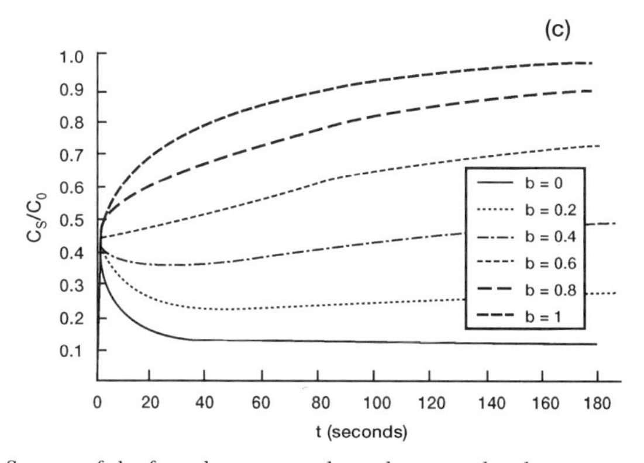

FIGURE 9.5 Influence of the fractal parameter b on the normalized antigen concentration near the fiber-optic surface, c_s/c_0, for a second-order reaction when nonspecific binding is (a) absent ($\alpha = 0$) and when it is present: (b) $\alpha = 0.01$; (c) $\alpha = 0.1$.

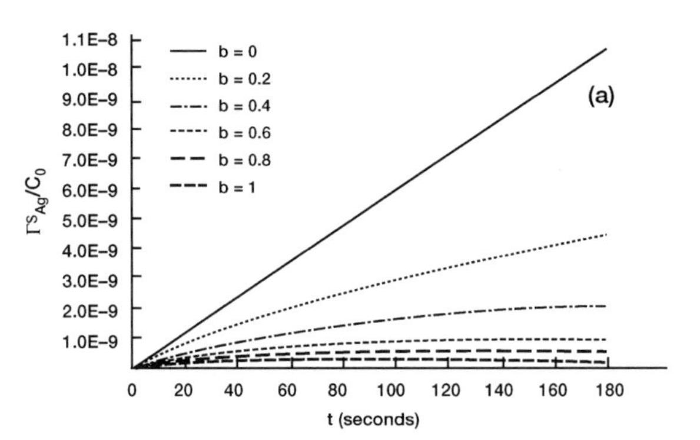

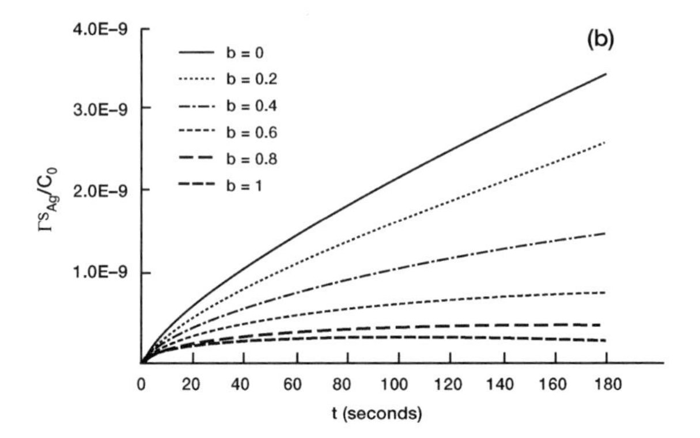

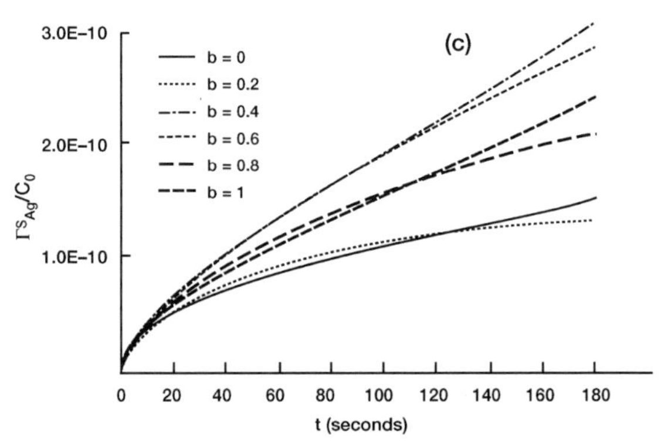

FIGURE 9.6 Influence of the fractal parameter b on the amount of antigen specifically bound to the antibody immobilized on the biosensor surface, Γ^s_{Ag}/c_0, for a second-order reaction when nonspecific binding is (a) absent ($\alpha = 0$) and when it is present: (b) $\alpha = 0.01$; (c) $\alpha = 0.1$.

biosensor surface. This has been shown for both the one-and-one-half- and the second-order case. Apparently, this should apply to reactions higher than first order. This type of analysis provides not only novel physical insights into the reactions occurring at the sensor surface but also a condition of optimum operation, in this case the degree of heterogeneity on the reaction surface. Such an analysis should assist in improving the sensitivity and stability of evanescent fiber-optic biosensors.

9.4. OTHER EXAMPLES OF INTEREST

We now discuss other examples (Sadana and Vo-Dinh, 1997), to be presented this time without the actual nonspecific binding step. The figures are similar in nature to Figs. 8.4 and 8.5. Derivations similar to those given in Chapter 8 may be obtained for the different order of reactions thus realized. To avoid repetition the derivations are not indicated here. However, the examples presented are important, and the reader can easily derive the necessary equations, if need be. (In fact, it may be a good exercise to do so).

There has been an increasing interest in the use of immunological techniques for the detection of environmental hazards and for biomedical applications (Vo-Dinh and Niessner, 1995). The design of antibody-targeted agents for a large class of chemical species—such as polycyclic-aromatic compounds (PACs)—could be an important development for biosensors. Such biosensors could be used to screen samples for their overall content of PACs rather than specific PACs.

We now show some reaction mechanisms that would be involved when an antibody is targeted to a group of antigens having multiple-antigen sites. This model is relevant to the situation in which the antibody is designed to have a paratope targeted to only a monocyclic aromatic—or part of a monocyclic ring. Such an antibody would be capable of recognizing not only one PAC but a family of PACs. Figure 9.7 shows such an antibody targeted to a family of PACs. (Note that to be relevant to this chapter, and as indicated above, one needs to add an additional nonspecific binding step in Fig. 9.7.) Multivalency for antibodies requires certain conditions. In general, antibodies are larger than antigens. Therefore, certain size and steric conditions must be fulfilled to allow more than one antibody to be attached to an antigen. This could occur for antigens with sufficiently large size or with antibodies specifically designed to have a small size or sterically favorable paratope geometry. Note that the combining site on the antibody should not be so large that it completely encloses the antigen (PAC, in this case). Note also that steric hindrance may be particularly significant if the binding pockets are especially deep. This has implications as far as nonspecific binding is concerned, as

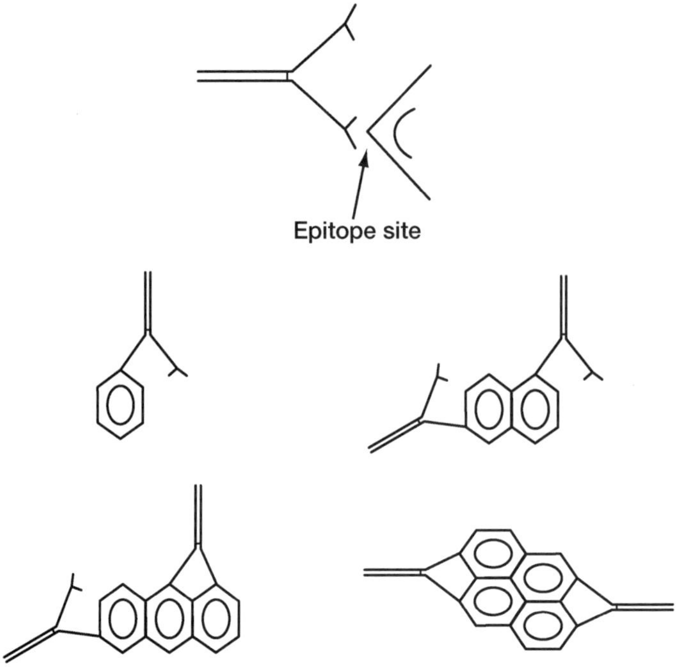

FIGURE 9.7 Schematic diagram of antibodies having paratopes targeted to the antigen series of polycyclic aromatic compounds.

mentioned at the beginning of this chapter. Apparently, in these cases, due to the steric hindrance, one may anticipate that nonspecific binding plays an increasingly important role, as compared to the cases with simple monovalent and divalent antigens. This will prove to be a difficult and challenging problem to engineers, chemists, and others who wish to design a PAC-combining site smaller than a PAC so that the site has a useful binding affinity while minimizing the deleterious effects of nonspecific binding. Perhaps an imaginative person could design systems consisting of parts of the antibody by cleaving and combining the appropriate epitopes. Note that the binding of multivalent antigens or antibodies to each other could have importance in the design of biosensor probes.

Another example where nonspecific binding may play an important role is shown in Fig. 9.8. Here we have the binding of the trivalent antigen in solution to the divalent antibody immobilized to the surface. Once again, the nonspecific binding step is not shown and must be included before one derives a reaction rate expression. (This is left as an exercise to the interested reader.) It is also of interest to analyze the reverse case, wherein the antibody

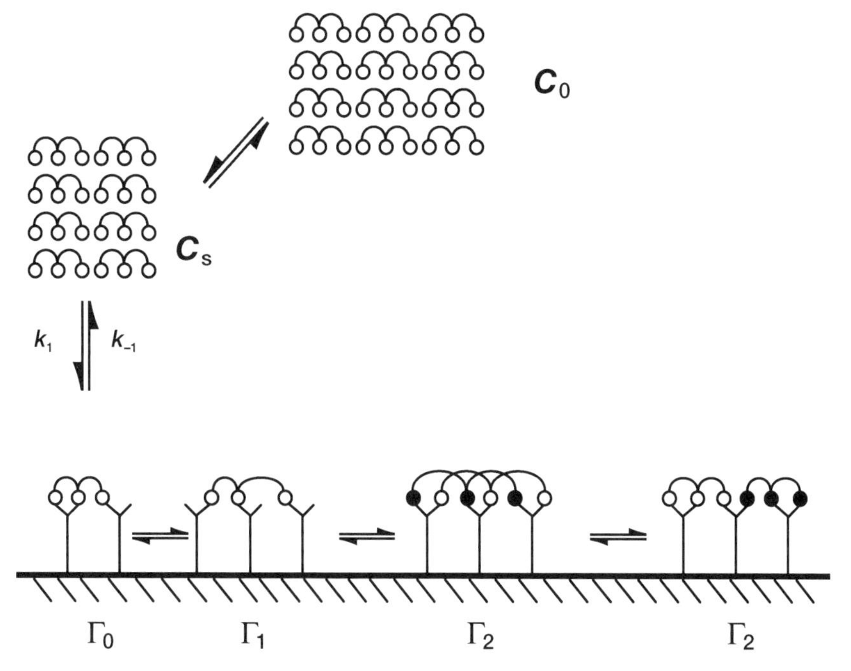

FIGURE 9.8 Elementary steps involved in the binding of trivalent antigen in solution to divalent antibody covalently attached to the surface.

is in solution and the antigen is covalently or noncovalently immobilized to the surface. The elementary steps involved in such a reaction are shown in Fig. 9.9. Once again, the nonspecific binding step is not shown but must be included before deriving a reaction rate expression. Figure 9.9a shows the elementary steps involved in the dual-step binding of divalent antibody in solution to monovalent antigen noncovalently or covalently attached to the surface. Figure 9.9b shows the elementary steps involved in the dual-step binding of divalent antibody in solution to trivalent antigen noncovalently or covalently attached to the surface.

Other suitable examples of antigen–antibody binding may also be possible. Be aware that all of the discussion presented in this chapter and in Chapter 8 should be applicable, with minor modifications if necessary, to analyte-receptor binding reactions in general.

PACs are one possible type of antigens that exhibit multivalencies and different binding sites for antibody binding. It is highly probable that as an antibody binds to an antigen-binding site, it either makes easier or constrains (owing to induced conformational changes on the molecule immobilized on

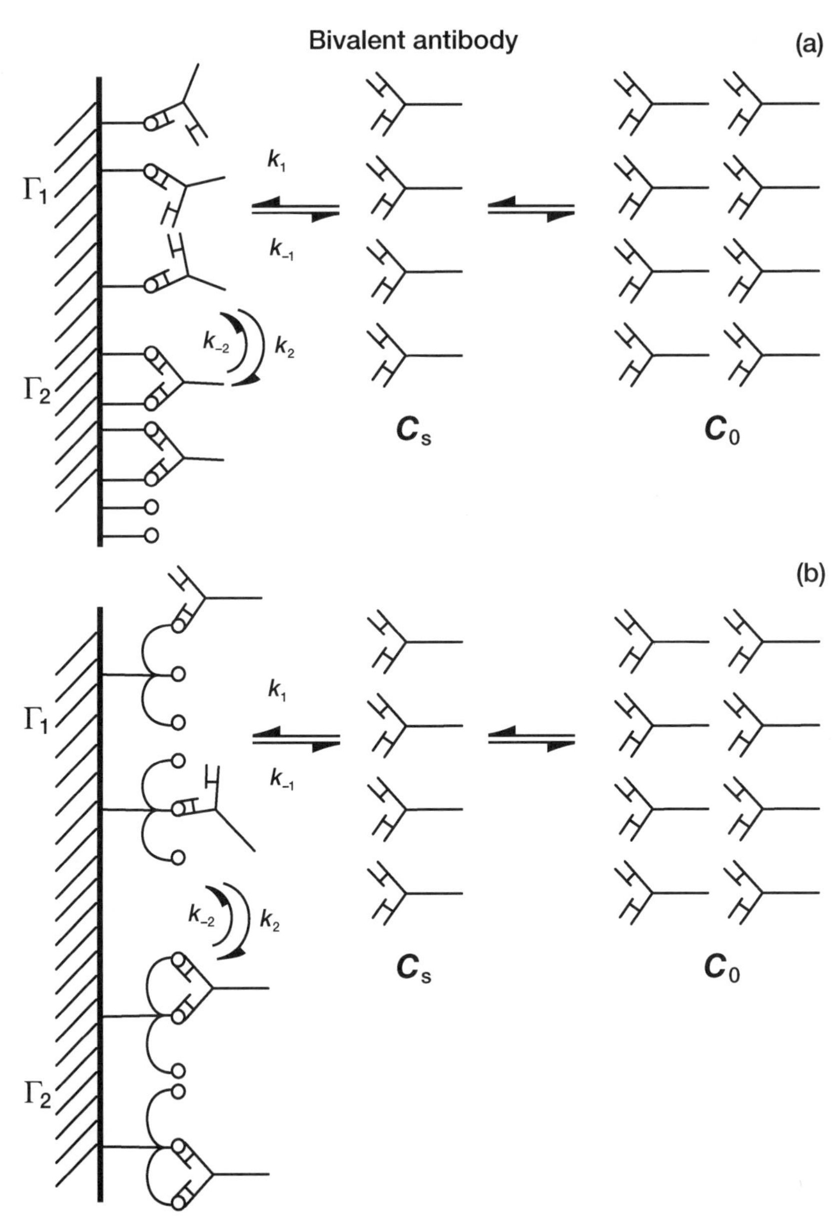

FIGURE 9.9 (a) Elementary steps involved in the dual-step binding of divalent antibody in solution to monovalent antigen noncovalently or covalently attached to the surface; (b) elementary steps involved in the dual-step binding of divalent antibody in solution to trivalent antigen non covalently or covalently attached to the surface.

the surface) binding of further antibody molecules to different binding sites on the same antigen molecule. This would lead to temporal binding rate coefficients, heterogeneities on the surface, and fractal reaction rate kinetics. All of these complexities and heterogeneities would contribute toward an increasing probability for and tendency toward nonspecific binding. Also, different PACs may bind with certain antibodies in more than one orientation. Furthermore, a population of antibodies may include some that bind to the same PAC in different orientations. This would lead to further heterogeneties and increasing disorder. These heterogeneities on the surface, as previously indicated, would lead to higher probabilities of nonspecific binding on the surface. It is reasonable to anticipate that higher-valency antigens would, in general, lead to higher levels of disorder on the surface. It would be of significant interest to relate the antigen valency to the fractal dimension on the surface, as well as how all of this may then affect the nonspecific binding on the surface. The eventual goal of the analysis, of course, is how one may improve the stability, sensitivity, reproducibility, and response time of biosensors.

9.5. CONCLUSIONS

The inclusion of nonspecific binding in the analysis of specific binding of the antigen in solution to the antibody immobilized on the surface provides a more realistic picture of what is happening on the surface, besides leading to temporal binding rate coefficients. Nonspecific binding may be present due to (1) the direct binding of the analyte (antigen or antibody, as the case may be) to the sensor surface, (2) the nonspecific (randomly oriented binding) of the antibody on the surface, and/or (3) the binding of other molecules to the sensing surface in multicomponent environments. A fractal analysis of the antigen–antibody binding reaction for the different reaction orders provides estimates of the changes in the rate and the amount of antigen in solution specifically bound to the antibody immobilized on the biosensor surface in the presence of nonspecific binding. We have suggested a method for determining the fractal parameter p (or $-b$) from experimental antigen–antibody binding kinetics data, which includes nonspecific binding, using regression analysis and Eq. (9.3). This demonstrates the applicability of the approach to real antibody surfaces.

For a first-order reaction, an increase in the fractal parameter b leads to a decrease in the amount of antigen in solution specifically bound to the antibody on the surface when nonspecific binding is absent or present. As expected, the presence of nonspecific binding leads to a decrease in the amount of antigen specifically bound to the antibody on the surface. The analysis provides quantitative estimates of the influence of nonspecific binding on the amount of antigen in solution specifically bound to the antibody immobilized on the surface. Note that in the ideal case nonspecific binding should be minimized. Nevertheless, it will be present to some degree in most (if not all) cases, and the model presented helps to estimate the derogatory influence NSB has on the biosensor performance.

For a one-and-one-half-order reaction and either when nonspecific binding is absent or for a very low value of α (equal to 0.01), an increase in the fractal parameter b leads to a decrease in the amount of antigen specifically bound to the antibody on the biosensor surface. However, for an α value of 0.1, the maximum rate and amount of antigen specifically bound to the antibody immobilized on the biosensor surface is obtained for fractal parameter b values of 0.2 and 0.4. This indicates that there is an optimum value or range of fractal parameter values that yield the maximum amount and rate of antigen in solution specifically bound to the antibody immobilized on the biosensor surface. Similar behavior was observed for the second-order reaction, where the fractal parameter value was 0.4. Apparently, some amount of heterogeneity is helpful in obtaining the optimum amount and rate of antigen in solution specifically bound to the antibody on the surface for the one-and-one-half-and second-order reactions. The analysis may presumably be extended to higher-order reactions, but no such optimum value of the fractal parameter was noted for the first-order reaction.

The fractal analysis of antigen–antibody binding kinetics in the presence of nonspecific binding helps provide novel physical insights into the conformational states of the antigen–antibody complex on the biosensor surface. Such an analysis should assist in controlling the reactions to advantage, as well as improve the sensitivity, stability, and reaction time of biosensors. It is recommended that further similar studies (theoretical as well as experimental) be carried out to further delineate the conformational states of either the antigen–antibody or the antigen complex on the surface. Further detailed studies of the different mechanisms of nonspecific binding and how they influence antigen–antibody kinetics for biosensor applications also would be of assistance. The analysis presented here is general enough to be applied to analyte-receptor reactions occurring on other than biosensor surfaces, such as membrane surfaces. Finally, increasing complexity of either the analyte or the receptor, such as multivalent antigens (for example, PACs) or antibodies, would presumably increase the probability of nonspecific binding. In this case, the influence of nonspecific binding, or at the very least estimates of its influence on specific binding, should be considered.

REFERENCES

Anderson, J., NIH Panel Review Meeting, Case Western Reserve University, Cleveland, OH, July 1993.

Angel, S. M., *Spectroscopy*, **2**, 38–46 (1987).

Chen, Z., "Influence of Nonspecific Binding on Antigen–Antibody Binding Kinetics: Biosensor Applications," Master's thesis, University of Mississippi, MS, 1994.

Cuypers, P. A., Willems, G. M., Kop, J. M., Corsel, J. W., Jansen, M. P., and Hermens, W. T., in *Proteins at Interfaces: Physicochemical and Biochemical Studies*, J. L. Brash and T. A. Horbett, eds., Washington, DC: American Chemical Society, pp. 208–211, 1987.

Dambrot, S. M., *Bio/Technology* **10**, 120–120 (1993).

De Gennes, T. G. and Bann, J. G., *Radiat. Phys. Chem.* **22**, 193–196 (1982).

Dewey, T. G. and Bann, J. G., *Biophys. J.* **63**, 594–598 (1982).

Di Cera, E., *J. Chem. Phys.* **95**, 5082–5086 (1991).

Giaver, I., *J. Immunol.* **116**, 766–771 (1976).

Giona, M., *Chem. Eng. Sci.* **47**, 1503–1515 (1992).

Havlin, S., in *The Fractal Approach to Heterogeneous Chemistry: Surfaces, Colloids, and Polymers*, D. Avnir, ed., New York: Wiley, pp. 251–269, 1989.

Kopleman, R., *Science* **241**, 1620–1626 (1988).

Le Brecque, M., *Mosaic* **23**, 12–15 (1992).

Liebovitch, L. S., Fischbarg, J., Koniarek, J. P., Todorova, I., and Wang, M., *Math. Biosci.* **84**, 37–68 (1987).

Liebovitch, L. S. and Sullivan, J. M., *Biophys. J.* **52**, 979–988 (1987).

Li, H., Chen, S., and Zhao, H., *Biophys. J.* **58**, 1313–1320 (1990).

Lowe, C. R., *Biosensors* **1**, 3–16 (1985).

Mathewson, P. R. and Finley, J. W., eds., *Biosensor Design and Protocols*, ACS Symposium Series 511, Washington, DC: American Chemical Society, pp. ix–xi, 1992.

Nygren, H., *J. Immunol. Meth.* **114**, 107–114 (1988).

Nygren, H., *Biophys. J.* **65**, 1508–1512 (1993).

Nygren, H. and Stenberg, M., *J. Colloid & Interface Sci.* **107**, 560–566 (1985).

Nyikos, L. and Pajkossy, T., *Electrochim. Acta* **34**, 171–179 (1986).

Pfeifer, P., Avnir, D., and Farin, D. J., *J. Colloid & Interface. Sci.* **103**, 112–123 (1984).

Place, J. F., Sutherland, R. M., and Dahne, C. P., *Biosensors* **1**, 321–353 (1985).

Sadana, A. and Beelaram, A., *Biotech. Progr.* **10**, 291–298 (1994).

Sadana, A. and Beelaram, A., *Biosen. & Bioelectr.* **10**, 301–316 (1995).

Sadana, A. and Madagula, A., *Biotech. Progr.* **9**, 259–266 (1993).

Sadana, A. and Madagula, A., *Biosen. & Bioelectr.* **9**, 45–55 (1994).

Sadana, A. and Sii, D., *J. Colloid & Interface. Sci.* **151**, 166–177 (1992a).

Sadana, A. and Sii, D., *Biosen. & Bioelectr.* **7**, 559–568 (1992b).

Sadana, A. and Vo-Dinh, T., *Appl. Biochem. & Biotech.*, **67**, 1 (1997).

Scheller, F. W., Hintsche, R., Pfeiffer, D., Schubert, F., Reidel, K., and Kindervater, R., *Sensors & Actuators*, **4**, 197–206 (1991).

Sigmaplot, Scientific Graphing Software, User's manual, Jandel Scientific, San Rafael, CA, 1993.

Stenberg, M. and Nygren, H., *Anal. Biochem.* **127**, 183–192 (1982).

Stenberg, M., Stiblert, L., and Nygren, H., *J. Theor. Biol.* **120**, 129–142 (1986).

Vo-Dinh, T., and Neissner, R., eds., "Environmental Monitoring and Hazardous Waste Site Remediations," Bellingham, Washington: SPIE Publishers, 1995.

Wise, D. L. and Wingard Jr., L. B., eds., *Biosensors with Fiberoptics,* New York: Humana Press, pp. vii–viii, (1991).

CHAPTER 10

Fractal Dimension and Hybridization

10.1. INTRODUCTION

DNA–RNA hybridization wherein there is base pairing of a single-strand nucleic acid probe to a second complementary sequence contained in a second immobilized DNA or RNA target has recently gained increasing importance due to its applications in gene identification, gene mapping, DNA sequencing, and medical diagnostics. Waring (1981) indicates that DNA-binding agents possess antitumor, antiviral, or antimicrobial activities. Piehler *et al.* (1997) note that antitumor drugs often exert their influence by interfering with DNA function. Therefore, the interaction of low molecular weight compounds with nucleic acids is increasingly being studied in both biophysical and biochemical investigations. Chiu and Christopoulos (1996) emphasize that the strong and specific interaction of the two complementary nucleic acid strands is the basis of hybridization assays. They indicate that hybridization assays exhibit the potential to transform laboratory medicine as well as clinical testing. As with all assay procedures, there is continual increasing pressure to enhance the speed, economy, and reliability of the testing procedures. The performance of these genosensors will be significantly enhanced if more physical insights are obtained into each of the procedures or steps that are involved in the entire assay.

A very basic step in the hybridization procedure is the immobilization of the denatured nucleic acid target on a solid support. Both the immobilization and the denaturation of the nucleic acid target on a solid support will lead to a heterogeneous surface. In other words, no matter how careful one is with the experimentation, there will be a certain degree of heterogeneity on the surface. This will significantly influence the binding of the complementary strands and, subsequently, the performance of the hybridization assay procedure.

External diffusion limitations play a role in the analysis of immunoassays (Giaver, 1976; Eddowes, 1987/1988; Bluestein *et al.*, 1991; Place *et al.*, 1991) and will significantly affect the performance of hybridization assays. The influence of diffusion in immunoassays has been analyzed to some extent (Stenberg *et al.*, 1986; Nygren and Stenberg, 1985; Stenberg and Nygren, 1982; Sadana and Sii, 1992a, 1992b). Stenberg *et al.* (1986) have analyzed in great detail the effect of external diffusion on solid-phase immunoassay when the antigen is immobilized to a solid surface and the antibodies are in solution. These authors noted that diffusion plays a significant part when high concentrations of antigens (or binding sites) are immobilized on the surface.

Kopelman (1988) indicates that surface diffusion-controlled reactions that occur on clusters (or islands) are expected to exhibit anomalous and fractal-like kinetics. These fractal kinetics exhibit anomalous reaction orders and time-dependent rate (for example, binding) coefficients. Fractals are disordered systems, and the disorder is described by nonintegral dimensions (Pfeifer and Obert, 1989). The time-dependent adsorption rate coefficients observed experimentally (Cuypers *et al.*, 1987; Nygren and Stenberg, 1990) may also be due to nonidealities or heterogeneity on the surface. For hybridization assays, the immobilization of the denatured nucleic acid will lead to heterogeneities on the surface. This is a good example of a disordered system for which a fractal analysis is appropriate. In addition the DNA/RNA hybridization reaction on the solid surface is a good example of a low-dimension reaction system in which the distribution tends to be less random (Kopelman, 1988), and a fractal analysis would provide novel physical insights into the diffusion-controlled reactions occurring at the surface. Furthermore, Matuishita (1989) indicates that the irreversible aggregation of small particles occurs in many natural processes, such as polymer science, material science, immunology, etc. Also, when too many parameters are involved in a reaction, which is the case for these hybridization reactions on the solid surface, the fractal dimension may be a useful global parameter. Finally, Lee and Lee (1995) emphasize that in all heterogeneous systems, the geometry of the environment has a major impact on the reaction rate and performance. The fractal dimension is an appropriate descriptor for these irregularities since these structures are fractal-like in nature. Since the performance of biosensors and immunoassays, in general, is constrained by chemical binding kinetics, equilibrium, and mass transport of the complementary DNA in solution to the biosensor (or immunoassay) surface, it is appropriate to pay particular attention to the design of such systems and to explore new avenues by which further knowledge may be obtained about these systems.

In this chapter, we use fractals to analyze the binding of a single-strand nucleic acid probe of defined sequence in solution to a complementary DNA or RNA sequence immobilized to a biosensor or other immunoassay surface. The analysis is performed on data available in the literature. Examples are presented wherein either a single- or a dual-fractal analysis is required to adequately describe the hybridization binding kinetics. The fractal analysis is one way by which one may elucidate the time-dependent binding rate coefficients and the heterogeneity that exists on the sensing surface.

10.2. THEORY

There are similarities in the binding of antigen in solution to antibody immobilized on a biosensor surface and the binding observed in hybridization kinetics. In hybridization binding, there is base pairing of a single-strand nucleic acid probe in solution to a second complementary sequence contained in a second immobilized DNA or RNA target. This applies to hybridization observed using biosensors or any other immunoassay procedure involving a solid surface. Here we present a method of estimating actual fractal dimension values for hybridization binding kinetics observed in biosensor applications and other immunoassay procedures.

10.2.1. VARIABLE BINDING RATE COEFFICIENT

The diffusion-limited binding kinetics of antigen (or antibody or substrate) in solution to antibody (or antigen or enzyme) immobilized on a biosensor surface has been analyzed within a fractal framework (Sadana and Beelaram, 1994; Sadana *et al.*, 1995). One of the findings, for example, is that an increase in the surface roughness, or fractal dimension, leads to an increase in the binding rate coefficient. Furthermore, experimental data presented for the binding of HIV virus (antigen) to anti-HIV (antibody) immobilized on a surface displays a characteristic ordered disorder (Anderson, 1993). This indicates the possibility of a fractal-like surface. It is obvious that such a biosensor system (wherein either the antigen or the antibody is attached to the surface), along with its different complexities—including heterogeneities on the surface and in solution, diffusion-coupled reactions, time-varying adsorption or binding rate coefficients, etc.—can be characterized as a fractal system. In the previous chapters, the fractal analysis has been utilized to model analyte-receptor and analyte-receptorless (protein) systems. In this chapter, we extend these ideas to the ssRNA–ssDNA or ssDNA–ssDNA binding observed in hybridization kinetics.

Single-Fractal Analysis

Havlin (1989) indicates that the diffusion of a particle (in our case, a single-strand nucleic acid probe, $ssDNA_{soln}$) from a homogeneous solution to a solid surface (second complementary sequence contained in a second DNA- or RNA-coated surface, $ssRNA_{surf}$) where it reacts to form a hybrid product (dDNA) is given by

$$[\mathrm{dDNA}] \sim \begin{cases} t^{(3-D_f)/2} = t^p & t < t_c \\ t^{1/2} & t > t_c, \end{cases} \tag{10.1a}$$

where dDNA represents the stable double-strand DNA formed. An intermediate and partially hybridized DNA, DNA_{interm}, may be present, but for now we will ignore that. Here, D_f is the fractal dimension of the surface. Equation (10.1a) indicates that the concentration of the product, dDNA(t), in the reaction ssDNA + ssDNA→dDNA on a solid fractal surface scales at short and intermediate time frames as dDNA $\sim t^p$, with the coefficient $p = (3 - D_f)/2$ at short time frames and $p = 1/2$ at intermediate time frames.

Dual-Fractal Analysis

The single-fractal analysis can be extended to include two fractal dimensions. In this case, the product (double-strand DNA, dDNA) concentration on the immunosensor surface is given by

$$[\mathrm{dDNA}] \sim \begin{cases} t^{(3-D_{f1})/2} = t^{p_1} & t < t_1 \\ t^{(3-D_{f2})/2} = t^{p_2} & t_1 < t < t_2 = t_c \\ t^{1/2} & t > t_c. \end{cases} \tag{10.1b}$$

10.3. RESULTS

Abel *et al.* (1996) used a fiber-optic biosensor for the detection of 16 mer oligonucleotides in hybridization assays. The authors immobilized a biotinylated capture probe on the biosensor surface using either avidin or streptavidin. Fluorescence was utilized to monitor the hybridization with fluorescein-labeled complementary strands. These authors indicate that the capability of DNA and RNA fragments to bind selectively to complementary arranged nucleotides at other nucleic acids is the basis for *in vitro* tests.

Figure 10.1 shows the binding of 16*CFl (complementary oligonucleotide) in a 10 nM solution to 16*B (oligonucleotide) immobilized via sulfosuccinimidyl-6-(biotinamido)hexanoate (NHS-LC-biotin) and streptavidin to a biosensor. Both chemical (Fig. 10.1a) and thermal (Fig. 10.1b) regeneration was employed. In this case, a single-fractal analysis is sufficient to adequately describe the binding kinetics. Table 10.1a shows the values of the binding rate coefficients and the fractal dimensions obtained in these cases. The parameter values presented in Table 10.1 (and Table 10.2) were obtained from a regression analysis using Sigmaplot (1993) to model the data using Eqs. (10.1a) and (10.1b), wherein (dDNA) $= kt^p$ (single-fractal analysis) and (dDNA) $= kt^{(p_1 \text{ or } p_2)}$ (dual-fractal analysis). The k, D_f, k_1, k_2, D_{f_1}, and D_{f_2} values presented in Table 10.1 (and Table 10.2) are within 95% confidence limits. For example, for the binding of 10 nM 16*CFl (oligonucleotide) to 16*B immobilized via NHS-LC-biotin and streptavidin during chemical regeneration, the k value is 86.53 ± 3.21. The 95% confidence limit indicates that 95% of the k values will fall between 83.32 and 89.74. Note that as one goes from the chemical to the thermal regeneration cycle, k increases by about 15.5%—from 86.53 to 100, and D_f increases by about 13.4%—from 1.2112 to 1.3942.

Table 10.1 Influence of Different Parameters on Fractal Dimensions and Binding Rate Coefficients for Different Analyte-Receptor Hybridization Kinetics: Single-Fractal Analysis

Analyte in solution/receptor on surface	k	D_f	Reference
(a) 10 nM 16*CF1 (oligonucleotide)/16*B immobilized via NHS-LC-biotin and streptavidin (chemical regeneration)	86.53 ± 3.21	1.211 ± 0.026	Abel *et al.*, 1996
10 nM 16*CF1 (oligonucleotide)/16*B immobilized via NHS-LC-biotin and streptavidin (thermal regeneration)	100.0 ± 6.7	1.394 ± 0.046	Abel *et al.*, 1996
10 nM 16*CF1 (oligonucleotide)/16*B immobilized via NHS-LC-biotin and avidin	1338 ± 31	1.227 ± 0.021	Abel *et al.*, 1996
(b) Streptavidin-biotinylated aequorin conjugates; target DNA/capture probe (immobilized through digoxigenin–antidigoxigenin interaction); detection probe biotinylated	109.5 ± 2.17	2.043 ± 0.037	Galvan and Christopoulos, 1996
(c) 1 ng/100 μL rRNA from *B. thermospacta*/labeled DNA	0.0224 ± 0.0007	2.940 ± 0.010	Fliss *et al.*, 1995

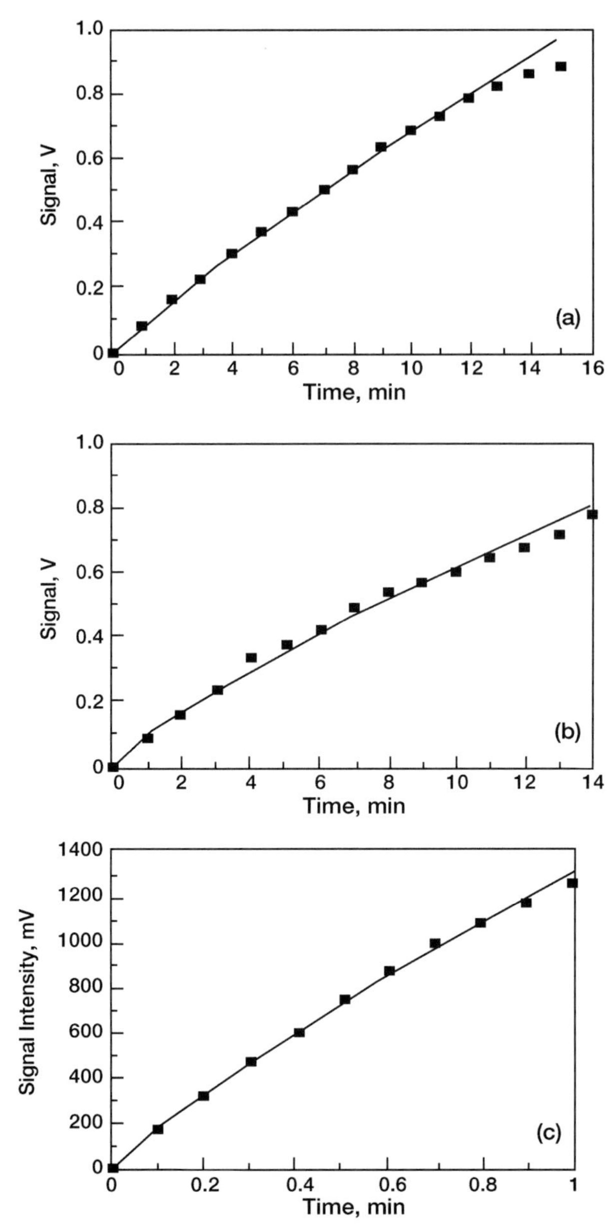

FIGURE 10.1 Hybridization of 10 nM 16*CFl (oligonucleotide) in solution to 16*B immobilized via NHS-LC-biotin and streptavidin to a biosensor surface (Abel *et al.*, 1996). (a) Chemical regeneration; (b) thermal regeneration. (c) Binding of 10 nM 16*CFl (oligonucleotide) in solution to 16*B immobilized via NHS-LC-biotin and avidin to a biosensor surface.

Table 10.2 Influence of Different Parameters on Fractal Dimensions and Binding Rate Coefficients for Different Analyte-Receptor Hybridization Kinetics: Dual-Fractal Analysis

Analyte in solution/receptor on surface	k	k_1	k_2	D_f	D_{f_1}	D_{f_2}	Reference
(a) Complementary ssDNA (cDNA)/ single-strand deoxyribonucleic acid (ssDNA) thymidylic acid icosanucleotides on derivatized quartz optical fiber	13.79 ± 4.61	29.13 ± 26.06	8.676 ± 0.944	1.000 ± 0.389	1.545 ± 1.12	0.4672 ± 0.399	Piunno *et al.*, 1995
(b) BCR-ABL mRNA characteristic of chronic myelogenous leukemia; mRNA (target) DNA/captured on streptavidin-coated wells and hybridized to the RNA probe	6.979 ± 1.356	3.174 ± 0.446	38.24 ± 0.44	2.793 ± 0.237	2.195 ± 0.335	3.000 ± 0.150	Radovich *et al.*, 1995
(c) Target DNA (200 bp) from the BCR-ABL mRNA from K652 cells/capture probe (immobilized through digoxi genin–antidigoxigenin interaction); detection probe biotinylated	1486 ± 434	633.9 ± 219.4	9243 ± 5	1.766 ± 0.303	1.226 ± 0606	2.594 ± 0.002	Radovich *et al.*, 1995
(d) 0.01 ng/100 μL rRNA from *L. Monocytogenes*/labeled DNA	0.1740 ± 0.029	0.1258 ± 0.016	0.7283 ± 0.047	2.230 ± 0.098	2.061 ± 0.120	2.760 ± 0.121	Fliss *et al.*, 1995
0.5 ng/100 μL rRNA from *L. Monocytogenes*/labeled DNA	0.1639 ± 0.053	0.1693 ± 0.004	1.666 ± 0.048	1.966 ± 0.174	1.674 ± 0.389	2.825 ± 0.043	Fliss *et al.*, 1995
1 ng/100 μL rRNA from *L. Monocytogenes*/labeled DNA	0.5254 ± 0.075	0.3871 ± 0.119	1.666 ± 0.109	2.363 ± 0.083	2.198 ± 0.058	2.796 ± 0.097	Fliss *et al.*, 1995

Abel *et al.* also analyzed the binding kinetics of 10 nM 16*CFl in solution to 16*B immobilized on a biosensor surface. The only difference between this case and the previous two cases is that for the immobilization step the authors used avidin instead of streptavidin. Figure 10.1c shows that a single-fractal analysis is adequate to model the binding kinetics. The values of k and D_f are given in Table 10.1a. In this case, the binding rate coefficient is much higher than in the previous two cases. The fractal dimension value of 1.2722 is between the two values obtained in the previous two cases. In this case, the fractal dimension analysis provides a quantitative estimate of the degree of heterogeneity that exists on the biosensor surface for each of the DNA hybridization assays. More data is required to establish trends or predictive equations for the binding rate coefficient in terms of analyte concentration in solution or the estimated fractal dimension of the biosensor surface under reaction conditions.

It is appropriate to indicate that the reliability of the data is rather limited. This is because both in the preceding example and in the examples that follow in this chapter we analyze a variety of different immobilization chemistries, different DNA densities/orders, and different detection chemistries. We have attributed the rate of signal development simply to hybridization rates, but this is an over simplification; we do not identify any limitations and analyze and correct for it in our analysis. For example, detection may be driven by hybridization, which in turn may be detected by the partitioning of a marker (for example, intercalation). At present, we do not offer a correction for this.

Also, there may be nonselective adsorption of the complementary nucleotide. Our analysis here does not include this nonselective adsorption. We do recognize that in some cases this may be a significant component of the adsorbed material and that the rate of association, which is of a temporal nature, would depend on surface availability. Furthermore, this adsorbed material could then have a probability of selective hybridization different from that observed for direct hybridization from solution. At present, our model does not allow for this nonselective adsorption and the subsequent rate of association. Accommodating the nonselective adsorption into the model would lead to an increase in the degree of heterogeneity on the surface, since by its very nature nonspecific adsorption is more heterogeneous than specific adsorption. This would lead to higher fractal dimension values since the fractal dimension is a direct measure of the degree of heterogeneity that exists on the surface. Future analysis of hybridization binding data may include this aspect in the analysis, which would be exacerbated by the presence of the inherent external diffusion limitations.

Furthermore, we do not present any independent proof or physical evidence for the existence of fractals in the analysis of these hybridization systems except to indicate that it has been applied in other areas and that it is a convenient means to make quantitative the degree of heterogeneity that exists on the surface. Thus, in all fairness, and as previously indicated, this is but one possible way by which to analyze this hybridization binding data.

Galvan and Christopoulos (1996) developed a hybridization assay using aequorin as a reporter molecule. These authors indicate that the target DNA was hybridized simultaneously with a capture probe and a detection probe. The detection probe was biotinylated, and the capture probe was immobilized on the wells by a digoxgenin–antidigoxigenin interaction. Aequorin covalently attached to streptavidin or biotinylated aequorin–streptavidin complexes were utilized to determine the hybrids. The authors used this assay procedure to detect mRNA for prostate-specific antigen (PSA).

Figure 10.2a shows the analysis of the biotinylated aequorin by HPLC using a size exclusion column. In this protocol, the streptavidin was first allowed to bind to biotinylated aequorin. Thereafter, the streptavidin–aequorin complex was allowed to bind to the hybrids. A single-fractal analysis

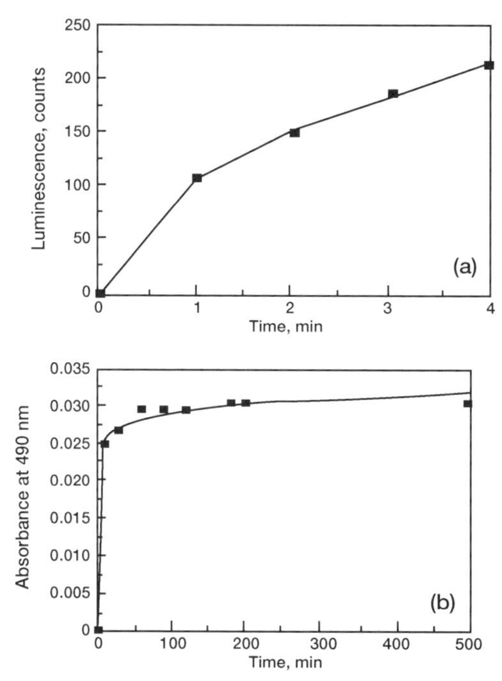

FIGURE 10.2 (a) Hybridization of target DNA in solution to capture probe and detection probe immobilized on a HPLC surface (Galvan and Christopoulos, 1996). (b) Binding of hybridized 1 ng/100 μl RNA from *B. thermospacta* in solution to labeled DNA on an immunosensor (Fliss *et al.*, 1995).

is sufficient to adequately describe the binding kinetics. Table 10.1b shows the values k and D_f. The analysis of much more data of this type is required before any reasonable statement (except that a single-fractal analysis is apparently sufficient) can be made with regard to the degree of heterogeneity that exists on the surface and its influence on the binding rate coefficient. Galvan and Christopoulos indicate that high nonspecific binding is a major sensitivity-limiting factor in this protocol. Nonspecific binding leads to increasing heterogeneity on the surface, and this contributes to the value of the fractal dimension, D_f, that exists on the surface.

Fliss *et al.* (1995) developed a liquid-phase hybridization assay for the detection of *Listeria* species. The authors utilized this anti-DNA–RNA approach to detect *Listeria* in pure cultures and in innoculated meat and meat products. The difference between this study and the studies previously mentioned is that the hybrid between the 784-bp DNA probe (specific for the genus *Listeria*) and a target RNA is first formed in solution. Thereafter, this hybrid binds to monoclonal antibody or antisera raised against hybrid nucleic acids in different immunoassays. These authors emphasize that their approach utilizes the adavantages of both nucleic acid probes and immunological methods. They were able to optimize the hybridization of the rRNA and a specfic DNA probe in solution. Thereafter, they could effectively develop an immunoassay to quantitatively detect the specific DNA–RNA hybrids formed.

Target RNA from *B. thermosphacta* was hybridized in solution at 69°C. Then, the hybridized nucleic acid was detected by labeled and immobilized DNA. A double-sandwich enzyme-linked immunosorbent assay procedure was utilized to quantitatively assess the hybrids formed. Figure 10.2b shows the analysis of the binding curve obtained using $1\,\text{ng}/100\,\mu\text{l}$ of the target RNA. A single-fractal analysis is sufficient to adequately describe the binding kinetics. Table 10.1c shows the values of k and D_f. Only a single piece of data was available from this reference. Once again, more data of this type is required before any reasonable statement can be made with regard to the degree of heterogeneity that exists on the surface and its influence on the binding rate coefficient. Nevertheless, the analysis of even a single piece of data does indicate the extent of heterogeneity that exists on the surface by providing a value of the fractal dimension, D_f (equal to 2.9403). This is a very high value of the fractal dimension (since the maximum value of D_f is 3). This indicates that there is a very high level of heterogeneity that exists on the surface, at least for this case.

Piunno *et al.* (1995) developed a fiber-optic DNA sensor for fluorometric nucleic acid detection. These authors indicate that their goals were to develop a biosensor that could provide results in minutes and to use PCR amplification techniques to provide the high sensitivity required. They covalently immobilized single-strand deoxyribonucleic acid (ssDNA) thymidyloic acid icosanucleotides (dT_{20}) on the surfaces of derivatized quartz optical fibers to develop an optical biosensor. They were then able to hybridize complementary ssDNA (cDNA) or ssRNA (cRNA) from solution using these covalently immobilized oligomers on the biosensor. Figure 10.3a shows the analysis of the binding curves obtained using both a single-fractal [Eq. (10.1a)] and a dual-fractal [Eq. (10.1b)] analysis. Clearly, a dual-fractal analysis is required in this case to provide a reasonable fit. Table 10.2a provides the values of the binding rate coefficients and the fractal dimensions obtained from both single- and dual-fractal analysis. Since the dual-fractal analysis clearly provides a better fit, the single-fractal analysis is not analyzed further. For the dual-fractal analysis, note that a decrease in the fractal dimension value by about 69.8%—from $D_{f_1} = 1.5458$ to $D_{f_2} = 0.4672$—leads to a decrease in the value of the binding rate coefficient by a factor of about 2.6—from $k_1 = 29.127$ to $k_2 = 8.6761$. The D_{f_2} value, which is rather low, indicates that there is a Cantor-like dust on the surface.

Radovich *et al.* (1995) utilized the polymerase chain reaction (PCR) to prepare a DNA template suitable for the direct synthesis of RNA probes. These authors prepared RNA probes specific for the BCR-ABL mRNA, which is characteristic of chronic mylogenous leukemia. Fluorometric hybridization assays were utilized. The amplified (by PCR) product—is the target DNA—was captured onto streptavidin-coated wells. Thereafter, it was hybridized to the RNA probe. Figure 10.3b shows the binding of the mRNA (target) DNA from the BCR-ABL mRNA. In this case, the capture probe was immobilized by the digoxigenin–antidigoxigenin interaction, and the detection probe was biotinylated. Once again, a dual-fractal analysis is required to provide a reasonable fit. The values of the binding rate coefficients and the fractal dimensions obtained from Eqs. (10.1a) and (10.1b) are shown in the Table 10.2b. Only the dual-fractal analysis is analyzed further since it provides a better fit. In this case, note that as the reaction (binding) proceeds, the fractal dimension increases by 36.6%—from $D_{f_1} = 2.1952$ to $D_{f_2} = 3.0$, and the binding rate coefficient increases by a factor of about 12.4—from $k_1 = 3.1745$ to $k_2 = 38.239$. Recall that 3 is the highest value that the fractal dimension can have. This indicates that for this case there is a very high degree of heterogeneity that exists on the reaction surface. Once again, the analysis of more data of this type is required before further reasonable statements can be made about the degree of heterogeneity that exists on the surface and its influence on the binding rate coefficient.

Chiu and Christopoulos (1996) developed a hybridization assay using an expressible DNA encoding firefly luciferase as a label. In this analysis, the target DNA (200 bp) is denatured and hybridized simultaneously with two oligonucleotides. One of the probes is immobilized in microtiter wells by the digoxigenin–antidigoxigenin interaction. The other probe is biotinylated. The

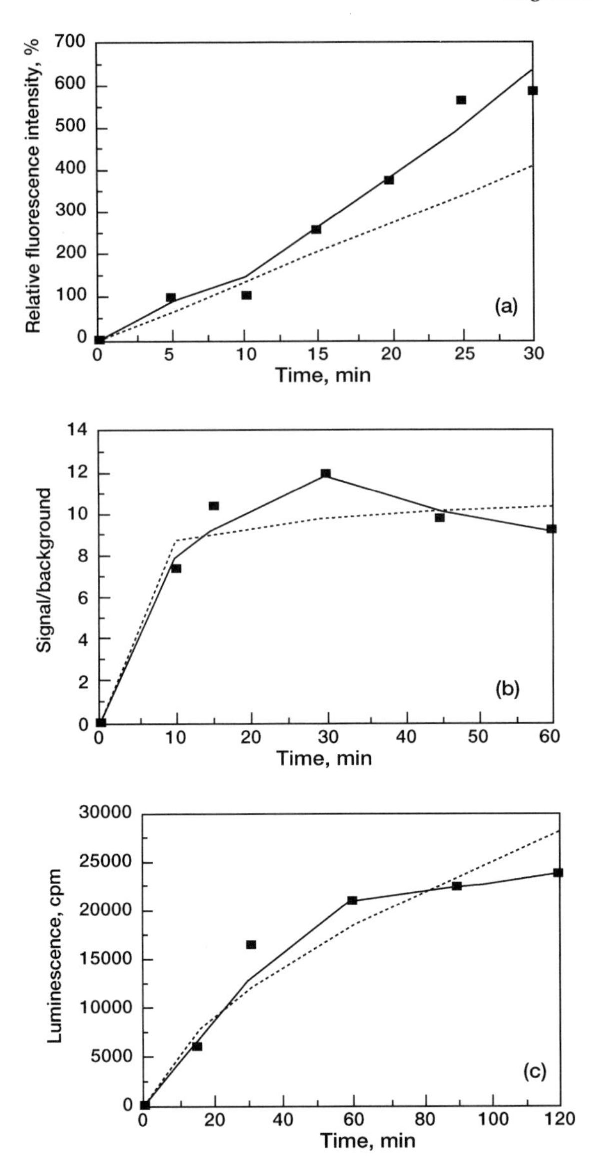

FIGURE 10.3 (a) Hybridization of complementary ssDNA (cDNA) in solution to single-strand deoxyribonucleic acid (ssDNA) thymidylic acid icosanucleotides on derivatized quartz optical fiber (Piunno *et al.*, 1995). (- - -, single-fractal analysis; —, dual-fractal analysis). [This key applies to parts (a), (b), and (c).] (b) Hybridization of BCR-ABL mRNA characteristic of chronic mylegenous leukemia in solution to RNA probe immobilized on a DNA template (Radovich *et al.*, 1995). (c) Hybridization of target DNA (200 bp) from the BCR-ABL mRNA from K652 cells in solution to oligonucleotides immobilized on an immunosensor (Chiu and Christopoulos, 1996).

hybrids are reacted with the streptavidin–luciferase reaction after the formation of the hybrids. Figure 10.3c shows the binding of the target (200 bp) from the BCR-ABL mRNA from K652 cells. Once again, a dual-fractal analysis is required to provide a reasonable fit. The values of the binding rate coefficients and the fractal dimensions obtained from Eqs. (10.1a) and (10.1b) are shown in Table 10.2c. For the dual-fractal analysis, an increase in the fractal dimension by a factor of about 2.11—from $D_{f_1} = 1.2264$ to $D_{f_2} = 2.5948$—leads to an increase in the binding rate coefficient by a factor of about 14.7—from $k_1 = 633.96$ to $k_2 = 9342.19$. In this case, the binding rate coefficient is rather sensitive to the fractal dimension or the degree of heterogeneity that exists on the biosensor surface.

Fliss *et al.* (1995) have also utilized target RNA from *L. monocytogenes* in their liquid-phase hybridization assay for the detection of *Listeria* species. The procedure is the same as presented before in this book, except that now the target RNA is from *L. monocytogenes* whereas previously it was from *B. thermosphacta.* Figure 10.4a–c shows the analysis of the binding curve obtained using 0.01 to 1 ng/100 μl of the target RNA in solution. A dual-fractal analysis is required to adequately describe the binding kinetics. Once again, for each of the three target concentrations in solution, note that an increase in the fractal dimension from D_{f_1} to D_{f_2} leads to an increase in the binding rate coefficient from k_1 to k_2 (see Table 2d).

From the data presented in Table 10.2, no reasonable trends were seen for the binding rate coefficient, k_2, and the fractal dimensions, D_{f_1} and D_{f_2}. However, the binding rate coefficient, k_1, increased with an increase in the target rRNA concentration in the range 0.01 to 1 ng/100 μl. In this rRNA concentration range, k_1, may be given by

$$k_1 = (0.2809 \pm 0.1814)(\text{rRNA})^{0.1868 \pm 0.1418}. \tag{10.2a}$$

Figure 10.5a shows that more data points are required to provide a better fit. There is some deviation, as indicated by the error estimates for both the coefficient and the exponent terms. Nevertheless, Eq. (10.2a) does indicate an increase in the binding rate coefficient, k_1, as the rRNA concentation in solution increases in the concentration range analyzed. The fractional exponent dependence of the binding rate coefficient on the rRNA concentration in solution lends support to the fractal nature of the system.

Figure 10.5b shows the dependence of the binding rate coefficient, k_1, on the fractal dimension, D_{f_1}. From the data presented in Table 2, the binding rate ceofficient, k_1, is given by:

$$k_1 = (0.0545 \pm 0.0581) D_{f_1}^{1.9385 \pm 1.6656}. \tag{10.2b}$$

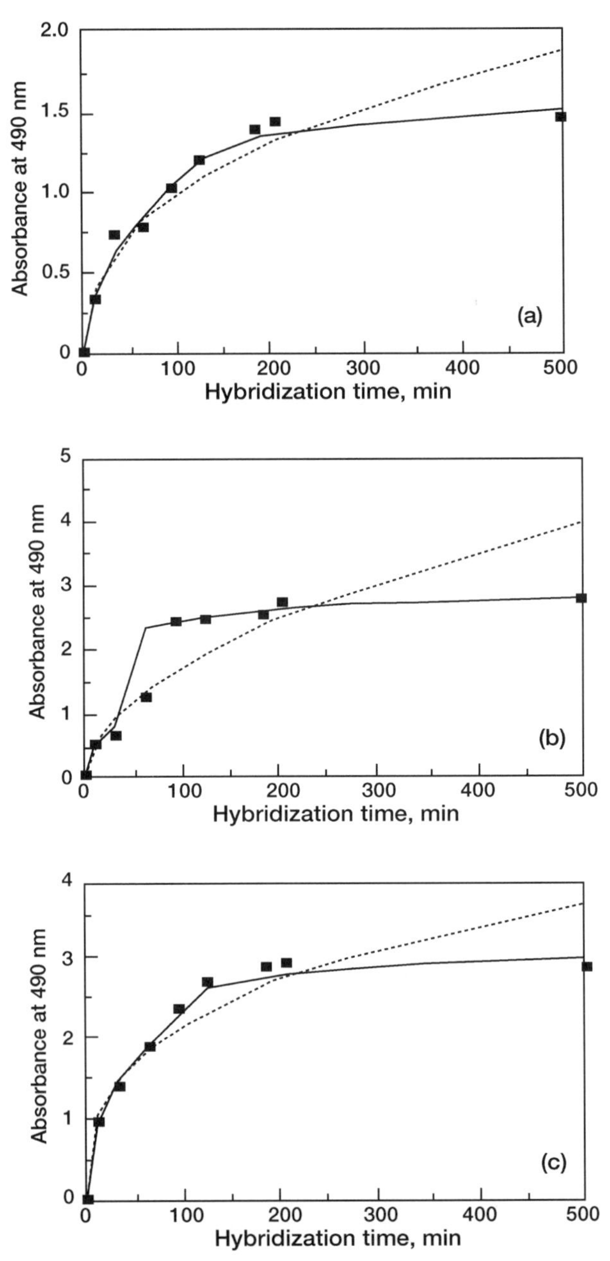

FIGURE 10.4 Hybridization of rRNA from *L. monocytogenes* in solution to labeled DNA immobilized on an immunosensor (Fliss *et al.*, 1995). (a) 0.01 ng/100 μl rRNA; (b) 0.5 ng/100 μl rRNA; (c) 1 ng/100 μl rRNA.

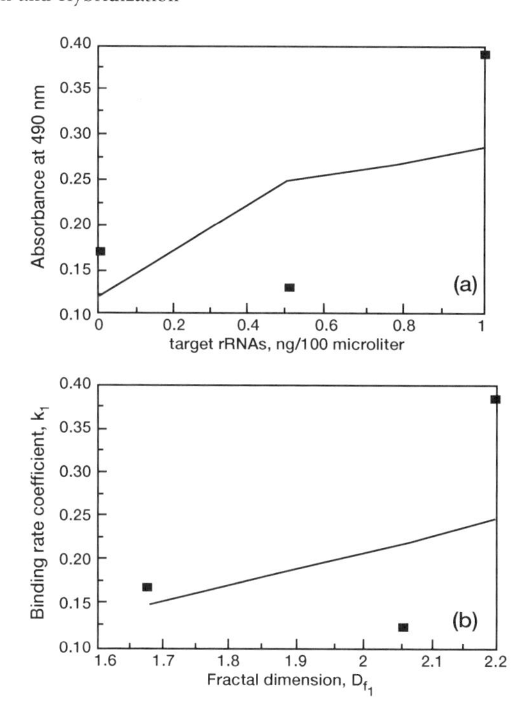

FIGURE 10.5 (a) Influence of the rRNA concentration (in ng/μl) in solution on the binding rate coefficient, k_1, for the hybridization of rRNA from *L. monocytogenes* in solution to labeled DNA immobilized on an immunosensor (Fliss *et al.*, 1995). (b) Influence of the fractal dimension, D_{f_1}

Once again, more data points are required to provide a better fit. There is quite a bit of deviation, which is indicated by the error estimates of the coefficient and the exponent terms. Nevertheless, Eq. (10.2b) is of value since it provides a quantitative measure of the influence of the heterogeneity on the surface on the binding rate coefficient, k_1. The binding rate coefficient, k_1, is sensitive to the fractal dimension, D_{f_1}. Furthermore, the exponent dependence of the binding rate coefficient on the fractal dimension lends further support to the fractal nature of the system.

Bier *et al.* (1997) analyzed the reversible binding of DNA nucleotides to immobilized DNA targets. Streptavidin or avidin was coupled to the surface and used as a bridge for immobilization of the DNA. The authors emphasize that nucleic acids determine complementary strands and are useful in clinical diagnostics, environmental monitoring, and hygiene. They analyzed the binding of 13 mer to 24 mer and 13 mer templates. These authors emphasize that biomolecular interaction analysis (BIA) is a useful means for the determination of analyte-receptor binding. The basis is the use of evanescent field technology in several transducers. For example, one may determine changes in refractive index on a waveguide surface. This leads to clear discrimination between unbound and bound species.

Figure 10.6a shows the binding of 1540-13 target oligonucleotide in

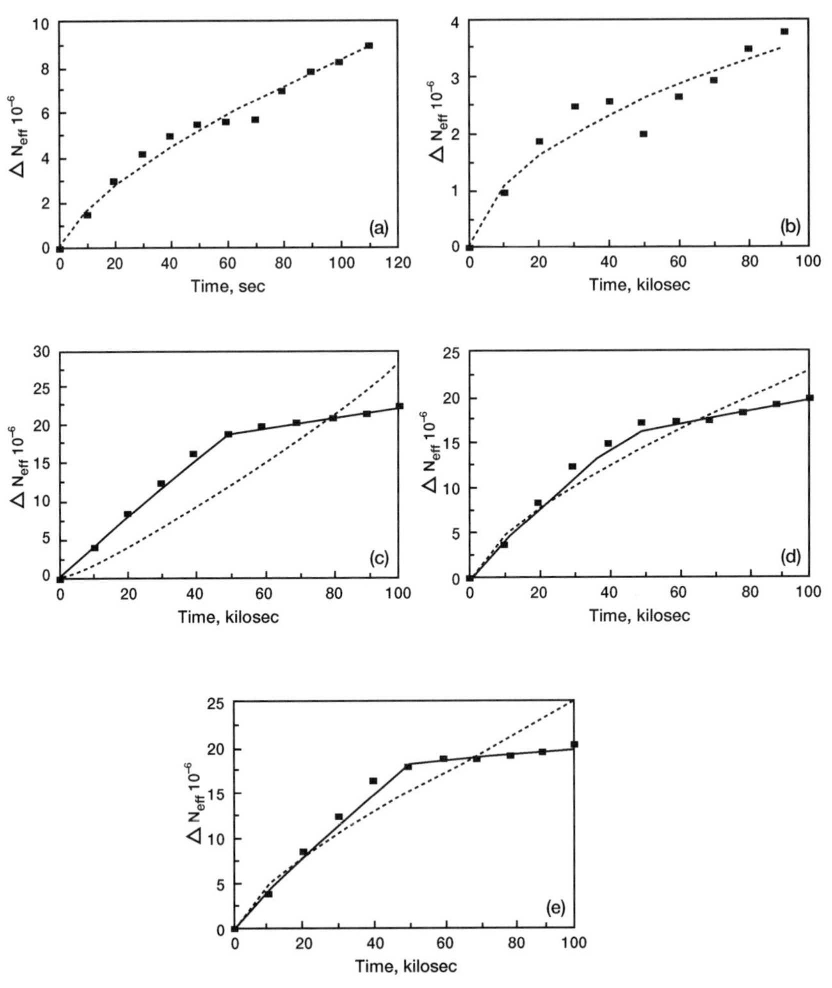

FIGURE 10.6 Hybridization of different 13 mer target oligonucleotides with different matching nucleotides to the 5′-mer-biot-TAG CTA TGG AAT TCC TAG GCA (biot-UP24TE) template immobilized on a grating coupler (Bier *et al.*, 1997) (a) 1540–13; (b) 1539–13; (c) 1543–13; (d) 1541–13; (e) 1542–13. (- - -, single-fractal analysis); —, dual-fractal analysis; ΔN_{eff}, the change in the refractive index measured by the grating coupler).

solution to a 5′-mer-biot-TAG CTA TGG AAT TCC TCG TAG GCA (biot-UP24TE) template immobilized on a grating coupler. The biot-UP24TE used as a template contained the EcoRI restriction sequence in the middle position. The authors selected the other parts of the sequence in an arbitrary but balanced manner, to avoid any self-complementarity. For the data in Fig. 10.6a, the first seven nucleotides were matching. There is excessive scattering in the experimental points in this case. Nevertheless, a single-fractal analysis is sufficient to adequately describe the binding kinetics. The k value (0.36 ± 0.04) is within 95% confidence limits. These values are precise enough that a dual-fractal analysis is not required. The excessive scatter is primarily responsible for the 0.32 to 0.40 range. It is possible in this case that the hybridization mechanism is being controlled not only by diffusion but also by some other means. Table 10.3 shows the values of the binding rate coefficients and the fractal dimensions obtained. The parameter values presented in Table 10.3 (and Table 10.4) were obtained from regression analysis using Sigmaplot (1993) to model the data using Eqs. (10.1a) and (10.1b). Once again, the values presented in Table 10.3 (and Table 10.4) are within 95% confidence limits.

Table 10.3 Influence of Different 13 mers on Binding Rate Coefficients and Fractal Dimensions for Hybridization Binding Kinetics Using a Grating Coupler (Bier *et al.*, 1997)

Analyte (13 mer) in solution/receptor (template) on surface	k	D_f	k_1	k_2	D_{f_1}	D_{f_2}
1540-13/biot-UP24TE, 5 to 10 nt* matching	0.36 ± 0.04	1.37 ± 0.08	na	na	na	na
1539-13/biot-UP24TE 5 to 10 nt matching	0.34 ± 0.06	1.95 ± 0.16	na	na	na	na
1543-13/biot-UP24TE, 5 to 10 nt matching	0.09 ± 0.03	0.50 ± 0.25	0.46 ± 0.05	8.09 ± 0.09	1.10 ± 0.12	2.36 ± 0.05
1541-13/biot-UP24TE, 5 to 10 nt matching	1.06 ± 0.19	1.66 ± 0.15	0.58 ± 0.08	5.64 ± 0.01	1.26 ± 0.17	2.46 ± 0.09
1542-13/biot-UP24TE, 5 to 10 nt matching	0.99 ± 0.20	1.61 ± 0.16	0.51 ± 0.06	9.86 ± 0.12	1.17 ± 0.15	2.69 ± 0.06
1543-13/biot-UP24TE, 10 nt matching	0.28 ± 0.07	0.96 ± 0.18	0.09 ± 0.01	7.09 ± 0.05	0.26 ± 0.13	2.43 ± 0.02
1544-13/biot-UP24TE, 12 nt matching and one central nt mismatch	0.24 ± 0.05	0.85 ± 0.16	0.09 ± 0.08	5.78 ± 0.09	0.23 ± 0.11	2.29 ± 0.05

*Nucleotide.

Table 10.4 Binding Rate Coefficients and Fractal Dimensions for Different Hybridization Binding Analyte Receptor Systems

Analyte in solution/receptor on surface	k	D_f	k_1	k_2	D_{f_1}	D_{f_2}	Reference
22 mer UP22H (4 ng/μl)/5′-biotinylated DNA-oligomer of 24 bases	0.11 ± 0.02	0.62 ± 0.31	0.05 ± 0.01	1.90 ± 0.03	0.06+0.39	1.93 ± 0.12	Bier and Scheller 1996
22 mer (UP22H) (40 ng/μl)/5′-biotinylated DNA-oligomer (complementary)	1.99 ± 0.25	1.64 ± 0.25	na	na	na	na	Bier and Scheller 1996
Target DNA/biotinylated mKTH 1301 and 1302 DNA and 125I-labeled plasmid pKTH 1300 DNA	1765 ± 246	2.10 ± 0.18	na	na	na	na	Syvant *et al.* 1986
Oligonucleotide 6/single-strand DNA (outer sequence)	167 ± 27.9	1.60 ± 0.17	84.6 ± 1.15	1350 ± 18.6	1.07 ± 0.03	2.73 ± 0.05	Nilsson *et al.* 1995

Figure 10.6b shows the binding of 1593-13 target oligonucleotide to the biot-UP24TE template immobilized on a grating coupler. In this case, the first six nucleotides were matching. Again, there is excessive scattering in the experimental points in this case. Nevertheless, once again, a single-fractal analysis is sufficient to adequately describe the binding kinetics. Table 10.3 shows the values of the binding rate coefficients and the fractal dimension obtained. In this case, the k value was 0.34 ± 0.06. The 95% confidence limit indicates that 95% of the k values will lie between 0.28 and 0.40. Thus these values are precise enough and a dual-fractal analysis is not required. The excessive scatter is primarily responsible for the 0.28 to 0.40 range. It is quite possible that the hybridization mechanism is controlled by some other means also. Note that as the number of matching nucleotides increases from six (1539-13) to seven (1540-13), the fractal dimension, D_f, decreases by about 29.7%—from 1.95 to 1.37—and the binding rate coefficient, k, increases by about 5.8%—from 0.34 to 0.36. In this case, a decrease in the fractal dimension (or a decrease in the degree of heterogeneity on the surface) leads to a slight increase in the binding rate coefficient. No explanation is presently offered for this behavior, except that a decreasing surface heterogeneity leads to an increase in the binding rate coefficient. The authors were careful to avoid steric hindrance by selecting sequences that would hybridize at the distal end rather than at the immobilized end. This decreasing or increasing surface roughness leads to a temporal binding rate coefficient.

Figure 10.6c shows the binding of the target nucleotide 1543-13 to the biot-UP24TE template immobilized on a grating coupler. Ten nucleotides were matching in this case. Here, a dual-fractal analysis is required to provide an adequate fit for the binding curve. The values of the binding rate coefficients, k, k_1, and k_2, and the fractal dimensions, D_f, D_{f_1}, and D_{f_2}, for a single- and a dual-fractal analysis are given in Table 10.3. Similarly, Fig. 10.6d shows the binding curve of the target 1541-13 to the biot-UP-24TE template immobilized on a grating coupler. In this case, eight nucleotides were matching. The values of the binding rate coefficients, k, k_1, and k_2, and the fractal dimensions, D_f, D_{f_1}, and D_{f_2}, are given in Table 10.3. Note that for a dual-fractal analysis when one goes from ten nucleotides matching (1543-13) to eight nucleotides matching (1541-13), the binding rate coefficient, k_1, and the fractal dimensions, D_{f_1}, and D_{f_2}, all exhibit increases. Only k_2 exhibits a decrease in value.

Figure 10.6e shows the binding of the target 1542-13 to the biot-UP24TE template immobilized on the grating coupler. In this case, nine nucleotides were matching. Again, a dual-fractal analysis was required to provide an adequate fit. The values of the binding rate coefficients, k, k_1, and k_2, and the fractal dimensions, D_f, D_{f_1}, and D_{f_2}, are given in Table 10.3. Compare Fig. 10.6c and 10.6e and note that when one goes from ten nucleotides matching

(1543-13) to nine nucleotides matching (1543-13), the binding rate coefficients, k_1, and k_2, and the fractal dimensions, D_{f_1} and D_{f_2}, all exhibit increases.

Figure 10.7a shows the binding of the target 1543-13 to the biot-UP24TE template immobilized on a grating coupler. In this case, ten nucleotides were matching. A dual-fractal analysis is required to provide an adequate fit. The values of the binding rate coefficients, k, k_1, and k_2, and the fractal dimensions, D_f, D_{f_1}, and D_{f_2}, are given in Table 10.3. Compare the results shown in Figs. 10.6c and 10.7a, both of which involve the target 1543-13. In one case, there were nine nucleotides matching (Fig. 10.6c), and in the other ten (Fig. 10.7a). In this case, an increase by 1 in the number of matching nucleotides leads to a decrease in the values of k_1, k_2, and D_{f_1}. Only D_{f_2} exhibits an increase.

Finally, Figure 10.7b shows the binding of the target 1544-13 to the biot-UP24TE template immobilized on a grating coupler. In this case, twelve

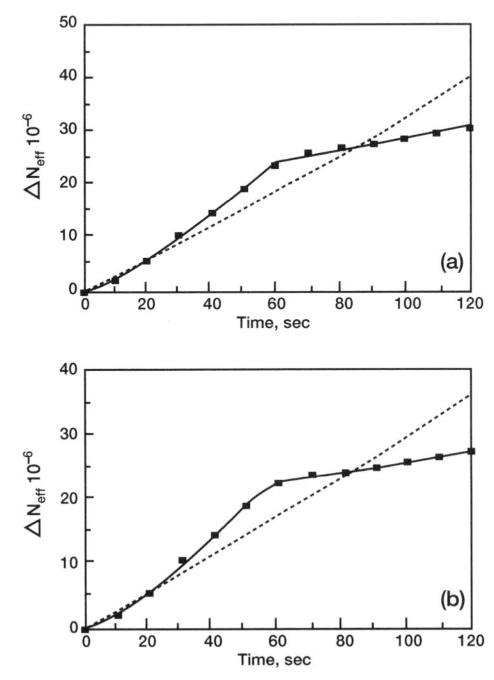

FIGURE 10.7 Hybridization of 13 mer target oligonucleotides with different matching nucleotides to the biot-UP24TE template immobilized to the grating coupler (Bier *et al.*, 1997). (a) 1543–13, 10 nucleotides matching; (b) 1544–13, 12 nucleotides matching. (- - -), single-fractal analysis; — dual-fractal analysis; ΔN_{eff}, the change in the refractive index measured by the grating coupler.)

nucleotides were matching, and there was one central mismatch. Once again, a dual-fractal analysis is required to provide an adequate fit. The values of the binding rate coefficients, k, k_1, and k_2, and the fractal dimensions, D_f, D_{f_1}, and D_{f_2}, are given in Table 10.3. Compare the results of Figs. 10.7a and 10.7b. As one goes from ten nucleotides matching (1543-13) to twelve nucleotides matching and one central mismatch (1544-13), the binding rate coefficients, k_1 and k_2, and the fractal dimensions D_{f_1} and D_{f_2}, all exhibit increases. From the data presented in Table 10.3, note that there is an increase in the binding rate coefficients, k_1 and k_2, with an increase in the fractal dimensions, D_{f_1} and D_{f_2}, respectively. For the data presented in Table 10.3 for the different targets, k_1, is given by

$$k_1 = (0.4293 \pm 0.0215) D_{f_1}^{1.11 \pm 0.03}. \tag{10.3}$$

Figure 10.8 indicates that the fit is quite reasonable considering that different targets are utilized. The binding rate coefficient is, however, not very sensitive to the fractal dimension, D_{f_1}, or the degree of heterogeneity that exists on the surface. This is because of the low value of the exponent. The fractional exponent dependence of the binding rate coefficient lends support to the fractal nature of the system.

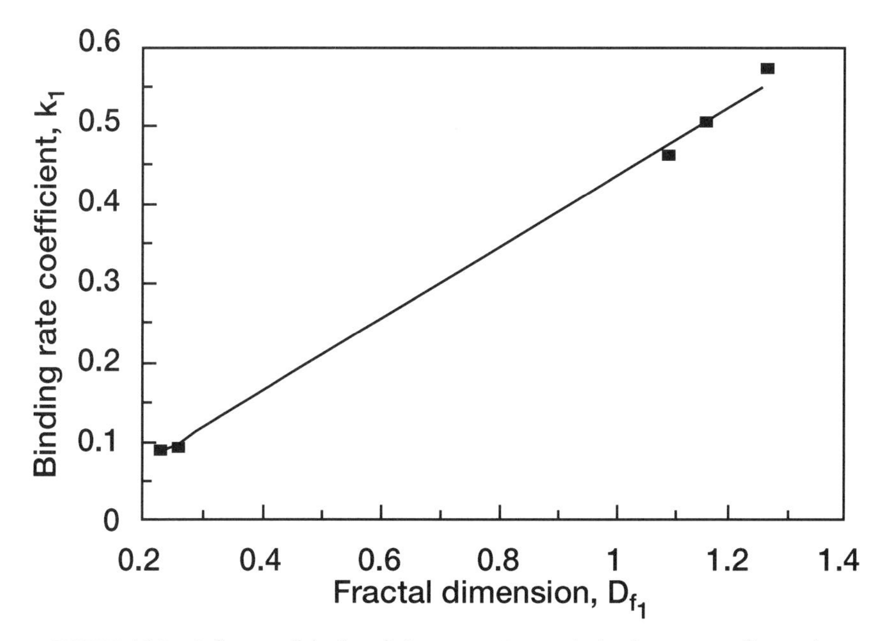

FIGURE 10.8. Influence of the fractal dimension, D_{f_1} on the binding rate coefficient, k_1.

Bier and Scheller (1996) also analyzed the binding of 4 ng/μl of a 22 mer UP22H to a 5′-biotinylated DNA-oligomer of 24 bases immobilized on a grating coupler. Figure 10.9a shows that the binding needs to be described by a dual-fractal analysis because the "flex" exhibited by the binding rate cannot be modeled adequately by a single-fractal analysis. Only smooth binding curves are adequately described by a single-fractal analysis. In this case, one cannot eliminate experimental uncertainty. The values of the binding rate coefficients, k, k_1, and k_2, and the fractal dimensions, D_f, D_{f_1}, and D_{f_2}, are given in Table 10.4. Figure 10.9b shows the binding of 40 ng/μl of the 22 mer UP22H to the 5′-biotinylated DNA-oligomer of 24 bases immobilized to a grating coupler. In this case, a single-fractal analysis is sufficient since the binding curve is smooth and the value of k obtained is 1.99 ± 0.25. The k values, with 95% confidence limits, will lie between 1.74 and 2.24. Note that

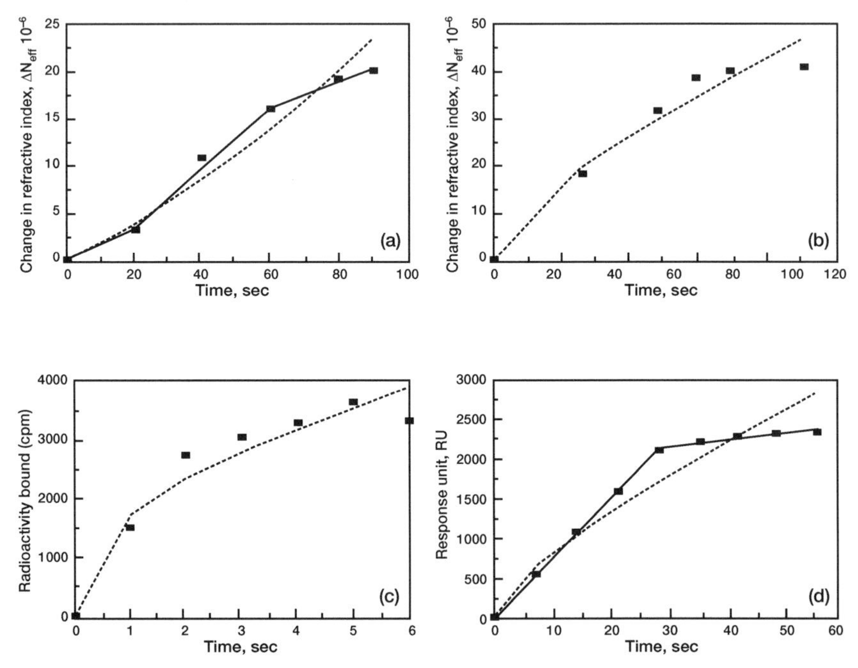

FIGURE 10.9 Hybridization of different analyte-receptor systems. (a) 4 ng/μl 22 mer UP22H to 5′-biotinylated DNA-oligomer of 24 bases immobilized on a grating coupler (Bier and Scheller, 1996); (b) 40 ng/μl 22 mer UP22H to 5′-biotinylated DNA-oligomer of 24 bases immobilized on a grating coupler (Bier and Scheller, 1996); (c) target DNA to a biotinylated mKTH 1301 and 1302 DNA and ^{125}I-labeled plasmid pKTH 1300 DNA immobilized on agarose using streptavidin (Syvanen *et al.*, 1986); (d) oligonuleotide 6 to single-strand DNA immobilized to a surface (outer sequence) (Nilsson *et al.*, 1995).

an increase in the analyte (22 mer UP22H) concentration leads to a change in the mechanism of binding since at the higher concentration only a single-fractal analysis is required (as compared to the dual-fractal analysis required at the lower concentration). No explanation is offered at present for this change in the mechanism, although experimental uncertainty may be a cause.

Syvanen *et al.* (1986) analyzed the binding of a target DNA to a biotinylated mKTH 1301 and 1302 DNA and ^{125}I-labeled plasmid pKTH 1300 DNA immobilized on agarose using streptavidin. These authors analyzed the hybridization of the nucleic acids by affinity-based hybrid collection. They indicate that their procedure is quite sensitive and has a detection limit of less than 1 attomole. Figure 10.9c shows that a single-fractal analysis is sufficient to adequately describe the binding kinetics. The values of k, and D_f, are given in Table 10.4. Only a single example of binding kinetics was available. Thus, no comparison is possible for this case. The k value obtained is 1745 ± 246, which indicates that within 95% confidence limits the k value will lie between 1499 and 1991. The visual fit is reasonable. A better fit could be obtained by using a dual-fractal analysis (four-parameter model).

Nilsson *et al.* (1995) monitored DNA manipulations using biosensor technology. These authors utilized a biosensor based on surface plasmon resonance (SPR) for the detection of changes in the refractive index with time on a sensor surface. The unit for measurement is the resonance unit (RU), which is proportional to the mass of the molecules bound to the surface. The authors utilized the high-affinity streptavidin–biotin system to immobilize DNA fragments to the sensor surface. This system was utilized to monitor DNA strand separation, DNA hybridization, and enzymatic modifications.

The influence of the relative distance from the streptavidin/dextran surface on the hybridization kinetics was analyzed by Nilsson *et al.* (1995) by noting the binding to different regions of the immobilized DNA strand. Figure 10.9d shows the binding of oligonucleotide 6 to the single-strand DNA immobilized on the surface. Oligonucleotide 6 hybridized to the sequence located at the 3′ end, the outermost sequence. A dual-fractal analysis is required to adequately describe the binding kinetics. The values of the binding rate coefficients, and the fractal dimensions, are given in Table 10.4. It is interesting to compare the values of the binding rate coefficients and the fractal dimensions obtained when the oligomer binds either to the outer (oligomer 6) or to the middle (oligomer 4) sequence. As one goes from oligomer 6 to oligomer 4 there is an increase in the fractal dimension, D_{f_1}, by about 29.3%—from 1.07 to 1.38. This leads to an increase in the binding rate coefficient, k_1, by about 43%—from 84.6 to 121. The values of the fractal dimension, D_{f_2}, for both cases are close to each other (2.73 and 2.75). However, the k_2 value for oligonucleotide 6 (equal to 1350) is higher by about 5.5% than for oligonucleotide 4 (equal to

1279). One might reasonably say that later on in the reaction (when D_{f_2} is involved), there is a very small difference in the degree of heterogeneity that exists on the surface for these two cases, which leads to very small changes in the binding rate coefficient.

Liu *et al.* (2000) utilized a molecular beacon DNA sensor to analyze the binding of complementary oligonucleotides and 1-base mismatch oligonucleotides in solution. The advantage of their technique is that no dye-labeled target molecule or intercalation agent is required. These authors indicate that hairpin-shaped oligonucleotides that report the presence of specific nucleic acids are utilized as molecular beacons. Figure 10.10a shows the curves obtained using Eq. (10.1a) for the binding of 30 nM complementary oligonucleotides in solution to a biotinylated ssDNA (molecular beacon) immobilized on an ultra-small optical fiber probe. A dual-fractal analysis is required to adequately describe the binding kinetics.

Table 10.5 shows (a) the values of k and D_f for a single-fractal analysis, and (b) the values of k_1, and k_2, D_{f_1}, and D_{f_2} for a dual-fractal analysis. The values of the binding rate coefficients and the fractal dimensions presented

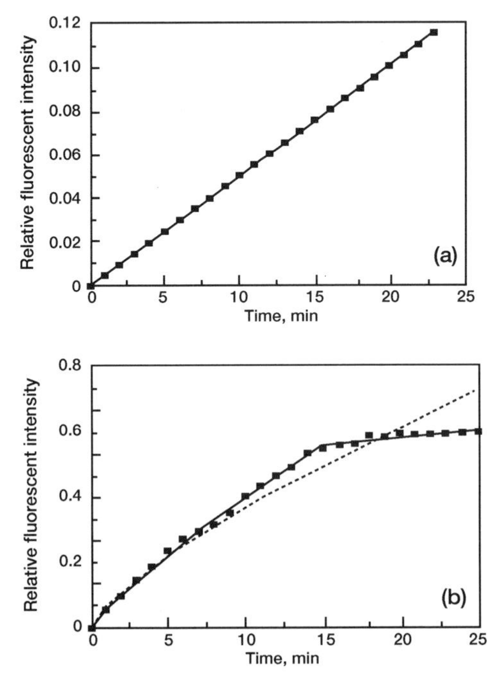

FIGURE 10.10 Binding of oligonucleotide in solution to biotinylated ssDNA molecular beacon immobilized on an SPR biosensor surface (Liu *et al.*, 2000) (a) 30 nM complementary oligonucleotide; (b) 1-base mismatch oligonucleotide.

Table 10.5 Influence of Noncomplementary Oligonucleotide and 1-Base Mismatch Oligonucleotide on the Binding Rate Coefficients and Fractal Dimensions for a Hybridization Reaction Using an Immobilized Molecular Beacon Optical Fiber Biosensor (Liu *et al.*, 2000)

Analyte in solution/receptor on surface	k	D_f	k_1	k_2	D_{f_1}	D_{f_2}
30 nM complementary oligonucleotide/ biotinylated ssDNA molecular beacon	0.066 ± 0.01	1.52 ± 0.05	0.058 ± 0.002	0.389 ± 0.006	1.32 ± 0.03	2.74 ± 0.06
30 nM 1-base mismatch oligonucleotide/ biotinylated ssDNA molecular beacon	0.005 ± 0.00	2.00 ± 0.00	na	na	na	na

in Table 10.5 were obtained from a regression analysis using Sigmaplot (1993) to model the experimental data using Eq. (10.1a), wherein (analyte · receptor) = kt^p for the binding step. The binding rate coefficient values presented in the table are within 95% confidence limits. For example, for the binding of 30 nM complementary oligonucleotide in solution to a biotinylated ssDNA molecular beacon immobilized on the optical fiber k_1 is equal 0.058 ± 0.002. The 95% confidence limit indicates that 95% of the k_1 values will lie between 0.056 and 0.060. This indicates that the values presented are precise and significant. The curves presented in the figures are theoretical curves.

Figure 10.10b shows the curves obtained for the binding of 30 nM 1-base mismatch oligonucleotide in solution to a biotinylated ssDNA immobilized on the ultra-small optical fiber probe. In this case, a single-fractal analysis is adequate to describe the binding kinetics. There is a change in the binding mechanism as one goes from the binding of complementary oligonucleotide (Fig. 10.10a) to the binding of the 1-base mismatch oligonucleotide (Fig. 10.10b), since a dual-fractal and a single-fractal analysis, respectively, are required to describe the binding kinetics.

It is surprising to note that the binding of a 1-base mismatch oligonucleotide in solution to the molecular beacon biosensor requires only a single-fractal (simple) mechanism whereas the binding of a complementary oligonucleotide in solution to the molecular beacon biosensor requires a dual-fractal (complex) mechanism. At the outset, it would appear that this should be the other way around. However, no explanation is offered at present for this.

Wink *et al.* (1999) have analyzed the binding of plasmid DNA in solution to a cationic polymer immobilized on a surface plasmon resonance (SPR) biosensor. Figure 10.11a shows the binding of 2 μg/ml plasmid DNA in solution to a nonthiolated poly(L-lysine) polymer layer immobilized on an SPR biosensor surface. A dual-fractal analysis is required to adequately describe the binding kinetics. The values of the binding rate coefficient, k, and the fractal dimension, D_f, for a single-fractal analysis, as well as the binding rate coefficients, k_1 and k_2, and the fractal dimensions, D_{f_1} and D_{f_2}, for a dual-fractal analysis are given in Table 10.6a. Note that for the dual-fractal analysis as the fractal dimension for binding increases from D_{f_1} to D_{f_2}, the binding rate coefficient increases from k_1 to k_2. A 30.8% increase in the fractal dimension—from $D_{f_1} = 2.14$ to $D_{f_2} = 2.80$—leads to an increase in the binding rate coefficient by a factor of 10.47—from $k_1 = 10.6$ to $k_2 = 111$. In other words, an increase in the degree of heterogeneity on the surface as the reaction proceeds leads to an increase in the binding rate coefficient.

Figure 10.11b shows the binding of 4 μg/ml of plasmid DNA in solution to a nonthiolated poly(L-lysine) polymer layer immobilized on an SPR biosensor

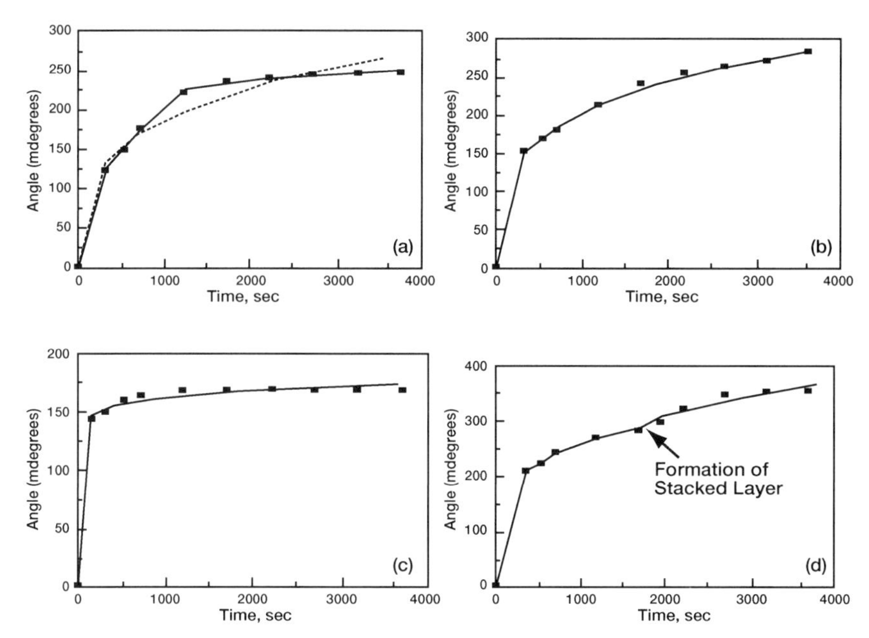

FIGURE 10.11 Binding of plasmid DNA (in μg/ml) in solution to nonthiolated poly(L-lysine) immobilized on an SPR biosensor surface (Wink *et al.*, 1999). (a) 2; (b) 4; (c) 8; (d) 10.

surface. A single-fractal analysis is sufficient to adequately describe the binding kinetics. The values of the binding rate coefficients and the fractal dimensions are given in Table 10.6a. Note that an increase in the plasmid DNA (analyte) concentration in solution from 2 to 4 μg/ml leads to a "simpler" binding mechanism. At the lower (2 μg/ml) analyte concentration a dual-fractal analysis is required, whereas at the higher (4 μg/ml) analyte concentration a single-fractal analysis is sufficient.

Figure 10.11c shows the binding of 8 μg/ml of plasmid DNA in solution to a nonthiolated poly(L-lysine) polymer layer immobilized on an SPR biosensor surface. Once again, a single-fractal analysis is sufficient to adequately describe the binding kinetics. The values of the binding rate coefficients and the fractal dimensions are given in Table 10.6a. Figure 10.11d shows the binding of 10 μg/ml of plasmid DNA in solution to a nonthiolated poly(L-lysine) polymer layer immobilized on an SPR biosensor surface. Here too, a single-fractal analysis is sufficient to adequately describe the binding kinetics. The values of the binding rate coefficients and the fractal dimensions are given in Table 10.6a. In this case, however, due to the high concentration of analyte in solution, there is a formation of a "stacked layer." Wink *et al.* (1999 clearly indicate the presence of a stacked layer. During the initial period, (at t close to 0), the binding is adequately described by a single-

Table 10.6. Influence of Plasmid DNA Concentration on Binding Rate Coefficients and Fractal Dimensions for Its Binding to (a) nonthiolated poly(L-lysine) and (b) poly[2-(dimethylamino) ethylmethacrylate] (p-DMAEMA) Immobilized on a Surface Plasmon Resonance Biosensor (Wink *et al.*, 1999)

Analyte in solution/receptor on surface	k	D_f	k_1	k_2	D_{f_1}	D_{f_2}
(a) 2 μ/ml plasmid DNA/nonthiolated poly (L-lysine)	26.5 ± 2.10	2.43 ± 0.06	10.6 ± 0.28	111 ± 1.49	2.14 ± 0.05	2.80 ± 0.03
4 μg/ml plasmid DNA/nonthiolated poly (L-lysine)	36.5 ± 0.81	2.50 ± 0.02	na	na	na	na
8 μg/ml plasmid DNA/nonthiolated poly (L-lysine)	116.1 ± 2.73	2.90 ± 0.01	na	na	na	na
10 μg/ml plasmid DNA/nonthiolated poly (L-lysine)	65.2 ± 0.92	2.59 ± 0.02	$k_{stack} = 34.9$	na	$D_{f,\ stack} = 2.60$	na
(b) 2 μg/ml plasmid DNA/5%-thiolated p-DMAEMA	3.40 ± 0.26	2.38 ± 0.06	na	na	na	na
6 μg/ml plasmid DNA/5%-thiolated p-DMAEMA	10.4 ± 0.34	2.43 ± 0.03	na	na	na	na
10 μg/ml plasmid DNA/5%-thiolated p-DMAEMA	44.7 ± 0.80	2.61 ± 0.01	na	na	na	na

fractal analysis. After this initial phase is over, there is a distinct formation of a stacked layer. During the stacked layer, the binding rate coefficient, k, is lower in value (equal to 34.9) than in the initial phase (equal to 65.2). However, the fractal dimension during the stacked phase (equal to 2.60) is almost the same as the fractal dimension during the initial phase (equal to 2.59).

In other words, in this case, even though the fractal dimension remains almost the same during the two phases (initial and stacked), there is a 46.47% decrease in the binding rate coefficient from the initial to the stacked phase of binding. This is of interest since it provides a possible means for manipulating (in this case, decreasing) the binding rate coefficient. Only a single example is provided, and more data needs to be analyzed using the stacked layer to see whether one may manipulate binding rate coefficients in desired directions utilizing one or even two (or more) stacked layers. Note that since the binding of the initial layer (at t close to 0) and that of the stacked layer may be described by a single-fractal analysis, this indicates that there is, at least, some similarity in the binding mechanism.

Figure 10.12a shows the binding of 2 μg/ml plasmid DNA in solution to 5%-thiolated pDMAEMA immobilized on an SPR biosensor surface (Wink *et al.*, 1999). A single-fractal analysis is adequate to describe the binding kinetics. The values of k and D_f are given in Table 10.6b. Figures 10.12b and 10.12c show the binding of 6 and 10 μg/ml DNA in solution to 5%-thiolated pDMAEMA immobilized on an SPR biosensor surface. Again, a single-fractal analysis is adequate to describe the binding kinetics, and the values of the binding rate coefficient and the fractal dimensions are given in Table 10.6b.

Figure 10.13a and Table 10.6 show that as the plasmid concentration in solution increases from 2 to 10 μg/ml the binding rate coefficient, k, increases. In the 2–10 μg/ml plasmid concentration range, k is given by

$$k = (1.050 \pm 0.702)[\text{plasmid DNA}]^{1.51 \pm 0.44}. \qquad (10.4a)$$

The binding rate coefficient is mildly sensitive to the plasmid DNA concentration in solution in the range analyzed and exhibits close to a one-and-one-half-order dependence. More data points are required to more firmly establish this equation. Nevertheless, Eq. (10.4a) is of value since it provides a quantitative indication of how the binding rate coefficient, changes with plasmid concentration in solution.

Figure 10.13b and Table 10.6 show that as the plasmid DNA concentration in solution increases from 2 to 10 μg/ml, the fractal dimension, D_f increases. In the 2–10 μg/ml plasmid DNA concentration range, D_f is given by

$$D_f = (2.28 \pm 0.08)[\text{plasmid DNA}]^{0.05 \pm 0.03}. \qquad (10.4b)$$

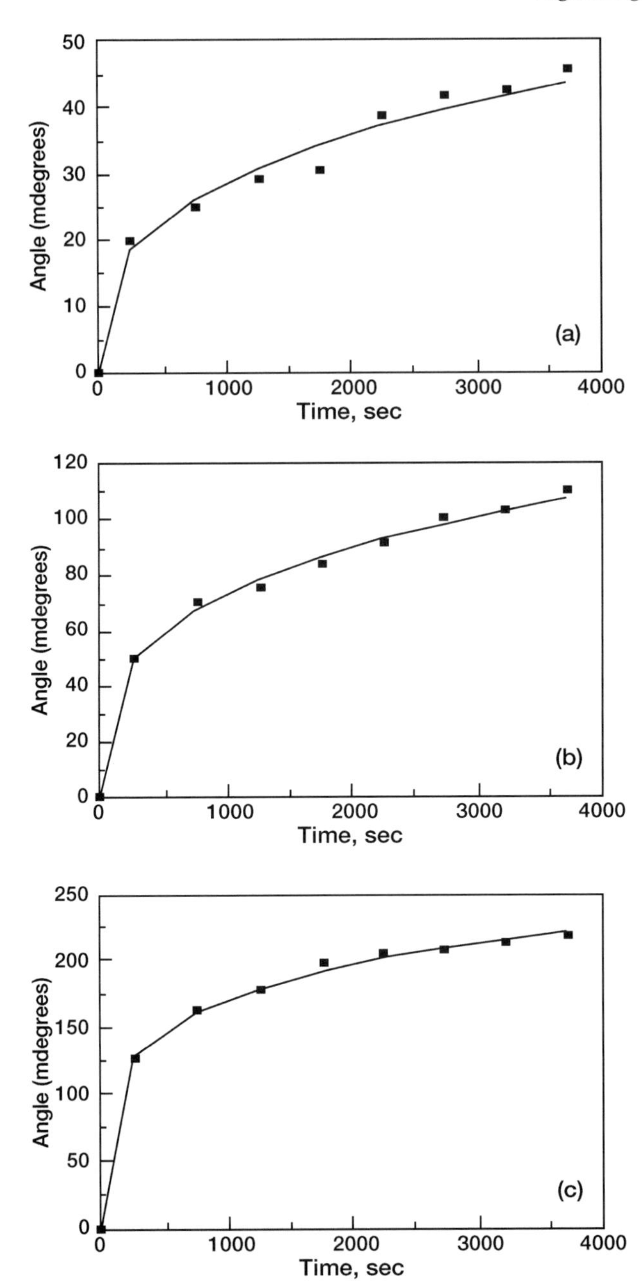

FIGURE 10.12 Binding of plasmid DNA (in μg/ml) in solution to 5%-thiolated poly(L-lysine) immobilized on an SPR biosensor surface (Wink *et al.*, 1999). (a) 2; (b) 6; (c) 10.

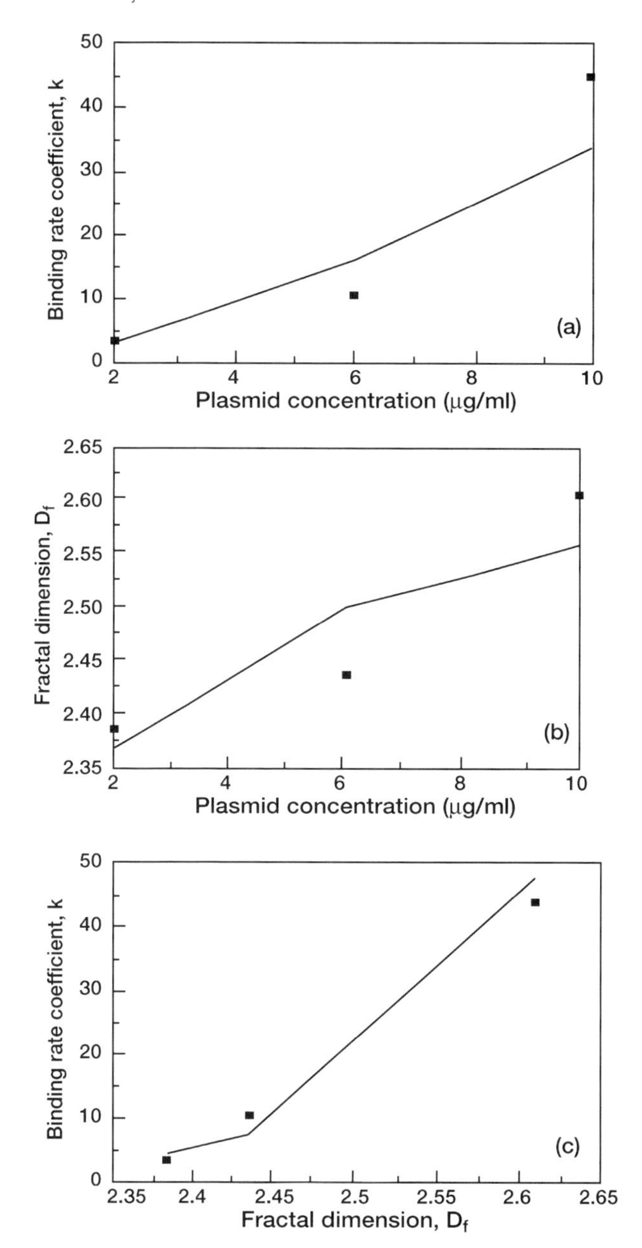

FIGURE 10.13 (a) Increase in the binding rate coefficient, k, with an increase in the plasmid DNA concentration; (b) increase in the fractal dimension, D_f, with an increase in the plasmid DNA concentration; (c) increase in the binding rate coefficient, k, with an increase in the fractal dimension, D_f.

The fractal dimension, exhibits only a very low dependence on the plasmid DNA concentration in solution, increasing very slowly with an increase in the plasmid DNA concentration in solution.

Figure 10.13c and Table 10.6 show that k increases as D_f increases. For the data presented in Table 10.6, k is given by

$$k = (3.57\text{E} - 10 \pm 1.78\text{E} - 10) D_f^{26.73 \pm 6.05} \quad (10.4\text{c})$$

For the three data points presented in Figure 10.13c, the fit is quite reasonable. More data points would more firmly establish this equation. The binding rate coefficient, is extremely sensitive to the degree of heterogeneity that exists on the surface, as noted by the very high value of the exponent. Note also the very low value of the coefficient (equal to 3.57E-10).

Wink *et al.* (1999) also analyzed the reversibility of plasmid DNA–pDMAEMA complex formation. At a pH of 7.4, the plasmid DNA in solution bound to the 5%-thiolated pDMAEMA immobilized on the SPR biosensor surface. As the pH was changed to 8.8, the plasmid DNA–pDMAEMA complex on the SPR biosensor surface dissociated. The authors further showed that the plasmid DNA in solution could be bound to and dissociated from the pDMAEMA immobilized on the SPR biosensor surface by changing the pH of the solution from 7.4 to 8.8 and vice versa. They showed such data for two cycles.

Figure 10.14 shows that a dual-fractal analysis is required to adequately describe the binding kinetics. However, a single-fractal analysis is sufficient to describe the dissociation kinetics. The values of the binding rate coefficient, k, and the fractal dimension, D_f, for a single-fractal analysis; the binding rate coefficients, k_1 and k_2, and the fractal dimensions, D_{f_1} and D_{f_2}, for a dual-fractal analysis; and the dissociation rate coefficient, k_d, and the fractal dimension for dissociation, $D_{f,d}$, are given in Table 10.7. Note that as the fractal dimension for binding increases by about 70%—from 1.76 to a maximum value of 3.0—the binding rate coefficient, increases by a factor of 348.9. Also note that the fractal dimension for dissociation, $D_{f,d} = 1.31$, is smaller than both the fractal dimensions for binding ($D_{f_1} = 1.76$ and $D_{f_2} = 3.0$). Finally, it is of interest to present values of the affinity, K, defined by k_d/k: $K_1(= k_d/k_1) = 0.129$ and $K_2(= k_d/k_2) = 0.0026$.

Compare the binding of plasmid DNA in solution to nonthiolated and 5%-thiolated pDMAEMA immobilized on the sensor surface. From Table 10.6, we note that comparisons can be made for the 2 and 10 μg/ml plasmid DNA concentration in solution. The binding of 2 μg/ml plasmid DNA in solution to the nonthiolated pDMAEMA immobilized on the SPR biosensor surface requires a dual-fractal analysis to adequately describe the binding kinetics. The binding of 2 μg/ml plasmid DNA in solution to 5%-thiolated pDMAEMA

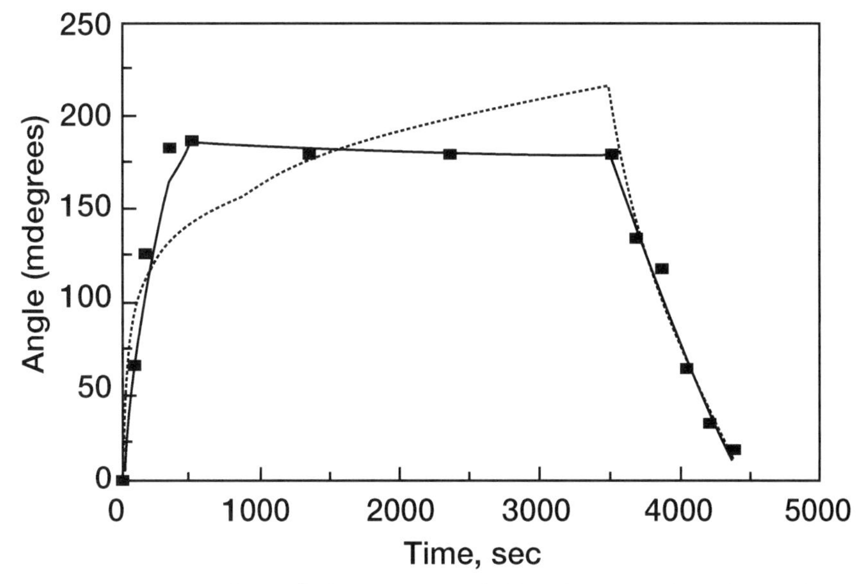

FIGURE 10.14 Reversibility of plasmid DNA-poly[2-(dimethylamino)ethylmethacrylate] (p-DMAEMA) complex binding on an SPR biosensor surface (Wink *et al.*, 1999) (--- single-fractal analysis; — dual-fractal analysis).

immobilized on the SPR biosensor surface requires a single-fractal analysis. This indicates that there is a change in the binding mechanism as one goes from the nonthiolated pDMAEMA immobilized in the SPR surface to the 5%-thiolated pDMAEMA immobilized on the SPR surface. That a dual-fractal analysis is required for the nonthiolated case indicates that in this case a complex binding mechanism is involved.

It is also of interest to compare the binding of 10 μg/ml plasmid DNA in solution to the nonthiolated as well as the thiolated case. For both cases, a single-fractal analysis is sufficient to adequately describe the binding kinetics when t is close to 0. Note that although the values of D_f are close to each other (2.59 for the nonthiolated and 2.61 for the thiolated case), the values of the binding rate coefficient are quite different from each other. The binding rate coefficient for the nonthiolated surface is higher than the binding to the thiolated surface by 44.8%. Note, however, that for the nonthiolated case there is a stacked layer of binding, as indicated earlier.

Table 10.7 **Influence of Reversibility of Plasmid DNA/poly[2-(dimethyl)ethylmethacrylate] (p-DMAEMA) Complex Formation on Binding, Dissociation, and Fractal Dimensions Using an SPR Biosensor (Wink *et al.*, 1999)**

Analyte in solution/receptor on surface	k	D_f	k_1	k_2	k_d	D_{f_1}	D_{f_2}	$D_{f,d}$
Plasmid DNA/5% p-DMAEMA	36.9 ± 12.6	2.57 ± 0.17	4.46 ± 0.83	217 ± 2.9	0.574 ± 0.08	1.76 ± 0.25	3.0 ± 0.0	1.31 ± 0.21

10.4. CONCLUSIONS

A single- and a dual-fractal analysis is presented for solid- and liquid-phase DNA hybridization utilized for biosensor or immunoassay purposes. The fractal analysis provides a quantitative indication of the state of disorder (fractal dimension) and the binding rate coefficient on the surface for the DNA–RNA binding systems analyzed. The fractal dimension value provides a quantitative measure of the degree of heterogeneity that exists on the biosensor or immunoassay surface. Initially, a single-fractal analysis was utilized to fit the DNA–RNA binding curve. This was done with the regression provided by Sigmaplot (1993). Only when the fit was not adequate was a dual-fractal analysis used. This was indicated by the regression analysis (sum of the $(\text{error})^2$) provided by Sigmaplot, where the error is the difference between the theoretical predicted value and the experimental value. This was further corroborated by visual inspection of the figures presented for fitting the data by a single- and a dual-fractal analysis. In addition, fractal dimensions for the dissociation step, $D_{f,\text{diss}}$, and dissociation rate coefficients, k_{diss}, were presented, providing a more complete picture of the analyte-receptor reactions occurring on the surface than can be seen with an analysis of the binding step alone, as done previously (Sadana, 1999). Besides, one may also use the numerical values for the rate coefficients for binding and the dissociation steps to classify the analyte-receptor biosensor system as moderate binding, extremely fast dissociation; moderate binding, fast dissociation; moderate binding, moderate dissociation; moderate binding, slow dissociation; fast binding, extremely fast dissociation; fast binding, fast dissociation; fast binding, moderate dissociation; or fast binding, slow dissociation.

For a single-fractal analysis and for the hybridization of 10 nM 16*CFl (oligonucleotide) to 16*B immobilized via NHS-LC-biotin and streptavidin using chemical and thermal regeneration (Abel *et al.*, 1996), an increase in D_f from 1.2112 (chemical regeneration) to 1.3942 (thermal regeneration) leads to an increase in k from 86.53 (chemical regeneration) to 100.0 (thermal regeneration). In other words, an increase in the fractal dimension or the degree of heterogeneity that exists on the biosensor surface leads to an increase in the binding rate coefficient. More data needs to be analyzed to see whether this holds for other hybridization reactions when they are utilized for diagnostic purposes. This has been observed previously for antibody–antigen and analyte-receptor binding reactions observed in biosensor reactions (Sadana and Sutaria, 1997).

For the examples where a dual-fractal analysis was used, an increase in the fractal dimension from D_{f_1} to D_{f_2} leads to an increase in the binding rate coefficient from k_1 to k_2. In other words, changes in the fractal dimension and in the binding rate coefficient are in the same direction. In these cases, an increase in the degree of heterogeneity on the reaction surface as the reaction progresses leads to an increase in the binding rate coefficient. There was, however, one exception to this. For the binding of complementary ssDNA in solution to single-strand DNA thymidylic acid icosanucleotides on derivatized quartz optical fiber, as the reaction progresses the binding rate coefficient and the fractal dimension both decrease. However, the changes exhibited by both the fractal dimension and the binding rate coefficient are still in the same direction. More such studies are required to further support the statement, that changes in the fractal dimension and the binding rate coefficient are exhibited in the same direction for hybridization reactions utilized as immunosensors or biosensors.

The parameter $K(=k_{diss}/k_{bind})$ value presented is of interest since it provides an indication of the stability, reusability, and regenerability of the biosensor. Also, depending on one's final goal, a higher or lower K value may be beneficial for a particular analyte-receptor system. During the binding of $8\,\mu$g/ml plasmid DNA in solution to a nonthiolated poly(L-lysine) polymer immobilized on an SPR biosensor surface, there was the distinct formation of a stacked layer (Wink *et al.*, 1999). Here, the fractal dimension remains almost the same during the initial phase (at t close to 0) ($D_f = 2.59$), and during the stacked phase ($D_f = 2.60$). However, the binding rate coefficient ($k = 34.9$) is 46.6% lower during the stacked phase as compared to the initial phase ($k = 65.2$). If this is indeed true, then perhaps this may exhibit the potential to decrease (or manipulate) the binding rate coefficient in a desired direction (in this case, decreasing). Much more data needs to be analyzed to determine whether this potential indeed exists.

This is apparently the first study where the binding rate coefficient is directly related to the degree of heterogeneity that exists on the surface for hybridization reactions. The analysis provides physical insights into the DNA–RNA reactions occurring on immunosensor or biosensor surfaces. In general, our analysis can be extended to other hybridization reactions occurring on different surfaces. The quantitative (predictive) expressions developed for the binding rate coefficient in terms of the analyte concentration and the fractal dimension for the detection of *Listeria* species should assist in the better control of biosensor performance parameters such as stability, selectivity, sensitivity, and response time. The development of more such equations for other hybridization systems would be valuable. In any case, many more detailed and precise studies are required to determine the influence of heterogeneity that exists on the immunosensor or biosensor surfaces on the binding rate coefficient where hybridization reactions are involved.

REFERENCES

Abel, A. P., Weller, M. G., Duveneck, G. L., Ehrat, M., and Widmer, H. M., *Anal. Chem.* **68**, 2905–2912 (1996).

Anderson, J., NIH Panel Review Meeting, Case Western Reserve University, Cleveland, OH, July 1993.

Bier, F. F., Kleinjung, F., and Scheller, F. W., *Sensors & Actuators* **38–39**, 78–82 (1997).

Bier, F. F., and Scheller, F. W., *Biosen. & Bioelectr.* **11**(6), 669–674 (1996).

Bluestein, B. I., Craig, M., Slovacek, R., Stundtner, L., Uricouli, C., Walziak, I., and Luderer, A., in *Biosensors with Fiberoptics* (D. Wise, L. B. Wingard Jr., eds.), Humana Press, New York, pp. 181–223, 1991.

Chiu, N. H. L. and Christopoulos, T. K., *Anal. Chem.* **68**, 2304–2308 (1996).

Cuypers, P. A., Willems, G. M., Kop, J. M., Corsel, J. W., Jansen, M. P., and Hermens, W. T., in *Proteins at Interfaces: Physicochemical and Biochemical Studies* (J. L. Brash and T. A. Horbett, eds.), American Chemical Society, Washington, DC, pp. 208–211, 1987.

De Gennes, P. G., *Radiat. Phys. Chem.* **22**, 193–196 (1982).

Eddowes, M. J., *Biosensors* **3**, 1–15 (1987/1988).

Fliss, R. St.-Laurent, M., Emond, E., Simard, R. E., Lemieux, R., Ettriki, A., and Pandian, S., *Appl. Microbiol. Biotech.* **43**, 717–724 (1995).

Galvan, B., Christopoulos, T. K. Bioluminescence Hybridization Assays Using Recombinant Aequorin. Application to the Detection of Prostate-Specific Antigen mRNA. *Anal. Chem.* 1996, **68**, 3545–3550.

Giaver, I., *J. Immunol.* **116**, 766–771.

Havlin, S., in *The Fractal Approach to Heterogeneous Chemistry: Surfaces, Colloids, Polymers* (D. Avnir, ed.), Wiley, New York, pp. 251–269, 1989.

Kopelman, R., *Science* **241**, 1620–1626 (1988).

Lee, C. K., and Lee, S. Y., *Surface. Sci.* **325**, 294–310 (1995).

Liu, X., Farmerie, W., Schuster, S., and Tan, W., *Anal. Biochem.* **283**, 175–181 (2000).

Matuishita, M., in *The Fractal Approach to Heterogeneous Chemistry: Surfaces, Colloids, and Polymers* (D. Avnir, ed.), Wiley, New York, pp. 161–179, 1989.

Nilsson, P., Persson, B., Uhlen, M., and Nygren, H. A., *Anal. Biochem.* **224**, 400–408 (1995).

Nygren, H. and Stenberg, M., *J. Colloid Interface. Sci.* **107**, 560–566 (1985).

Nygren, H. and Stenberg., M., *Biophys. Chem.* **38**, 67–75 (1990).

Nyikos, L. and Pajkossy, T., *Electrochim. Acta* **31**, 1347–1350 (1986).

Pfeifer, P. and Obert, M., in *The Fractal Approach to Heterogeneous Chemistry: Surfaces, Colloids, and Polymers* (D. Avnir, ed.), Wiley, New York, pp. 11–43, 1989.

Pfeifer, P., Avnir, D., and Farin, D. J., *Nature (London)* **308** (5956), 261–263 (1984a).

Pfeifer, P., Avnir, D., and Farin, D. J., *J. Colloid and Interface Sci.* **103**(1), 112–123 (1984b).

Piehler, P., Brecht, A., Gaulitz, G., Zerlin, M., Maul, C., Thiericke, R., and Grabley, M., *Anal. Biochem.* **249**, 94–102 (1997).

Piunno, P. A. E., Krull, U. J., Hudson, R. H. E., Damha, M. J., and Cohen, H., *Anal. Chem.* **67**, 2635–2643 (1995).

Place, J. F., Sutherland, R. M., Riley, A., and Mangan, C., in *Biosensors with Fiberoptics*; D. Wise, and L.B. Wingard, Jr., eds.), Humana Press, New York, pp. 253–291, 1991.

Radovich, P., Bortolin, S., and Christopoulos, T. K., *Anal. Chem.* **67**, 2644–2649 (1995).

Sadana, A., Alarie, J. P., Vo-Dinh, T., *Talanta* **42**, 1567–1574 (1995).

Sadana, A., and Beelaram, A., *Biotech. Progr.* **10**, 291–298 (1994).

Sadana, A., and Sii, D., *J. Colloid and Interface Sci.* **151**(1), 166–177 (1992a).

Sadana, A., and Sii, D., *Biosen. Bioelectr.* **7**, 559–568 (1992b).

Sadana, A. and Sutaria, M., *Biophys. Chem.* **65**, 29–44 (1997).
Sadana, A., *Biosen. & Bioelectr.*, **14**, 515–531 (1999).
Sigmaplot, Scientific Graphing Software, User's manual, Jandel Scientific, San Rafael, CA, 1993.
Stenberg, M. and Nygren, H. A., *Anal. Biochem.* **127**, 183–192 (1982).
Stenberg, M., Stiblert, L. and Nygren, H. A., *J. Theor. Biol.* **127**, 183–192 (1986).
Syvanen, A. C., Laaksonen, M., and Soderlund, H., *Nucleic Acids Res.* **14**(2), 5037–5048 (1986).
Waring, M., *Ann. Rev. Biochem.* **50**, 159–192 (1981).
Wink, T., de Beer, J., Hennick, W. E., Bult, A., and van Bennekom, W. P., *Anal. Chem.* **71**, 801–805 (1999).

CHAPTER 11

Fractal Dimension and Analyte-Receptor Binding in Cells

11.1. INTRODUCTION

The importance of providing a better understanding of the mode of operation of biosensors to improve their sensitivity, stability, and specificity has been emphasized both in the literature (Scheller *et al*., 1991) and throughout this book. A particular advantage of biosensors is that no reactant labeling is required. However, for the binding interaction to occur, one of the components must be bound or immobilized on a solid surface. This solid surface may be, for example, a biosensor or cell surface. This often leads to mass transfer limitations and subsequent complexities, as already indicated. There is a need to characterize the reactions occurring at the biosensor surface as well as other receptor-coated surfaces (such as cell surfaces) in the presence of the diffusional limitations inevitably present in these types of systems. It is our intention in this chapter as we further develop the knowledge on analyte-receptor binding kinetics for biosensor applications, to extend and apply that knowledge to provide insights into cellular analyte-receptor reactions.

Van Cott *et al*. (1994) emphasize that there is a critical need to develop serologic tools predictive of antibody function. This applies both to *in vitro* as well as to *in vivo* studies. For example, these authors note that antibodies directed toward the V3 loop of the envelope glycoprotein gp20 of HIV-1 is important due to its prevalence in natural infection and its ability to neutralize HIV-1 *in vitro*. Thus, the authors utilized surface plasmon resonance and biosensor technology to analyze the binding and dissociation kinetics of V3-specific antibodies with a biosensor matrix-immobilized recombinant-gp120. The biosensor-immobilized V3 peptides were found to mimic their conformational structure in solution. Fratamico *et al.* (1998) used the BIACORE biosensor for the detection of *Escherichia coli* 0157:H7 on antibodies that are reactive against this pathogen. If one binds suitable ligands to the cell sensor chip surface, the binding of the bacterial and human blood cells may be analyzed (BIAtechnology Note 103, 1994).

Recently, Byrnes and Griffin (1998) indicate that alphaviruses pose a significant threat to human health and cause a wide variety of diseases such as arthralgia, myalgia, and encephalitis. These authors emphasize that a better understanding of the cellular receptors used by the alphaviruses would provide a clearer insight into the pathogenesis of these viruses, perhaps leading to the design of effective ("live-attenuated") vaccines against them. The authors analyzed the binding of Sindbis virus to cell-surface heparan sulfate and found that glycosaminoglycan (GAG) heparan sulfate participates in the binding of Sindbis virus to cells. In its absence, the binding of this virus to the cell is diminished, although, it still does occur.

Kelly *et al.* (1999) analyzed the influence of a synthetic peptide adhesion epitope as an antimicrobial agent. These authors indicate that an early step in microbial infection is the adherence of binding of specific microbial adhesions to the mucosa of different tracts, such as oro-intestinal, nasorespiratory, or genitourinary. Utilizing a surface plasmon resonance biosensor, the authors attempted to inhibit the binding of cell-surface adhesion of *Streptococcus mutans* to salivary receptors *in vitro*. They used a synthetic peptide, p1025, which corresponded to residues 1025–1044 of the adhesion. The two residues Q1025 and E1037 that contributed to the binding were identified by site-directed mutagenesis. Kelly *et al.* indicate that this technique of utilizing peptide inhibitors of adhesion also may be employed to control other microorganisms in which adhesions are involved.

Teixera (1999) analyzed the influence of nanostructure and biomimetic surfaces on cell behavior. This author indicates the importance of ridges and grooves on cell surfaces for cell adhesion on examining the membranes in the cell lining of the cornea. Lauffenberger (1999) emphasized the importance of cellular diagnostics. He indicates the need for the development of design parameters for diagnostic procedures: Input–output relations are required wherein the input is the cellular analyte-receptor binding and the output is the cellular reaction. Protein surface characteristics (just like cellular surfaces) play an important role in binding, diffusion, association, and recognition of a ligand. Different techniques have been utilized to estimate or determine the roughness or the fractal dimension of surfaces. Some examples of the fractal dimension of proteins include bacterial serine protease A ($D_f = 2.09$), lysozyme ($D_f = 2.17$), and trypsin ($D_f = 2.62$). Gustaffson (1999) indicates that cells are exposed to a wide variety of chemicals in the environment such as hormones, metabolic intermediates, and compounds. The cells respond to these (ligands) by receptors on the surface. Some of these ligands have been identified. Other ligands for these receptors (orphans) have not been identified. Thus there is a need to identify the ligands for these (orphan) receptors since the receptors bind to ligands on the surface and move to the nucleus where they initiate changes. Gustaffson emphasizes the effort made by the pharmaceutical industry to identify the receptors and ligands that could be of help in the treatment of diseases.

Although we will emphasize cellular reactions occurring on biosensor surfaces in our analysis, the analysis is also applicable to ligand-receptor and analyte-receptorless systems for biosensor and other applications. The emphasis is to promote the understanding of cell-surface reactions. The fractal approach is not new and has been used previously in the studies of immunosensors and phenomena in membranes by Tam and Tremblay (1993). Theirs was an early attempt at a more extended application of fractal analysis to the investigation of analyte-receptor binding kinetics for biosensors with the eventual goal of providing a better understanding of these reactions on cell surfaces. In our analysis, performed on data available in the literature, similarities with immunoassay kinetics are also discussed whenever appropriate. We present of examples wherein either a single- or a dual-fractal analysis is required to adequately describe the analyte-receptor binding kinetics. The fractal analysis is one way by which to elucidate the time-dependent binding rate coefficients and the heterogeneity that exists on the biosensor or cell surface.

11.2. THEORY

Most of the theory and background material has been covered in the previous chapters and is not repeated here. Only the basic equations are provided for easy reference.

11.2.1. Single-Fractal Analysis

Havlin (1989) indicates that the diffusion of a particle (in our case, an analyte) from a homogeneous solution to a solid surface (receptor on a cellular surface or immobilized on a biosensor surface) where it reacts to form an analyte-receptor complex is given by

$$(\text{Analtye} \cdot \text{Receptor}) \sim \begin{cases} t^{(3-D_f)/2} = t^p & t < t_c \\ t^{1/2} & t > t_c, \end{cases} \tag{11.1a}$$

where (analyte · receptor) represents the binding complex formed on the surface. Equation (11.1a) indicates that the concentration of the product (analyte · receptor) on a solid fractal surface scales at short and intermediate time frame as (analyte · receptor) $\sim t^p$ with the coefficient $p = (3 - D_f)/2$ at short time frames and $p = \frac{1}{2}$ at intermediate time frames. This equation is associated with the short-term diffusional properties of a random walk on a fractal surface. Note that, in perfectly stirred kinetics on a regular (nonfractal) structure (or surface), k_1, is a constant; that is, it is independent of time. In other words, the limit of regular structures (or surfaces) and the absence of diffusion-limited kinetics leads to k_1 being independent of time. In all other situations, one would expect a scaling behavior given by $k_1 \sim k' t^{-b}$ with $-b = p < 0$. Also, the appearance of p different from 0 is the consequence of two different phenomena: the heterogeneity (fractality) of the surface and the imperfect mixing (diffusion-limited) condition.

Havlin indicates that the crossover value may be determined by $r_c^2 \sim t_c$. Above the characteristic length r_c, the self-similarity of the surface is lost. Above t_c, the surface may be considered homogeneous, and regular diffusion is now present. One may consider our analysis as an intermediate heuristic approach in that in the future one may also be able to develop an autonomous (and not time-dependent) model of diffusion-controlled kinetics in disordered media.

11.2.2. Dual-Fractal Analysis

The single-fractal analysis just presented can be extended to include two fractal dimensions. At present, the time ($t = t_1$) at which the first fractal dimension changes to the second fractal dimension is arbitrary and empirical. For the most part, it is dictated by the data analyzed and experience gained by handling a single-fractal analysis. In this case, the product (analyte · receptor) concentration on the biosensor surface is given by

$$(\text{Analyte} \cdot \text{Receptor}) \sim \begin{cases} t^{(3-D_{f_1})/2} = t^{p_1} & t < t_1 \\ t^{(3-D_{f_2})/2} = t^{p_2} & t_1 < t_1 < t_2 = t_c \\ t^{1/2} & t > t_c. \end{cases} \tag{11.1b}$$

11.3. RESULTS

First, it is appropriate to mention that the mathematical approach is straightforward. It is assumed that the fractal approach applies, which may be a limitation; it is really just possible one way to analyze the diffusion-limited binding kinetics assumed to be present in the cellular systems. The parameters thus obtained provide a useful comparison of the different situations. The cellular analyte-receptor binding reaction analyzed is a complex reaction, and the fractal analysis via the fractal dimension and the binding rate coefficient provide a useful lumped-parameter analysis of the diffusion-limited situation.

Kelly *et al.* (1999) analyzed the inhibitory binding of cell surface adhesion of *Streptococcus mutans* to salivary receptors *in vitro*. They utilized a synthetic peptide p1025 corresponding to residues 1025–1044 of the adhesion. Peptide p1025 contains two residues, Q1025 and E1037, that contribute to the binding. Figures 11.1a and 11.1b, show the binding of 250 nM peptide E1037A back-mutagenized to wild-type (E1037A/E) in solution to salivary agglutinin immobilized on a sensor chip. The binding was analyzed using a

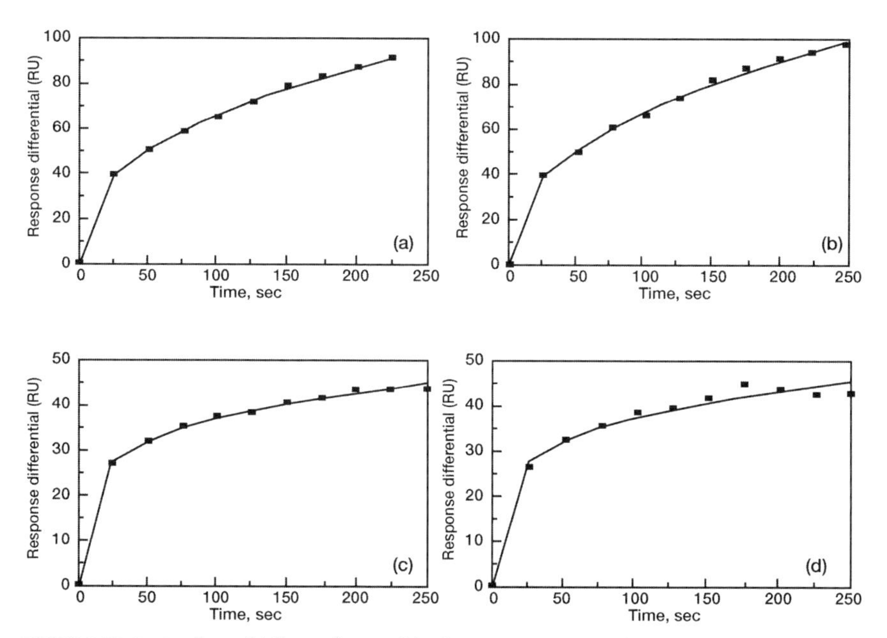

FIGURE 11.1 Binding of different forms of the fragment E1037 in solution to salivary agglutinin immobilized on a sensor chip. (a) and (b) 250 nM peptide E1037 back-mutagenized to wild-type fragment E1037A/E; (c) and (d) 250 nM E1037 mutagenized fragment (Kelly *et al.*, 1999).

surface plasmon resonance biosensor. In both cases, a single-fractal analysis is sufficient to adequately describe the binding kinetics. Table 11.1a shows the

Table 11.1 **Influence of Different Forms of Streptococcal Antigen (I/II) (SA I/II) (Wild-Type, Mutagenized, and Back-Mutagenized to Wild-Type) on the Binding Rate Coefficients and Fractal Dimensions for the Binding to Salivary Agglutinin Using a Surface Plasmon Resonance Biosensor (Kelly *et al.*, 1999)**

Analyte in solution/receptor on surface	Binding rate coefficient, k	Fractal dimension, D_f
(a) 250 nM E1037A back-mutagenized to wild-type (E1037A/E)/salivary agglutinin immobilized on sensor chip	11.6 ± 0.17	2.24 ± 0.01
250 nM E1037A back-mutagenized to wild-type (E1037A/E)/salivary agglutinin immobilized on sensor chip	11.5 ± 0.23	2.22 ± 0.02
250 nM E1037 mutagenized form of C-terminal fragment/salivary agglutinin immobilized on sensor chip	14.2 ± 0.23	2.59 ± 0.01
250 nM E1037 mutagenized form of C-terminal fragment/salivary agglutinin immobilized on sensor chip	14.31 ± 0.58	2.58 ± 0.06
(b) 500 nM wild-type fragment/salivary agglutinin immobilized on sensor chip	19.5 ± 0.70	2.17 ± 0.02
500 nM wild-type fragment/salivary agglutinin immobilized on sensor chip	13.1 ± 0.48	2.05 ± 0.044
500 nM wild-type mutant form Q1025 A/salivary agglutinin immobilized on sensor chip	0.32 ± 0.04	0.80 ± 0.12
(c) 500 nM C-terminal fragment in which E1037 mutagenized to E1037 A/salivary agglutinin immobilized on sensor chip	9.45 ± 0.26	2.01 ± 0.02
500 nM wild-type fragment/salivary agglutinin immobilized on sensor chip	16.6 ± 0.47	2.12 ± 0.03
500 nM wild-type fragment/salivary agglutinin immobilized on sensor chip	22.4 ± 0.81	2.20 ± 0.02
(d) 100 nM wild-type fragment/salivary agglutinin immobilized on sensor chip	5.60 ± 0.19	1.84 ± 0.03
200 nM wild-type fragment/salivary agglutinin immobilized on sensor chip	8.52 ± 0.112	1.892 ± 0.1026
300 nM wild-type fragment/salivary agglutinin immobilized on sensor chip	10.55 ± 0.15	1.91 ± 0.01
400 nM wild-type fragment/salivary agglutinin immobilized on sensor chip	15.9 ± 0.41	2.00 ± 0.02
400 nM wild-type fragment/salivary agglutinin immobilized on sensor chip	16.6 ± 0.36	2.00 ± 0.02
500 nM wild-type fragment/salivary agglutinin immobilized on sensor chip	24.3 ± 0.30	2.11 ± 0.01

values of k, and D_f obtained using Sigmaplot (1993) to fit the data. The equation

$$(\text{Analyte} \cdot \text{Receptor}) = kt^{(3-D_f)/2} \tag{11.2}$$

was used to obtain the values of k and D_f for a single-fractal analysis. The values of the parameters presented in the table are within 95% confidence limits. For example, the value of k reported for the binding of E1037A/E to salivary agglutinin is 11.6 ± 0.17. The 95% confidence limit indicates that 95% of the k values will lie between 11.4 and 11.8.

In all fairness, we realize that this is just one possible way by which to analyze this analyte-receptor binding data. One might justifiably argue that appropriate modeling may be achieved by using a Langmuir or other approach. The Langmuir approach was originally developed for gases (Thomson and Webb, 1968). Consider a gas at pressure, p, in equilibrium with a surface. The rate of adsorption is proportional to the gas pressure and to the fraction of uncovered surface. Adsorption will only occur when a gas molecule strikes a bare site. Researchers in the past have successfully modeled the adsorption behavior of analytes in solution to solid surfaces using the Langmuir model even though it does not conform to theory. Rudzinski *et al.* (1983) indicate that other appropriate liquid counterparts of the empirical isotherm equations have been developed. These include counterparts of the Freundlich (Dabrowski and Jaroniec, 1979), Dubinin–Radushkevich (Oscik *et al.*, 1976), and Toth (Jaroniec and Derylo, 1981) empirical equations. These studies, with their known constraints, have provided some restricted physical insights into the adsorption of adsorbates on different surfaces. The Langmuir approach may be used to model the data presented if one assumes the presence of discrete classes of sites. In our analysis, the results presented in Figs. 11.1a and 11.1b are repeat runs under the same conditions, which indicates the reproducibility of the experiments and our analysis procedure.

Figures 11.1c and 11.1d show the binding of the 250 nM E1037 mutagenized form of the C-terminal fragment in solution to salivary agglutinin immobilized on a sensor chip. Once again, these are repeat runs. A single-fractal analysis is sufficient to adequately describe the binding kinetics. The values of k and D_f are given in Table 11.1a. Note that as one goes from the mutagenized form (E1037A) to the back-mutagenized form (E1037A/E), both the fractal dimension, D_f, and the binding rate coefficient, k, decrease. One would have anticipated an increase in the binding rate coefficient on back-mutagenization. However, the binding rate coefficient is sensitive to the binding data at initial times (at t close to 0), which is reflected in the k values. The effect of the mutagenization is more clearly revealed at later times at or approaching saturation values. As expected, on comparing the saturation value (mutagenization) with the approaching saturation value (back-mutagenization), the value for mutagenization is much lower than for back-mutagenization. It seems that the saturation value is lower if the fractal dimension value is higher. At least, it appears so in this case.

Table 11.1a and Fig. 11.2 indicate that an increase in the fractal dimension, D_f, leads to an increase in the binding rate coefficient, k. An increase in D_f, by about 16.1%—from 2.23 to 2.59—leads to an increase in k, by about 24.3%—from a value of 11.5 to 14.3. For these runs for the binding of mutagenized and back-mutagenized forms of E1037, the binding rate coefficient, k, is given by

$$k = (3.69 \pm 0.02) D_f^{1.42 \pm 0.04}. \tag{11.3}$$

This equation predicts the binding rate coefficient values presented in Table 11.1a reasonably well. The availability of more data points would more firmly establish this equation, especially points between those presented and over a wider range. In this case, the binding rate coefficient is not very sensitive to the fractal dimension, as noted by the value of the exponent. There is an initial degree of heterogeneity that exists on surface, and this determines the value of k. It is this degree of heterogeneity on the surface that leads to the temporal binding rate coefficient. For a single-fractal analysis, it is assumed that this degree of heterogeneity remains constant during the reaction,

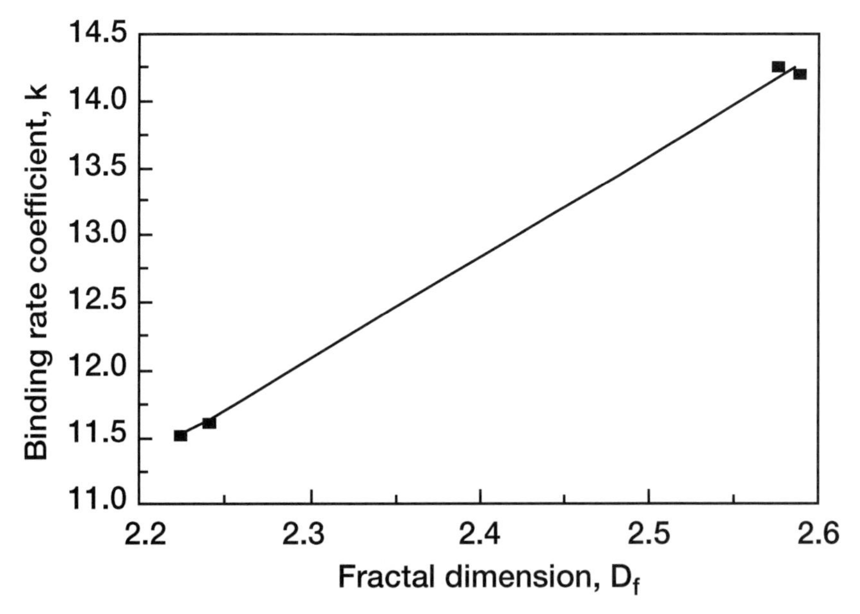

FIGURE 11.2 Influence of the fractal dimension, D_f, on the binding rate coefficient, k.

exhibiting a single D_f value. Note that in the present case an increase in D_f leads to an increase in k.

Figures 11.3a, and 11.3b show the binding of 500 nM wild-type fragment in solution to salivary agglutinin immobilized on a sensor chip. These are repeat runs under same conditions. In both cases, a single-fractal analysis is sufficient to adequately describe the binding kinetics. Table 11.1b shows the values of k, and D_f. In this case, there is a difference between the binding rate coefficient, k, and the fractal dimension, D_f, values obtained. Note, however, that even here an increase in D_f from a value of 13.1 to 19.5 leads to an increase in k from 2.05 to 2.17.

Figure 11.3c shows the binding of 500 nM of the mutant form Q1025 in solution to salivary agglutinin immobilized on a sensor chip. In this case too, a single fractal analysis is sufficient. The values of k and D_f, are given in Table 11.1b. Note that as one goes from the wild-type form to the mutant form (Q1025A) there is a significant decrease in both D_f and k. For example, an increase in D_f by a factor of 2.71—from 0.80 to 2.17—leads to an increase in k, by a factor of 60.9—from 0.32 to 19.5.

Table 11.1b and Fig. 11.4 indicate that an increase in the fractal dimension, D_f, leads to an increase in the binding rate coefficient. For these runs for the binding of the wild-type fragment and the mutant form (Q1025A), the binding rate coefficient is given by

$$k = (0.80 \pm 0.10)D_f^{4.02 \pm 0.15} \tag{11.4}$$

This equation predicts the k values presented in Table 11.1b reasonably well. There is some deviation in the data, which is reflected in the error estimate for the coefficient as well as in the exponent. The availability of more data points would more firmly establish this equation. Note the high exponent dependence of the binding rate coefficient, k, on the fractal dimension, D_f. This underscores that the binding rate coefficient is very sensitive to the degree of heterogeneity that exists on the surface. No theoretical explanation is offered to explain this high exponent dependence.

Figure 11.5a shows the binding of 500 nM of the mutant form of the fragment in which the E1037 is mutagenized to E1037A in solution to salivary agglutinin immobilized on a sensor chip. Once again, a single-fractal analysis is sufficient. The values of k, and D_f, are given in Table 11.1c. Figures 11.5b and 11.5c show the binding of 500 nM of the wild-type fragment in solution to salivary agglutinin immobilized on a sensor chip. Here again, a single fractal analysis is sufficient to adequately describe the binding kinetics. The values of k and D_f are given in Table 11.1c.

Table 11.1c and Fig. 11.6 indicate that an increase in D_f, leads to an increase in k. For these runs for the binding of the wild-type fragment and the

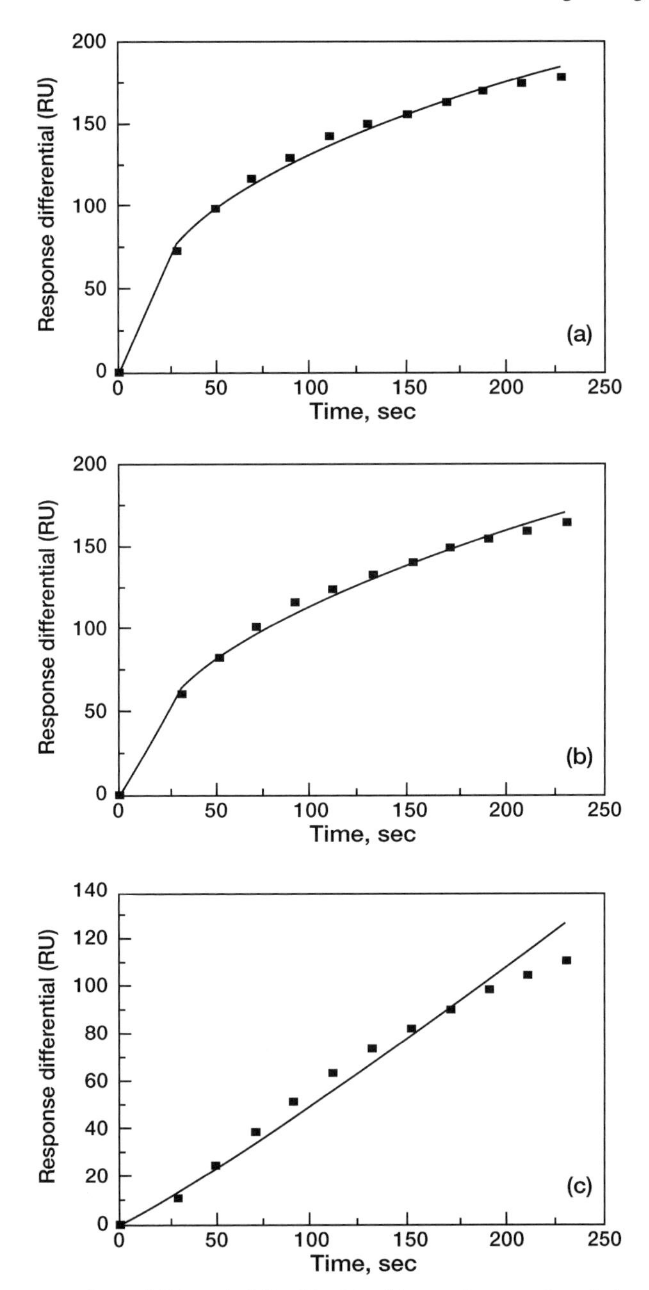

FIGURE 11.3 Binding of [(a) and (b)] 500 nM wild-type fragment in solution to salivary agglutinin immobilized on a sensor chip; (c) 500 nM of the mutant form Q1025 in solution to salivary agglutinin immobilized on a sensor chip (Kelly *et al.*, 1999).

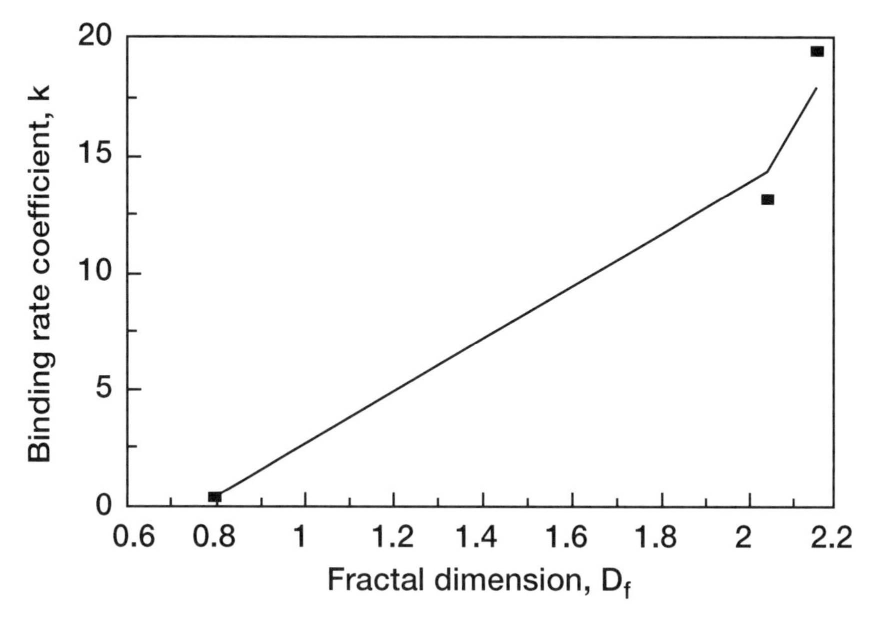

FIGURE 11.4 Influence of the fractal dimension, D_f, on the binding rate coefficient, k.

mutant form (E1037A), the binding rate coefficient is given by

$$k = (0.0099 \pm 0.0007) D_f^{9.85 \pm 1.12}. \tag{11.5}$$

This equation predicts the k values presented in Table 11.1c reasonably well. There is some deviation in the data, which is reflected in the error estimate for the coefficient as well as in the exponent. The availability of more data points would more firmly establish this equation. Note the very high value of the exponent. This, once again, underscores that the binding rate coefficient is very sensitive to the degree of heterogeneity that exists on the surface. Once again, and as indicated previously in this text, the data in Fig. 11.6 could very easily be reasonably represented by a linear function. The Sigmaplot (1993) program provides the nonlinear function. The lack of data points (only three data points are available) once again prevents a clearer model discrimination between a nonlinear and a linear representation.

Figures 11.7a–f show the binding of different concentrations (100 to 500 nM) of wild-type fragment in solution to salivary agglutinin immobilized on a sensor chip. In each case, starting from 100 nM with increasing increments of 100 nM to 500 nM, the binding curve may be described by a single-fractal analysis. The values of k and D_f for each case are given in Table 11.1d. The 400 nM wild-type fragment run was repeated once.

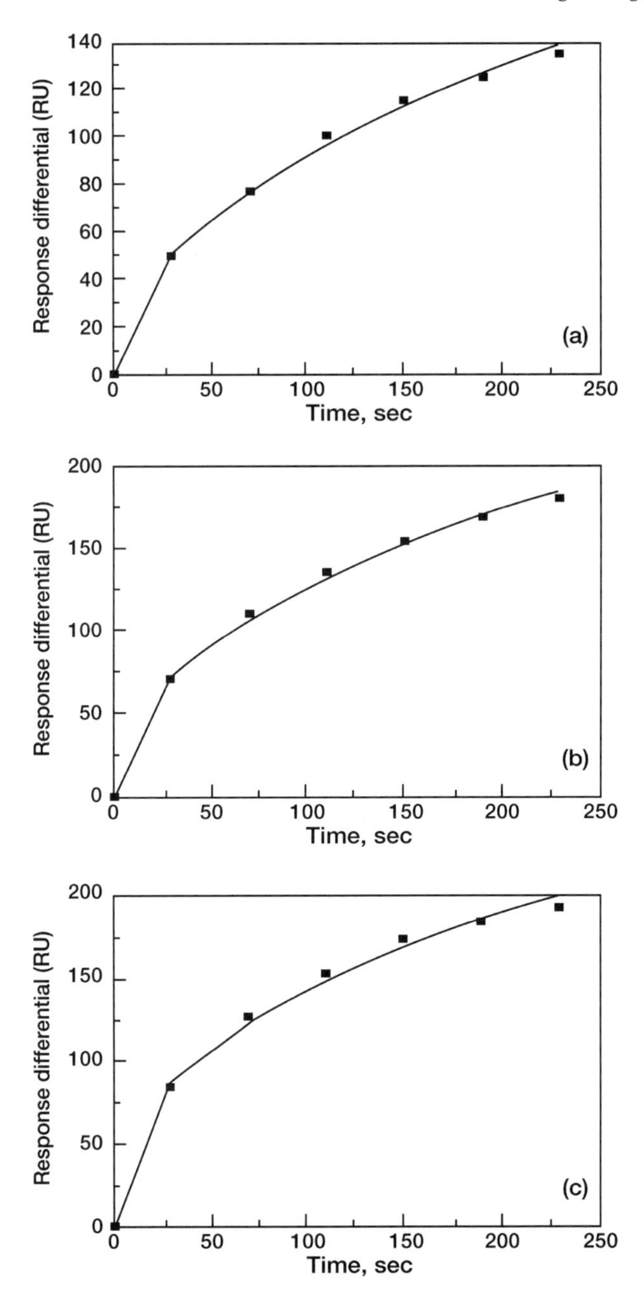

FIGURE 11.5 (a) Binding of 500 nM of the mutant form of the E1037 fragment in solution to salivary agglutinin immobilized on a sensor chip; (b) and (c) 500 nM wild-type fragment in solution to salivary agglutinin immobilized on a sensor chip (Kelly *et al.*, 1999).

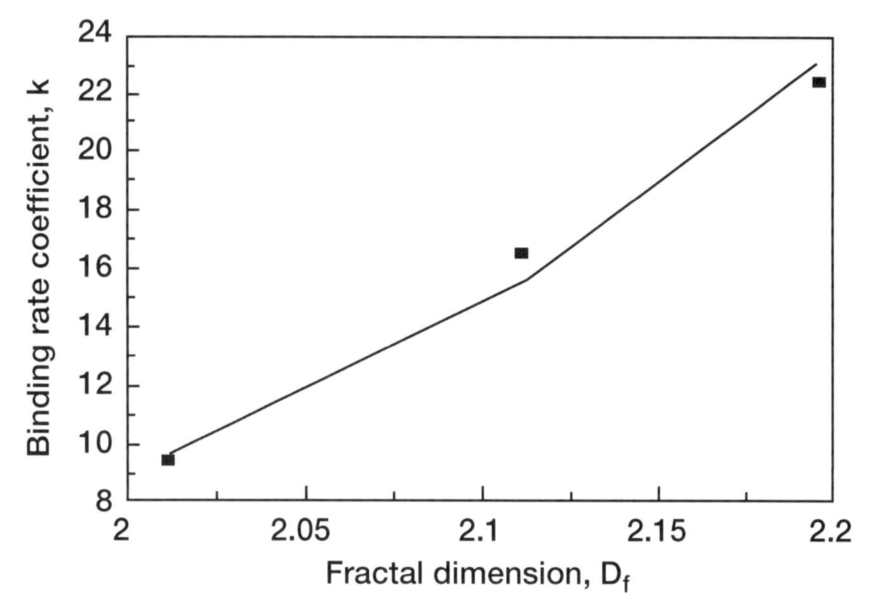

FIGURE 11.6 Influence of the fractal dimension, D_f, on the binding rate coefficient, k.

Table 11.1d and Fig. 11.8a indicate that an increase in the wild-type fragment concentration in solution leads to an increase in the binding rate coefficient, k. For the 100 to 500 nM wild-type fragment concentration in solution, the binding rate coefficient is given by

$$k = (0.097 \pm 0.016)[\text{wild-type fragment}]^{0.86 \pm 0.12}. \tag{11.6a}$$

This equation predicts the k values presented in Table 11.1d reasonably well. This is a nonlinear representation, although the exponent dependence is close to 1 (0.86). There is some scatter in the data, reflected in the error estimate for the coefficient as well as in the exponent. The availability of more data points would more firmly establish this equation. A better fit could presumably be obtained from an equation such as

$$k = \text{a}[\text{wild-type fragment}]^{b} + c[\text{wild type fragment}]^{d}. \tag{11.6b}$$

However, this would only involve more parameters. Here a, b, c, and d are parameters that need to be determined by regression. In this case, more data points would definitely be required.

Table 11.1d and Fig. 11.8b indicate that an increase in the wild-type fragment concentration in solution leads to an increase in the fractal

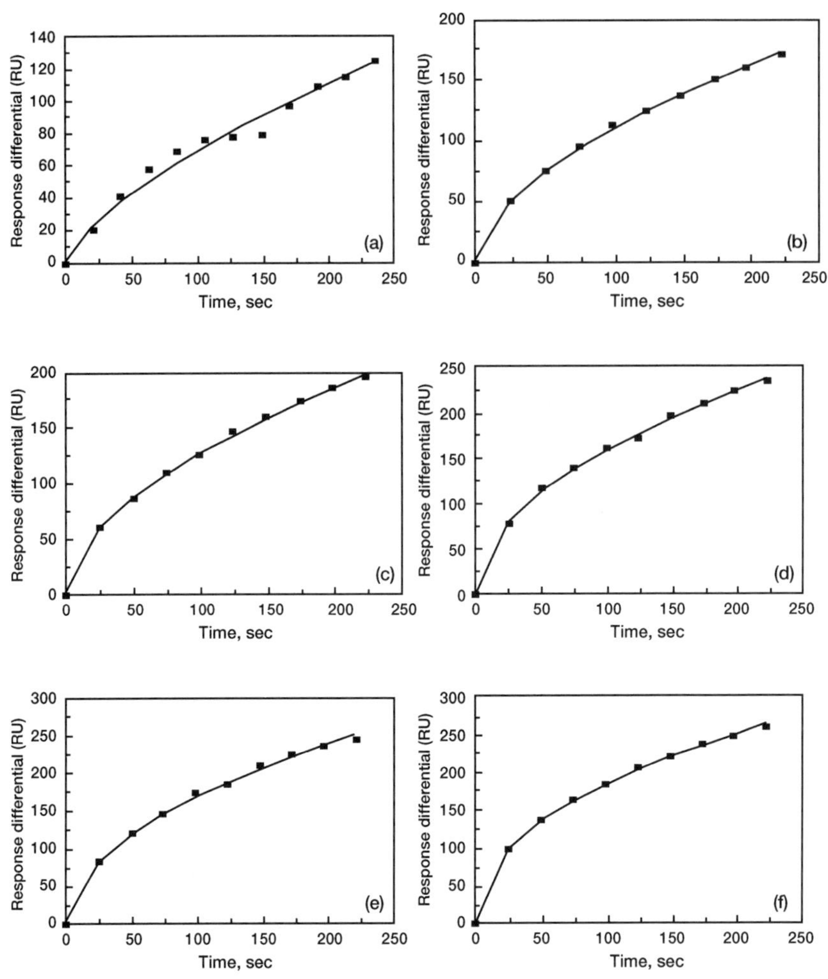

FIGURE 11.7 Binding of different concentrations (in nM) of wild-type fragment in solution to salivary agglutinin immobilized on a sensor chip (Kelly *et al.*, 1999). (a) 100; (b) 200; (c) 300; (d) 400; (e) 400; (f) 500.

dimension, D_f. For the 100 to 500 nM wild-type concentration in solution utilized, the fractal dimension is given by

$$D_f = (1.28 \pm 0.03)[\text{wild-type fragment}]^{0.075 \pm 0.019}. \qquad (11.6c)$$

This equation predicts the D_f values given in Table 11.1d reasonably well. The representation is nonlinear, even though the exponent dependence on the

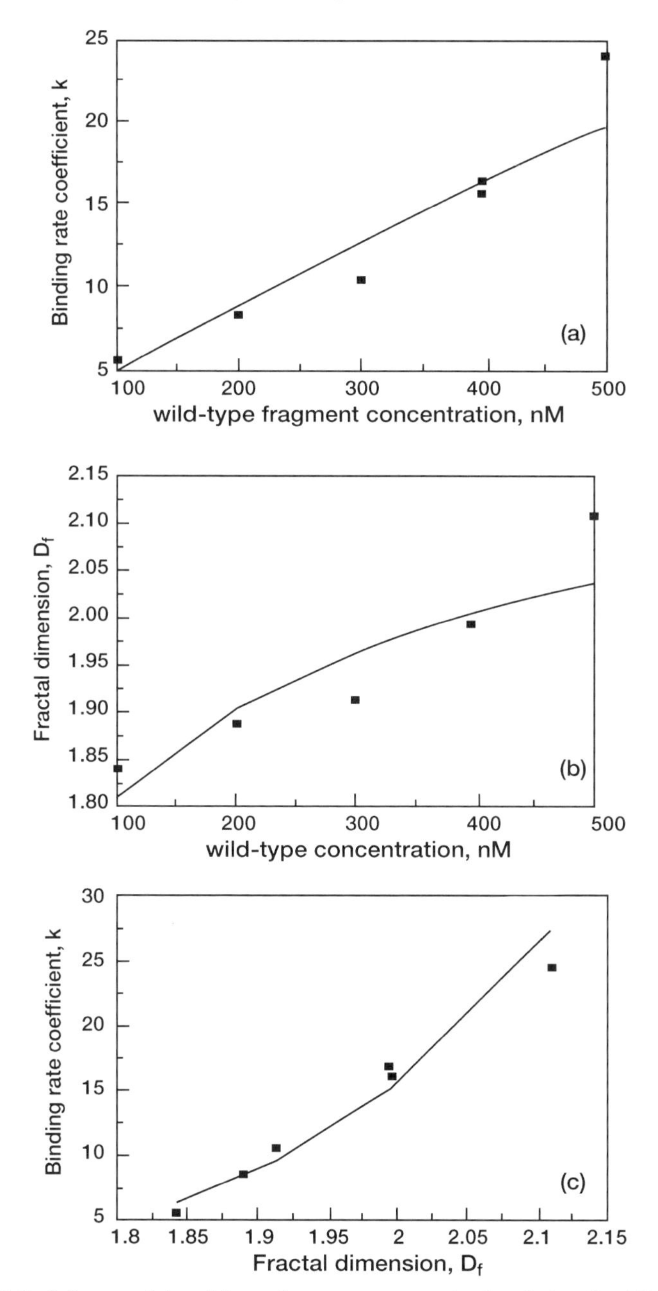

FIGURE 11.8 Influence of the wild-type fragment concentration in solution (in nM) on (a) the binding rate coefficient, k, and (b) the fractal dimension, D_f. (c) Influence of the fractal dimension, D_f, on the binding rate coefficient, k.

wild-type fragment concentration is rather low (0.075). The rather low exponent dependence exhibited is presumably responsible for the opposite curvature exhibited. There is deviation in the data, reflected in the error for the exponent. The availability of more data points would more firmly establish this equation. A better fit could be obtained if an equation of the form

$$D_f = a[\text{wild-type fragment}]^b + c[\text{wild-type fragment}]^d \tag{11.6d}$$

were used. However, this again would just introduce more parameters and data points would definitely be required. The coefficients *a*, *b*, *c*, and *d* need to be determined by regression.

Table 11.1d and Fig. 11.8c indicate that an increase in the fractal dimension, D_f, leads to an increase in the binding rate coefficient, *k*. For these runs for the binding of the wild-type fragment in the 100 to 500 nM concentration range in solution, the binding rate coefficient is given by

$$k = (0.011 \pm 0.001) D_f^{10.5 \pm 1.05}. \tag{11.6e}$$

This equation predicts the *k* values presented in Table 11.1d reasonably well. There is some deviation at the higher fractal dimension values. This is reflected in the error of the estimate for the coefficient as well as in the exponent. Note the very high exponent dependence of *k* on D_f. This, once again, underscores that the binding rate coefficient is very sensitive to the degree of heterogeneity that exists on the surface. Once again, no theoretical explanation is offered to explain this very high exponent dependence. One might very reasonably argue that the data presented in Fig. 11.8c could very well be effectively represented by a linear function in the narrow range of fractal dimension presented. No explanation is offered at present, except that the Sigmaplot (1993) provided the results. In cases such as these, apparently a better model discrimination program (between nonlinear and linear representations of the dependent variable dependence on the narrow range of independent variable presented) is apparently required.

Scroth-Diez *et al.* (1998) analyzed the fusion activity of transmembrane and cytoplasmic domain chimeras of the glycoprotein from influenza virus. The authors indicate that chimeric constructs enable one to analyze dependencies of protein-induced membrane fusion on specific amino acid sequences. In addition, the properties of transmembrane and cytoplasmic domain of the fusion protein are elucidated. Figure 11.9 shows the binding of R18-labeled human erythrocytes (RBCs) to different chimeric-construct-transfected CV-1 cells at different pH values. The details of the experimental procedure are available (Scroth-Diez *et al.*, 1998) and not repeated here. The

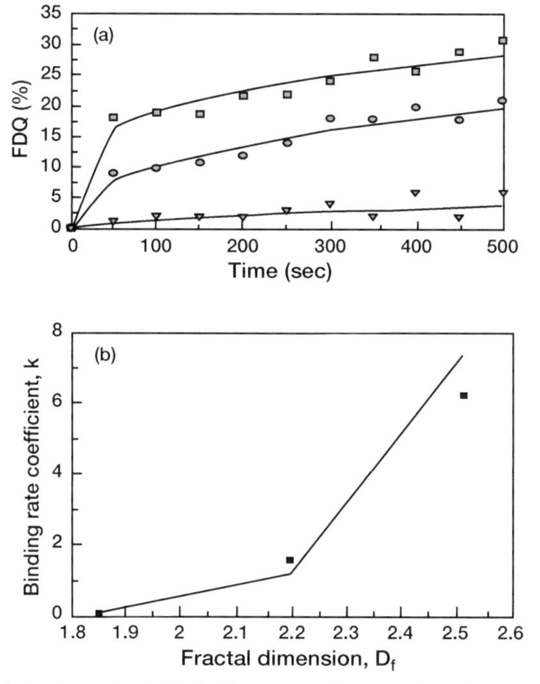

FIGURE 11.9 (a) Binding of R18-labeled human erythrocytes (RBCs) in solution to different chimeric-construct-transfected CV-1 cells at different pH values (Scroth-Diez *et al.*, 1998): ⊡ H/H/N at pH 5.0; ⊙ H/H/R at pH 5.0; ▽ H/H/N at pH 6.7. (b) Influence of the fractal dimension, D_f, on the binding rate coefficient, k.

binding kinetics were analyzed using fluorescence dequenching (FDQ). Figure 11.9a shows the binding of RBCs to (a) H/H/N at pH 5.0, (b) H/H/R at pH 5.0, and (c) H/H/N at pH 6.7. In each case, a single-fractal analysis is sufficient to adequately describe the binding kinetics. Table 11.2a shows the values of k, and D_f.

Table 11.2a and Fig. 11.9b indicate that an increase in the fractal dimension, leads to an increase in the binding rate coefficient. For the data presented in Table 11.2a, the binding rate coefficient is given by

$$k = (3.95\text{E}-05 \pm 1.61\text{E}-05)D_f^{13.2 \pm 1.57}. \tag{11.7}$$

This equation predicts the k values presented in Table 11.2a reasonably well. More data is required to more firmly establish this equation. Note that the binding rate coefficient is very sensitive to the fractal dimension. No

Table 11.2 Influence of Different Parameters on Fractal Dimensions and Binding Rate Coefficients for the Binding Kinetics of Fluorescence Dequenching (FDQ) of R-18-Labeled Human Erythrocytes (RBCs) to Different Chimeric-Transfected CV-1 Cells at Different pH values (Scroth-Diez *et al.*, 1998)

Analyte in solution/ receptor on surface	k	D_f	k_1	k_2	D_{f_1}	D_{f_2}
(a) RBC/H/H/N at pH 5.0	6.31 ± 0.52	2.52 ± 0.07	na	na	na	na
RBC/H/H/R at pH 5.0	1.65 ± 0.18	2.20 ± 0.09	na	na	na	na
RBC/H/H/N at pH 6.7	0.12 ± 0.06	1.85 ± 0.36	na	na	na	na
(b) RBC/H/H/H at pH 4.8	18.29 ± 0.86	2.69 ± 0.04	na	na	na	na
RBC/H/H/H at pH 5.8	2.49 ± 0.44	2.05 ± 0.15	0.56 ± 0.09	9.21 ± 0.46	1.38 ± 0.30	2.50 ± 0.12
RBC/H/F/H at pH 4.8	19.9 ± 0.90	2.74 ± 0.04	na	na	na	na
RBC/H/F/H at pH 5.8	0.59 ± 0.07	1.73 ± 0.10	na	na	na	na
RBC/H/H/F at pH 4.8	31.3 ± 0.50	2.84 ± 0.04	na	na	na	na
RBC/H/H/F at pH 5.8	4.58 ± 0.75	2.19 ± 0.14	1.23 ± 0.19	23.3 ± 1.02	1.62 ± 0.27	2.76 ± 0.11

theoretical explanation is offered to explain this very high exponent dependence.

Figure 11.10a shows the binding of R18-labeled RBCs to H/H/H-transfected CV-1 cells at pH 4.8 (Scroth-Diez *et al.*, 1998). Once again FDQ was used. A single-fractal analysis is sufficient, and the values of k and D_f are given in Table 11.2b. Figure 11.10b shows the binding kinetics of R-18-labeled RBCs to H/H/H-transfected CV-1 cells at pH 5.8. In this case, a dual-fractal analysis is required to adequately describe the binding kinetics. The values of the binding coefficient(s) and the fractal dimension(s) for a single- and a dual-fractal analysis are given in Table 11.2b. The values obtained indicate that there is a change in the binding mechanism as one changes the pH from 4.8 (single-fractal analysis) to 5.8 (dual-fractal analysis).

Figure 11.11a shows the binding of R-18-labeled RBCs to H/F/H-transfected cells at pH 4.8 and 5.8. In each case, a single-fractal analysis is sufficient. Figure 11.11b shows the binding of R-18-labeled RBCs to H/H/F-transfected CV-1 cells at pH 4.8. In this case too, a single-fractal analysis is

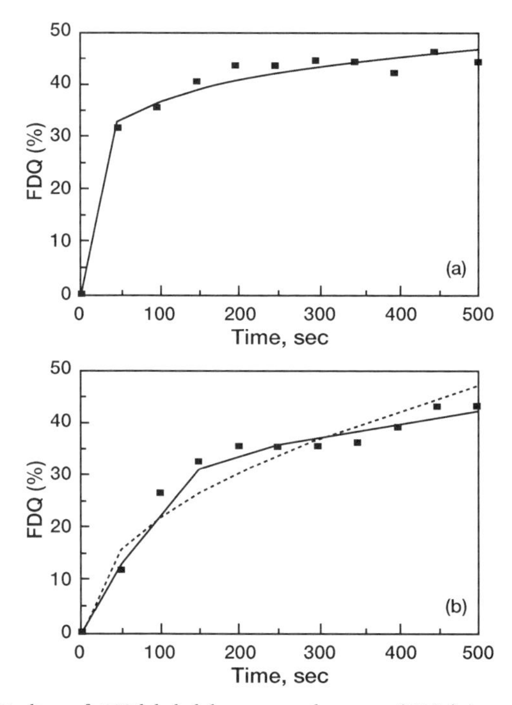

FIGURE 11.10 Binding of R18-labeled human erythrocytes (RBCs) in solution to H/H/H-transfected CV-1 cells at different pH values (Scroth-Diez *et al.*, 1998). (a) 4.8; (b) 5.8 (- - -, single-fractal analysis; —, dual-fractal analysis).

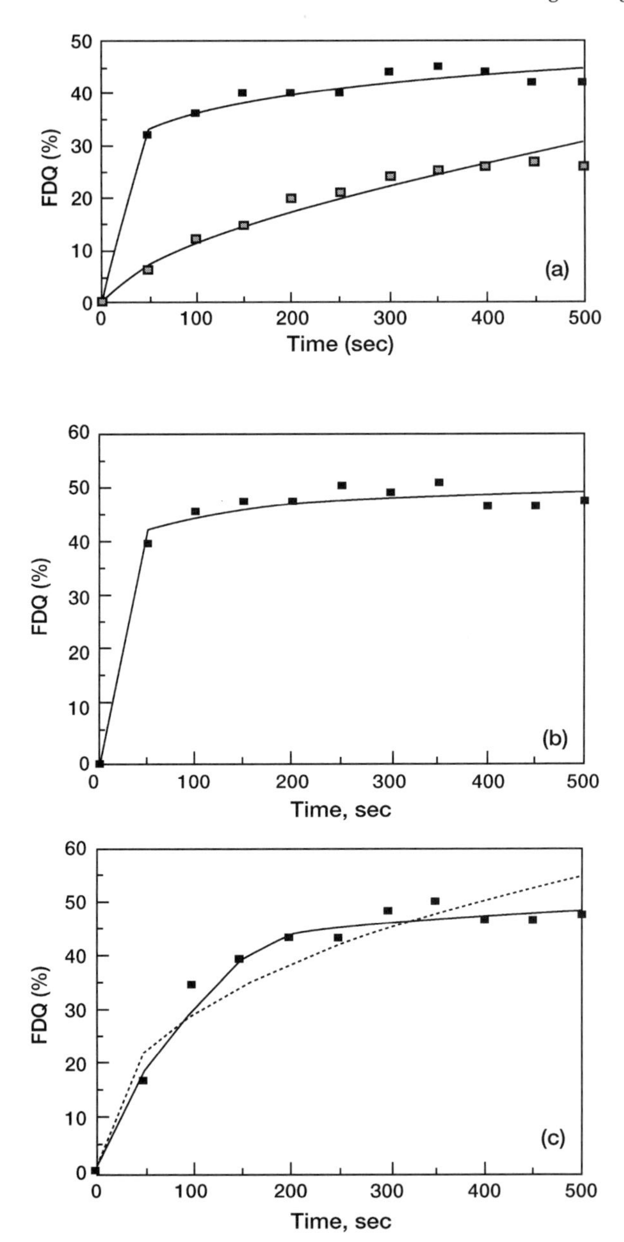

FIGURE 11.11. (a) Binding of R18-labeled RBCs in solution to H/F/H-transfected CV-1 cells at different pH values: ⊡ 4.8; ■ 5.8 (Scroth-Diez *et al.*, 1998). (b) Binding of R18-labeled RBCs to H/H/F-transfected CV-1 cells at pH 4.8. (c) Binding of R18-labeled RBCs to H/H/F-transfected CV-1 cells at pH 5.8 (- - -, single fractal analysis; —, dual-fractal analysis).

sufficient. Finally, Fig. 11.11c shows the binding of R18-labeled RBCs to H/H/F-transfected CV-1 cells at pH 5.8. In this case, a dual-fractal analysis is required to adequately describe the binding kinetics. The values of the binding rate coefficient(s) and the fractal dimension(s) for a single- and a dual-fractal analysis for all these cases are given in Table 11.2b.

For a single-fractal analysis, Table 11.2b and Fig. 11.12 indicate that an increase in the fractal dimension leads to an increase in the binding rate coefficient. For the data presented in Table 11.2b, the binding rate coefficient is given by

$$k = (0.0081 \pm 0.0007) D_f^{7.81 \pm 0.21}. \tag{11.8}$$

This equation predicts the k values presented in Table 11.2b reasonably well. More data is required to more firmly establish this equation. Note that the binding rate coefficient is very sensitive to the fractal dimension. No theoretical explanation is offered to explain this very high exponent dependence.

Rux *et al.* (1998) analyzed the binding of glycoprotein-D (gD) of herpes simplex virus (HSV) to herpes entry mediator (HveAt) using surface plasmon resonance. The authors have utilized the BIACORE biosensor since it directly measures analyte-receptor binding in real time without the use of labels. From

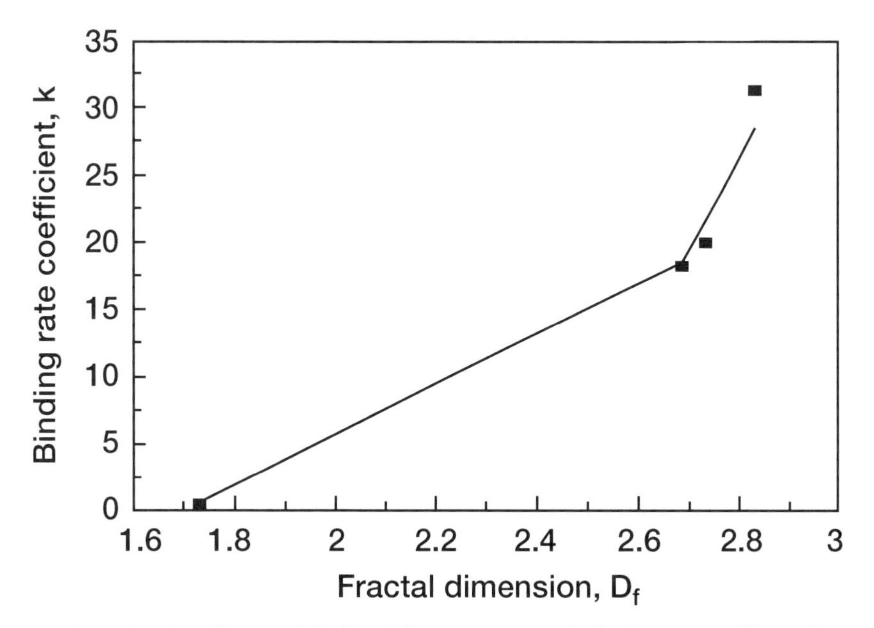

FIGURE 11.12 Influence of the fractal dimension, D_f, on the binding rate coefficient, k.

the analysis of binding data for monoclonal antibodies to distinct sites on the HSV gD that block HSV binding to HveAt (Nicola *et al.*, 1998), Rux *et al.* concluded that HveAt regulates HSV entry by acting as a receptor for the virus and that the principal ligand of HSV binding to HveAt is virion gD. These authors further indicate that a truncated form of a functional region IV variant, gD1(Δ290-299t), exhibited an increasing tendency to block virus entry and to bind to the HveAt. The authors further explained this increased affinity to block virus entry by analyzing the binding of gD1 variants to the HveAt.

Figure 11.13 shows the binding of 0.03 to 0.25 mM gD1(Δ277–299t) in solution to about 2000 resonance units (RU) of HveAt immobilized on a flow cell of a CM5 sensor chip via primary amines. In this case, a single-fractal analysis is sufficient to adequately describe the binding kinetics of 0.03 mM gD1(Δ277–299t). See Fig. 11.13a and Table 11.3a. However, a dual-fractal analysis is required to adequately describe the binding kinetics of 0.06 to 0.25 mM gD1(Δ277–299t) in solution. This indicates that there is a change in

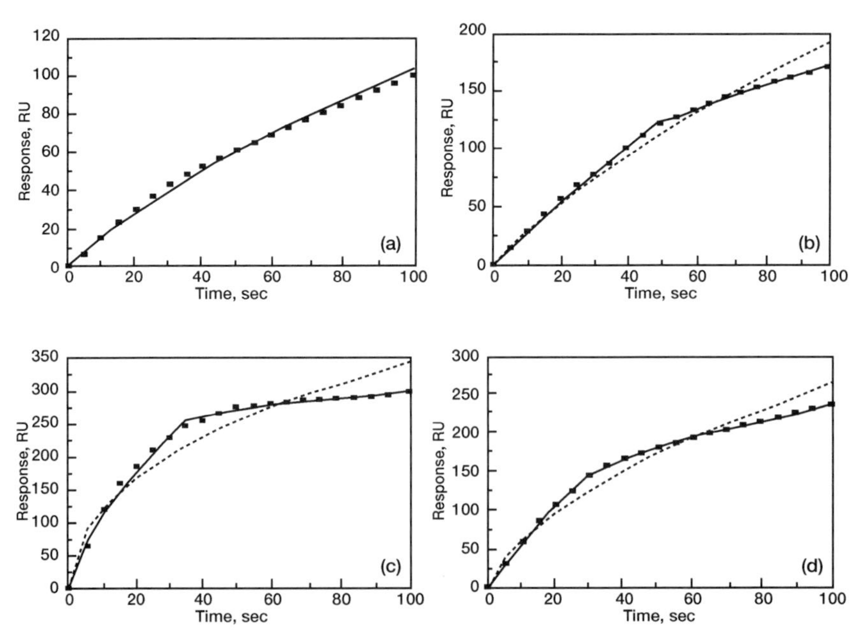

FIGURE 11.13 Binding of different concentrations (in mM) of variant of glycoprotein D [gD1(Δ277–299t)] of herpes simplex virus (HSV) to herpes entry mediator (HveAt) using surface plasmon resonance (Rux et al., 1998). (a) 0.3 (single-fractal analysis); (b) 0.6; (c) 0.125; (d) 0.25 (- - -, single-fractal analysis; —, dual-fractal analysis).

Table 11.3 Influence of Glycoprotein D Variants [gD1(Δ277-299t) and gD1(275t)] Concentration from the Herpes Simplex Virus on the Fractal Dimensions and Binding Rate Coefficients for Functional Region IV Binding to the Herpes Entry Mediator (HveAt) Coupled to a Flow Cell of a CM5 Sensor Chip (Rux *et al.*, 1998)

Analyte in solution/receptor on surface	k	D_f	k_1	k_2	D_{f_1}	D_{f_2}
(a) 0.03 mM gD1 (Δ277-299t)/HveAt coupled to a flow cell of a CM5 sensor chip	2.57 ± 0.15	1.39 ± 0.03	na	na	na	na
0.06 mM gD1 (Δ277-299t)/HveAt coupled to a flow cell of a CM5 sensor chip	5.11 ± 0.42	1.43 ± 0.04	3.96 ± 0.19	14.80 ± 0.13	1.25 ± 0.04	1.94 ± 0.02
0.125 mM gD1 (Δ277-299t)/HveAt coupled to a flow cell of a CM5 sensor chip	15.28 ± 1.69	1.75 ± 0.06	8.31 ± 0.35	36.76 ± 0.18	1.28 ± 0.06	2.17 ± 0.007
0.25 mM gD1 (Δ277-299t)/HveAt coupled to a flow cell of a CM5 sensor chip	45.08 ± 5.70	2.12 ± 0.07	26.55 ± 1.99	145.08 ± 1.56	1.73 ± 0.08	2.68 ± 0.02
(b)0.03 mM gD1 (275t)/HveAt coupled to a flow cell of a CM5 sensor chip	1.48 ± 0.11	1.29 ± 0.04	na	na	na	na
0.06 mM gD1 (275t)/HveAt coupled to a flow cell of a CM5 sensor chip	3.58 ± 0.04	1.44 ± 0.05	2.10 ± 0.04	9.50 ± 0.13	1.04 ± 0.02	1.90 ± 0.02
0.125 mM gD1 (275t)/HveAt coupled to a flow cell of a CM5 sensor chip	9.99 ± 1.34	1.72 ± 0.07	4.84 ± 0.31	27.27 ± 0.73	1.18 ± 0.08	2.20 ± 0.03
0.25 mM gD1 (275t)/HveAt coupled to a flow cell of a CM5 sensor chip	24.66 ± 4.29	2.01 ± 0.09	11.98 ± 1.42	87.97 ± 2.02	1.48 ± 0.12	2.63 ± 0.02
0.50 mM gD1 (275t)/HveAt coupled to a flow cell of a CM5 sensor chip	60.56 ± 8.26	2.36 ± 0.07	29.35 ± 2.97	149.35 ± 1.36	1.80 ± 0.13	2.80 ± 0.01

the binding mechanism as one goes from the lowest to the other (increasing) gD1(Δ277–299t) concentrations utilized.

For the higher concentrations (0.06 to 0.25 mM) of gD1(Δ277–299t) (Figs. 11.13b–11.13d) and where a dual-fractal analysis is required, note that an increase in the gD1(Δ277–299t) concentration in solution leads to an increase in the binding rate coefficients, k_1 and k_2, and an increase in the fractal dimensions, D_{f_1} and D_{f_2}. For example, an increase in the gD1(Δ277–299t) concentration by a factor of about 4.33—from 0.06 to 0.25 mM—leads to an increase in k_1 by a factor of 6.72—from 3.95 to 26.55—and in k_2 by a factor of 9.81—from 14.76 to 145.08. Also, there is a corresponding increase in D_{f_1} by a factor of 1.38—from 1.24 to 1.73—and in D_{f_2} by a factor of 1.32—from 1.94 to 2.68. Thus, the binding rate trends follow the fractal dimensionality trends. No explanation is presently offered for this behavior, except that an increasing roughness or heterogeneity on the surface leads to an increase in the binding rate coefficient.

Figure 11.14 shows the binding of 0.03 to 0.50 mM of the variant gD1(275t) concentration in solution to about 200 resonance units (RU) of HveAt immobilized on a flow cell of a CM5 sensor chip via primary amines. In this case too, both a single- and a dual-fractal analysis are required to adequately describe the binding kinetics (see Table 11.3b). A single-fractal analysis is adequate to describe the binding kinetics of 0.03 mM gD1(275t) (Fig. 11.14a) but a dual-fractal analysis is required to adequately describe the binding kinetics of 0.06 to 0.50 mM gD1(275t) (Figs. 11.14b–11.4e). Once again, this indicates that there is a change in the binding mechanism as one goes from the lowest to the other (increasing) gD1(275t) concentrations utilized.

For the higher concentrations of gD1(275t) utilized and where a dual-fractal analysis is required, note that an increase in the gD1(275t) concentration in solution leads to an increase in the binding rate coefficients, k_1 and k_2, and to an increase in the fractal dimensions, D_{f_1} and D_{f_2}. For example, an increase in the gD1(275t) concentration by a factor of 8.66—from 0.06 to 0.50 mM—leads to an increase in k_1 by factor of about 13.9—from 2.10 to 29.35—and in k_2 by a factor of 15.7—from 9.50 to 149.35. Also, there is a corresponding increase in D_{f_1} by a factor of 1.72—from 1.04 to 1.80—and in D_{f_2} by a factor of 1.27—from 2.20 to 2.80.

Figure 11.15a shows that dependence of the binding rate coefficient, k_1, on the fractal dimension, D_{f_1}. From the data presented in Table 11.3b, the binding rate coefficient is given by

$$k_1 = (1.93 \pm 0.28) D_{f_1}^{4.69 \pm 0.33}. \tag{11.9}$$

More data are required to provide a better and more reliable fit. Nevertheless,

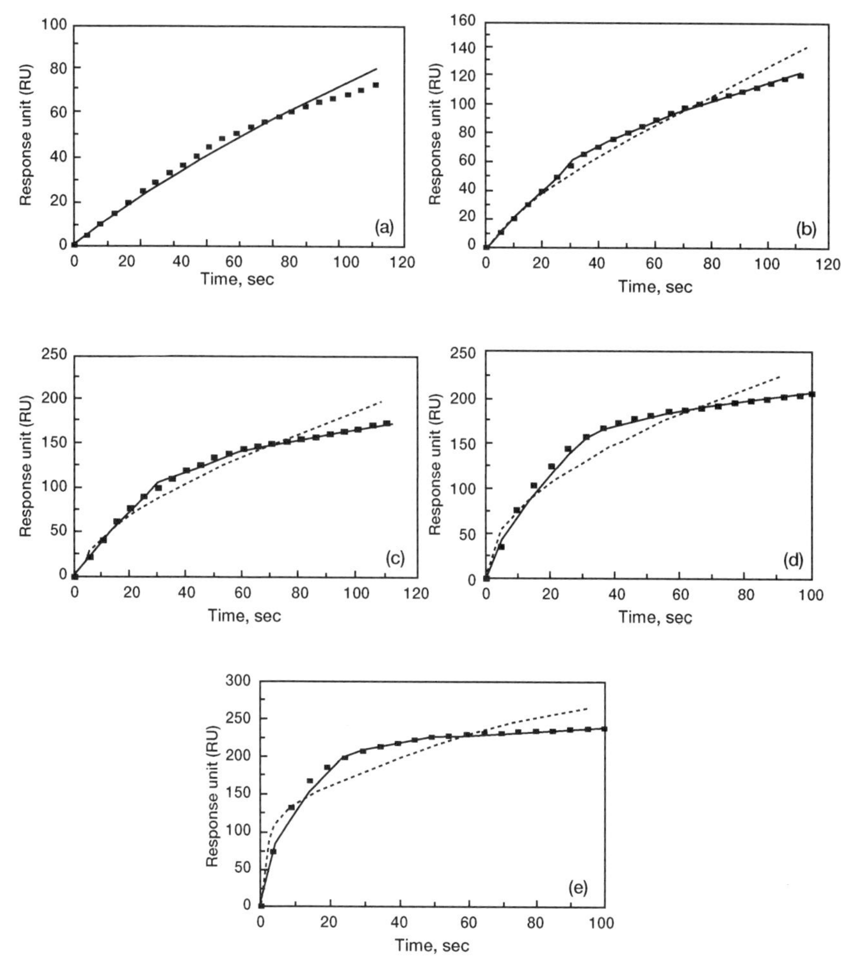

FIGURE 11.14 Binding of different concentrations (in mM) of variant of glycoprotein D [(gD1(275t)] of herpes simplex virus (HSV) to herpes entry mediator (HveAt) using surface plasmon resonance (Rux *et al.*, 1998). (a) 0.3 (single-fractal analysis); (b) 0.6; (c) 0.125; (d) 0.25; (e) 0.5 (- - -, single-fractal analysis; —, dual-fractal analysis).

the Eq. (11.9) is of value since it provides a quantitative measure of the influence of the heterogeneity on the surface on the binding rate coefficient, k_1. The binding rate coefficient, k_1, is sensitive to the fractal dimension, D_{f_1}, or the degree of heterogeneity that exists on the biosensor surface, as noted by the high value of the exponent.

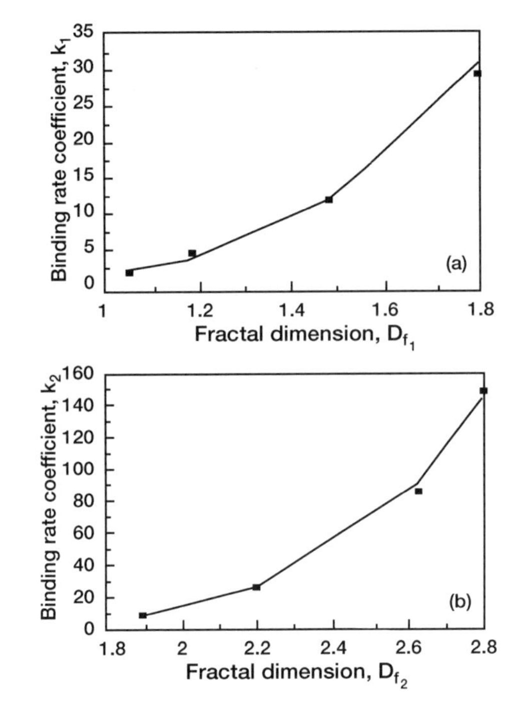

FIGURE 11.15 Influence of (a) the fractal dimension, D_{f_1} on the binding rate coefficient, k_1, and (b) the fractal dimension, D_{f_2} on the binding rate coefficient, k_2.

Figure 11.15b shows the dependence of the binding rate coefficient, k_2, on the fractal dimension, D_{f_2}. From the data presented in Table 11.3b, the binding rate coefficient is given by

$$k_2 = (0.11 \pm 0.005) D_{f_2}^{6.98 \pm 0.16}. \tag{11.10}$$

Again, more data points are required to provide a better and more reliable fit. However, this predictive equation is also of value since it provides a quantitative measure of the influence of the heterogeneity on the surface on the binding rate coefficient, k_2. The binding rate coefficient, k_2, is very sensitive to the degree of heterogeneity that exists on the biosensor surface, as noted by a very high exponent value.

Rella *et al.* (1996) used living phenol oxidizing *Bacillus stearothermphillus* microbial cells immobilized on a hydroxyethyl methacrylate membrane of an amperometric biosensor to detect phenol and related compounds. The authors emphasize that this sensor can be applied for the on-line detection and monitoring of phenols in industrial waste effluents. To minimize low enzyme stability, the authors selected a thermophile microorganism.

Figure 11.16a shows the binding of 25 mM chlorophenol in solution to the *Bacillus stearothermophillus* cells immobilized on a membrane of an amperometric biosensor. In this case, a single-fractal analysis is adequate to describe the binding kinetics. The values of k and D_f are given in Table 11.4. Figures 11.16b–11.16f show the binding of 25 mM 2-naphthol, o-cresol, p-cresol, phenol, and 4-methylcatechol, respectively, to the *Bacillus stearothermophillus* cells immobilized on the biosensor. In each case, a dual-fractal analysis is required to adequately describe the binding kinetics. The values of the binding rate coefficients and the fractal dimensions obtained for a single- and a dual-fractal analysis are given in Table 11.4.

Figures 11.16g and 11.16h show the binding of o-cresol and catechol, respectively, to the *Bacillus stearothermophillus* cells immobilized on the amperometric biosensor. In both cases, a dual-fractal analysis is required to adequately describe the binding kinetics. However, also in both cases, the binding curve exhibits a classical S-shaped curve or sigmoidal behavior. This sigmoidal behavior leads to a 0 value for D_f (for a single-fractal analysis) and for D_{f_1} (for a dual-fractal analysis). The positive error estimate in the value of the fractal dimension in each case is also given. A fractal dimension value less than 1 indicates that there are "holes" on the surface where there is no attachment of the receptor on the surface. Low values of the fractal dimension indicate that the surface exists as a Cantor-like dust. No explanation is presently offered to account for the 0 values estimated for the fractal dimensions, when either a single- or a dual-fractal analysis is used. The values of the binding rate coefficients and the fractal dimensions for a single- and a dual-fractal analysis are given in Table 11.4.

For the dual-fractal analysis examples analyzed (excluding the sigmoidal curves exhibited by o-cresol and catechol), the dependence of the binding rate coefficient, k_1 on the fractal dimension, D_{f_1}, is shown in Fig. 11.17a. From the data presented in Table 11.4, the binding rate coefficient is given by

$$k_1 = (0.05 \pm 0.04) D_{f_1}^{2.86 \pm 0.43} \tag{11.11}$$

Some of the scatter in the data may be attributed to the different phenols analyzed. More data are required to provide a better and more reliable fit. Nevertheless, the predictive equation is of value since it provides a quantitative measure of the influence of heterogeneity on the surface on the binding rate coefficient, k_1. The binding rate coefficient, k_1, is sensitive to the fractal dimension, D_{f_1}, or the degree of heterogeneity that exists on the biosensor surface, as noted by the high value of the exponent. Note that this equation applies to the different phenols whose binding to the *Bacillus*

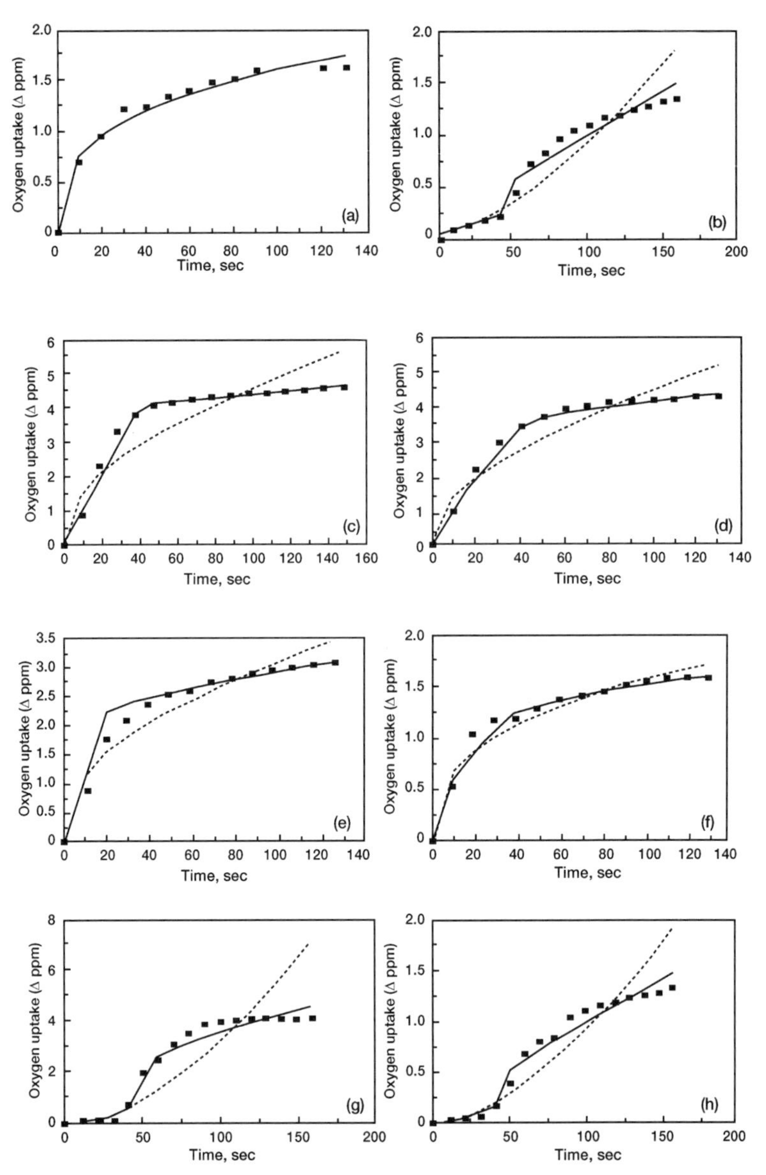

FIGURE 11.16 (a) Binding of 25 mM chlorophenol in solution to *Bacillus stearothermophillus* cells immobilized on a membrane of an amperometric biosensor (single-fractal analysis). [(b)–(f)] Binding of different phenols to *Bacillus stearothermophillus* cells immobilized on an amperometric biosensor. (b) 2-naphthol; (c) o-cresol; (d) p-cresol; (e) phenol; (f) 4-methylcatechol. [(g) and (h)] Binding of (g) o-cresol and (h) catechol to *Bacillus stearothermophillus* cells immobilized on an amperometric biosensor (sigmoid-kinetics) (- - -, single-fractal analysis; —, dual-fractal analysis) (Rella *et al.*, 1996).

Table 11.4 Influence of Different Compounds on the Fractal Dimensions and the Binding Rate Coefficients for Analyte-Receptor Reaction Kinetics on a *Bacillus stearothermophillus* Biosensor (Rella *et al.*, 1996)

Analyte (compound)/receptor	k	D_f	k_1	k_2	D_{f_2}	D_{f_1}
(a) *Single fractal analysis:* 25 μM 3-chlorophenol/phenol oxidizing *Bacillus Stearothermophillus* cells immobilized on hydroxy ethyl methacrylate membrane	0.36 ± 0.02	2.35 ± 0.04	na	na	na	na
(b) *Dual-fractal analysis:* 25 μM 2-naphthol/phenol oxidizing *Bacillus Stearothermophillus* cells immobilized on hydroxy ethyl methacrylate membrane	0.0014 ± 0.0004	$0.15 + 0.18$	0.0016 ± 0.0005	0.019 ± 0.002	$0.34 + 0.44$	1.28 ± 0.20
25 μM o-cresol/phenol oxidizing *Bacillus Stearothermophillus* cells immobilized on hydroxy ethyl methacrylate membrane	0.50 ± 0.13	2.04 ± 0.16	0.10 ± 0.02	3.24 ± 0.02	1.04 ± 0.27	2.87 ± 0.14
25 μM p-cresol/phenol oxidizing *Bacillus Stearothermophillus* cells immobilized on hydroxy ethyl methacrylate membrane	0.36 ± 0.07	1.89 ± 0.14	0.12 ± 0.02	1.68 ± 0.04	1.17 ± 0.23	2.59 ± 0.04
25 μM phenol/phenol oxidizing *Bacillus Stearothermophillus* cells immobilized on hydroxy ethyl methacrylate membrane	0.45 ± 0.06	2.17 ± 0.10	0.25 ± 0.03	1.37 ± 0.02	1.80 ± 0.17	2.67 ± 0.04
25 μM 4-methylcatechol/phenol oxidizing *Bacillus Stearothermophillus* cells immobilized on hydroxy ethyl methacrylate membrane	0.29 ± 0.04	2.27 ± 0.09	0.17 ± 0.03	0.53 ± 0.01	1.90 ± 0.24	2.54 ± 0.02
(c) *Dual-fractal analysis (zero fractal dimension sigmoid kinetics):* 25 μM o-cresol/phenol oxidizing *Bacillus Stearothermophillus* cells immobilized on hydroxy ethyl methacrylate membrane	0.0015 ± 0.0009	$0 + 0.33$	0.00057 ± 0.00005	0.28 ± 0.03	$0 + 1.16$	1.89 ± 0.17
25 μM catechol/phenol oxidizing *Bacillus Stearothermophillus* cells immobilized on hydroxy ethyl methacrylate membrane	0.00073 ± 0.00002	$0 + 0.21$	0.00073 ± 0.0003	0.12 ± 0.02	$0 + 0.62$	1.25 ± 0.19

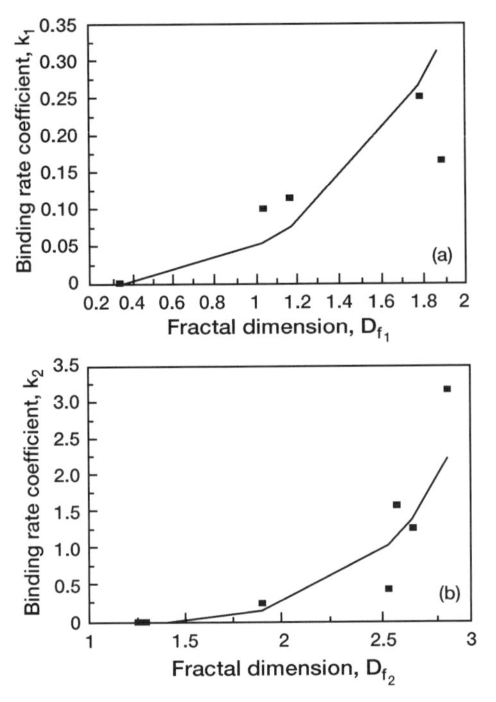

FIGURE 11.17 Influence of (a) the fractal dimension, D_{f_1}, on the binding rate coefficient, k_1, and (b) the fractal dimension, D_{f_2}, on the binding rate coefficient, k_2, for the binding of different phenols (Rella *et al.*, 1996).

stearothermophillus cells immobilized on the biosensor may be described by a dual-fractal analysis. This equation should also apply, within reason, to the binding of other phenols to this same microorganism.

The dependence of the binding rate coefficient, k_2, on the fractal dimension, D_{f_2}, is shown in Fig. 11.17b. From the data presented in Table 11.4, the binding rate coefficient is given by

$$k_2 = (0.0050 \pm 0.0026) D_{f_2}^{5.81 \pm 0.48}. \tag{11.12}$$

Once again, there is some scatter in the data and more data points would provide a better and more reliable fit. Some of the scatter may be attributed to the different phenols analyzed. Nevertheless, the predictive equation is of value since it provides a quantitative measure of the influence of heterogeneity on the surface on the binding rate coefficient, k_2. Again, the binding rate coefficient, k_2, is sensitive to the degree of heterogeneity that exists on the biosensor surface, as noted by the high value of the exponent dependence on D_{f_2}.

11.4. CONCLUSIONS

A fractal analysis of the binding of cellular analyte in solution to a (cellular) receptor immobilized on a biosensor surface provides a quantitative indication of the state of disorder (fractal dimension, D_f) and the binding rate coefficient, k, on the surface. The D_f value provides a quantitative measure of the degree of heterogeneity on the biosensor surface for the cellular analyte-receptor binding systems analyzed. Even though the cellular receptor is immobilized to a biosensor surface, it does provide insights into the cellular analyte-receptor binding reaction. Both a single- and a dual-fractal analysis were utilized to provide an adequate fit. This was done by the regression analysis provided by Sigmaplot (1993). The dual-fractal analysis was utilized to provide an adequate fit in spite of not presently having a molecular basis for it.

Quantitative expressions were developed for the binding rate coefficient as a function of the fractal dimension. The fractal dimension is not a classical independent variable, as indicated earlier, such as analyte concentration. Nevertheless, the expressions for the binding rate coefficient obtained as a function of the fractal dimension indicate the sensitivity of k_1 and k_2 on D_{f_1} and D_{f_2}, respectively. This is clearly brought out by the high-order and fractional dependence of the binding rate coefficient on the fractal dimension.

The binding rate coefficient has been directly related to the fractal dimension for cellular analyte-receptor reactions occurring on biosensor surfaces. Again, even though the reaction is occurring on a biosensor surface, the analysis provides physical insights into cellular analyte-receptor reactions. The quantitative expressions developed for the analyte-receptor systems should not only assist in enhancing biosensor performance parameters (such as sensitivity, selectivity, and stability), but also in understanding cellular analyte-receptor reactions in general. More such studies are required to determine whether the binding rate coefficient is indeed sensitive to the degree of heterogeneity that exists on the surface. For cellular analyte-receptor binding reactions, this provides an extra dimension of flexibility by which these reactions may be controlled. Cells may be induced to modulate their surfaces in desired directions. The analysis should encourage cellular experimentalists to pay increasing attention to the nature of the surface and how it may be manipulated to control cellular analyte-receptor binding reactions in desired directions.

REFERENCES

Anderson, J., NIH Panel Review Meeting, Case Western University, Cleveland, OH, July 1993.

BIAtechnology Note 103 (1994).

Bluestein, R. C., Diaco, R., Hutson, D. K., Neelkantan, N. V., Pankratz, T. J., Tseng, S. Y., and Vickery, E. K., *Clin. Chem.* **33**(9), 1543–1547 (1987).

Brynes, A. P. and Griffin, D. E., *J. Virology* 7349–7356 (1998).

Dabrowski, A. and Jaroniec, M., *J. Colloid & Interface Sci.* **73**, 475–482 (1979).

De Gennes, P. G., *Radiat. Phys. Chem.* **22**, 193–196 (1982).

Eddowes, M. J., *Biosensors* **3**, 1–15 (1987/1988).

Fischer, R. J., Fivash, M., Casas-Finet, J., Bladen, S., and McNitt, K. L., *Methods* **6**, 121–133 (1994).

Fratamico, P. M., Strobaugh, T. P., Medina, M. B., and Gehring, A. G., *Biotechnol. Techniq.* **12**, 571 (1998).

Giaver, I., *J. Immunol.* **116**, 766–771 (1976).

Glaser, R. W., *Anal. Biochem.* **213**, 152–158 (1993).

Gustaffson, J. A., *Science* (Washington, D.C.), **284**(5418), 1285–1286 (1999).

Havlin, S., in *The Fractal Approach to Heterogeneous Chemistry: Surfaces, Coloids, and Polymers* (D., Avnir, ed.), Wiley, New York, pp. 251–269, 1989.

Jaroniec, M. and Derylo, A., *Chem. Eng. Sci.* **36**, 1017–1019 (1981).

Kelly, C. G., Younson, J. S., Hikmat, B. Y., Todryk, S. M., Czisch, M., Haris, P. I., Flindall, I. R., Newby, C., Mallet, A. I., Ma, J. K. C., and Lehner, T., *Nature Biotech.* **17**, 42–47 (1999).

Kopelman, R., *Science* **241**, 1620–1626 (1988).

Lauffenberger, D., "New Tales from Cell Engineering; Cytokine Design, Gene Delivery, and Signal Transduction," Annual American Institute of Chemical Engineers Meeting, Dallas, Texas, October 31–November 5, 1999.

Markel, V. A., Muratov, L. S., Stockman, M. I., and George, T. F., *Phys. Rev. B* **43**(10), 8183–8188 (1991).

Nicola, A. V., Ponc de Leon, M., Xu, R., Hou, W., Whitbeck, J. C., Krumenacher, C., Montgomery, R. I., Spear, P. G., Eisenberg, R. J., and Cohen, G. H., *J. Virology* **72**, 3595 (1998).

Nygren, H. and Stenberg, M., *J. Colloid & Interface Sci.* **107**, 560–568 (1985).

Nyikos, L. and Pajkossy, T., *Electrochim. Acta* **31**, 1347–1350 (1986).

Oscik, J., Dabrowski, A., Jaroniec, M., and Rudzinski, W., *J. Colloid & Interface Sci.* **56**, 403–412 (1976)

Pajkossy, T. and Nyikos, L., *Electrochim. Acta* **34**, 171–179 (1989).

Pfeifer, P., Avnir, D., and Farin, D. J., *Nature (London)*, **308**(5956), 261–263 (1984a).

Pfeifer, P., Avnir, D., and Farin, D. J., *J. Colloid & Interface Sci.* **103**(1), 112–123 (1984b).

Pfeifer, P. and Obert, M., in *The Fractal Approach to Heterogeneous Chemistry: Surfaces, Colloids, and Polymers* (D. Avnir, ed.), and Wiley, New York, pp. 11–43, 1989.

Place, J. F., Sutherland, R. M., Riley, A., and Mangan, C., in *Biosensors with Fiberoptics* (D., Wise and L. B., Wingard Jr., eds.), Humana Press, New York, pp. 253–291, 1991.

Rella, R., Ferrara, D., Barison, G., Doretti, L., and Lora, S. *Biotechnol. Appl. Biochem.* **24**, 83 (1996).

Rudzinski, W., Lattar, L., Zajac, J., Wofram, E., and Paszli, J., *J. Colloid & Interface Sci.* **96**, 339–359 (1983).

Rux, A. H,. Willis, S. H., Nicola, A. V., Hou, W., Peng, C., Lou, H., Cohen, G.H., and Eisenberg, R. J., *J. Virology* **72**(9), 7091 (1998).

Sadana, A. and Sii, D., *J. Colloid & Interface Sci.* **151**(1), 166–177 (1992a).

Sadana, A., Alarie, J. P., and Vo-Dinh, T., *Talanta* **42**, 1567–1574 (1995).

Sadana, A. and Beelaram, A., *Biotech. Progr.* **9**, 45–55 (1994).

Sadana, A. and Beelaram, A., *Biosen. & Bioelectr.* **10**, 1567–1574 (1995).

Sadana, A. and Vo-Dinh, T., *Biotech. Progr.* **14**, 782–790 (1998).

Scheller, F. W., Hintsche, R., Pfeifer, P., Schubert, D., Reidel, K., and Kindevater, R., *Sensors & Actuators* **4**, 197–207 (1991).

Scroth-Diez, B., Ponimaskin, E., Reverey, H., Schmidt, M. F. G., and Herrmann, A., *J. Virology* **72**(1), 133–141 (1998).

Sigmaplot, Scientific Graphing Software, User's manual, Jandel Scientific, San Rafael, CA, (1993).

Sorenson, C. M. and Roberts, G. C., *J. Colloid & Interface Sci.*, **186**, 447–453 (1977).

Stenberg, M. and Nygren, H. A., *Anal. Biochem.* **127**, 183–192 (1982).

Stenberg, M., Stiblert, L., and Nygren, H. A., *J. Theor. Biol.* **120**, 129–142 (1986).

Tam, C. M. and Tremblay, A. Y., *Desalination* **1–3**, 77–92 (1993).

Teixira, A. I., "Effect of Nanostructure and Biomimetic Surfaces on Cell Behavior," Annual American Institute of Chemical Engineers Meeting, Dallas, Texas, October 31–November 5, (1999).

Thomson, S. J. and Webb, G., *Heterogeneous Catalysis*, Wiley, New York, p. 23, 1968.

Van Cott, T. C., Bethke, F. R., Polonis, V. R., Gorny, M. K., Zolla-Pazner, S., Redfield, R. R., and Birx, D. L., *J. Immunol.* **153**, 449–458 (1994).

CHAPTER 12

Surface Plasmon Resonance Biosensors

12.1. INTRODUCTION

An optical technique that has gained increasing importance in recent decades is the surface plasmon resonance (SPR) technique (Nylander *et al.*, 1982; Lukosz, 1991). This is particularly due to the development of the BIACORE biosensor, which is based on the SPR method and has found increasing industrial usage. Lofas *et al.* (1991) has reviewed the use of SPR for biospecific interaction analysis. Sigal *et al.* (1996) note that it is a particularly useful technique for analyzing processes occurring near or at the surfaces. In the SPR technique when the angle of incident electromagnetic radiation is larger than the critical angle, total reflection occurs and a nonradiative evanescent field is generated (Wink *et al.*, 1998). These authors indicate that at the resonance angle, a minimum in reflectivity is observed. This angle depends on the wavelength of incident light, the thickness of the gold layer (a thin metal layer is used to enhance the evanescent field), the dielectric properties of the glass, the gold, and the medium adjacent to the gold. In this technique, one monitors the refractive index of the evanescent field layer (which will change as the analyte-receptor binding interaction occurs on the surface).

If the interface between two media is coated with a thin layer of metal and the light is monochromatic and p-polarized (i.e., the electric component is parallel to the plane of incidence), a sharp "shadow," or intensity dip, appears in the reflected light at a specific angle. This phenomenon is the surface plasmon resonance (SPR), and the incident light angle is the SPR angle. SPR arises through the interaction of the evanescent wave created by total internal reflection with the delocalized surface electrons present in the thin metal film at the interface with the medium of lower refractive index. This interaction excites collective resonant oscillation to the electrons, or plasmons. The evanescent wave is amplified as a result of this plasmon resonance.

A change in the refractive index will lead to a shift in the resonance angle. The lower limit of detection by the SPR method is determined by the accuracy of measuring the shift in the resonance angle. According to Wink *et al.*, a disadvantage of the SPR method is that it does not detect low levels of concentrations of analytes. This is especially true for smaller molecules since they lead to smaller shifts in the resonance angle. Thus, an important area of investigation is to enhance the sensitivity of the SPR biosensor by analyzing different strategies or by providing a better understanding of the analyte-receptor binding interaction.

The change in the SPR signal—the SPR response—is directly related to the change in the surface concentration of the molecules. The refractive index (RI) is continuously monitored, detected as a change in the resonance angle in the detected volume (O'Shannessy *et al.*, 1995). In a sensorgram, the y axis is the resonance angle and is indicated in resonance or response units (RUs). Typically, a response of 1000 RU (or a 1 kiloRU) corresponds to a shift of 0.1° in the resonance angle, which in turn represents a change in the surface protein concentration of about $1\,ng/mm^2$, or in the bulk refractive index of about 10^{-3}. Typically, the total range covered by the SPR detector is 3°, corresponding to about 30,000 RU.

The reflected light angle that leads to SPR is mainly determined by the following three parameters.

(a) Metal film properties (thickness, optical constants, uniformity, microstructure).
(b) Wavelength of the incident light.
(c) Refractive index of the media on either side of the metal film.

Modern SPR devices can detect about $0.5\,ng/cm^2$ surface concentration, 0.1 nm film thickness or 10^{-12} M. One must take care while preparing samples, buffers, and protein solutions to achieve this accuracy level. Any inhomogeneity, impurities, gas bubbles, or temperature fluctuations will lead to an increase in the noise. The SPR sensitivity to analyte concentration depends strongly on the affinity constants and specificity of the immobilized biomolecules on the surface as well a: their surface concentration, packing density orientation, and denaturation level (Silin and Plant, 1997).

Today's sensors tend to be costly, cumbersome, and specialized (Service, 1997). This author indicates that it would be helpful to develop new sensors that are based on dirt-cheap starting materials. Such sensors could be effectively used as low-cost detectors for medical diagnostics, industrial monitoring, and environmental testing. The BIACORE biosensor is almost prohibitively expensive, and a cost reduction would significantly expand its user base as well as the number and types of analyte that could be used for possible detection.

Considering the importance of the SPR in biospecific interaction analysis, and particularly its increasing usage in industrial settings, it is worthwhile to further examine the role of surface roughness on the speed of response, specificity, stability, and sensitivity of SPR biosensors. Our intent is to provide physical insights into the binding interaction and thereby eventually help minimize the cost of the SPR biosensor. We will analyze both analyte-receptor and analyte-receptorless (protein adsorption) binding reactions with examples using both single- and dual-fractal analysis. The noninteger orders of dependence obtained for the binding rate coefficient on the fractal dimension will further reinforce the fractal nature of these analyte-receptor and analyte-receptorless binding systems.

12.2. THEORY

Since detailed theory was developed in previous chapters, we offer only a brief version of the theory here. We present a method of estimating actual fractal dimension values for analyte-receptor and analyte-receptorless (protein) binding systems utilized in surface plasmon resonance biosensors.

Experimental data presented for the binding of HIV virus (antigen) to the anti-HIV (antibody) immobilized on a surface displays a characteristic ordered disorder (Anderson, 1993). This indicates the possibility of a fractal-like surface. It is obvious that such a biosensor system (wherein either the antigen or the antibody is attached to the surface), along with its complexities (which include heterogeneities on the surface and in solution, diffusion-coupled reaction, time-varying adsorption or binding rate coefficients, etc.), can be characterized as a fractal system. Considering the complexities of the SPR biosensor surface and the clustering of the gold particles that is inevitable on these surfaces, a fractal analysis of the analyte-receptor binding is also appropriate here. As indicated in earlier chapters, the diffusion of reactants toward fractal surfaces has been analyzed (Sadana and Madagula, 1994; Sadana and Sutaria, 1997a, 1997b). Also, Havlin (1989) has briefly reviewed and discussed these results.

12.2.1. Single-Fractal Analysis

Havlin (1989) indicates that the diffusion of a particle (analyte) from a homogeneous solution to a solid surface (receptor-coated surface) where it reacts to form a product (analyte-receptor complex; analyte-receptor) is given by

$$(\text{Analyte} \cdot \text{Receptor}) \sim \begin{cases} t^{(3-D_{\mathrm{f,bind}})/2} = t^{p} & (t < t_c) \\ t^{1/2} & (t > t_c). \end{cases} \tag{12.1a}$$

Here, D_{f} is the fractal dimension of the surface. Equation (12.1a) indicates that the concentration of the product analyte • receptor(t) in the reaction analyte + receptor→analyte • receptor on a solid fractal surface scales at short and intermediate frames as analyte • receptor $\sim t^{p}$, with the coefficient $p = (3 - D_{\mathrm{f}})/2$ at short time frames and $p = \frac{1}{2}$ at intermediate time frames. This equation is associated with the short-term diffusional properties of a random walk on a fractal surface. The appearance of p different from 0 is the consequence of two different phenomena: the heterogeneity (fractality) of the surface and the imperfect mixing (diffusion-limited) condition. Havlin (1989) indicates that the crossover value may be determined by $r_c^2 \sim t_c$. Above the characteristic length, r_c, the self-similarity property disappears and regular diffusion is now present. For our analysis, t_c is arbitrarily chosen.

12.2.2. Dual-Fractal Analysis

The single-fractal analysis can be extended to include two fractal dimensions. At present, the time ($t = t_1$) at which the first fractal dimension changes to the second fractal dimension is arbitrary and empirical. For the most part, it is dictated by the data analyzed and experience gained by handling a single-fractal analysis. A smoother curve is obtained in the transition region if care is taken to select the correct number of points for the two regions. In this case, the product (analyte • receptor complex) concentration on the SPR surface is given by

$$(\text{Analyte} \cdot \text{Receptor}) \sim \begin{cases} t^{(3-D_{\mathrm{f}_1,\mathrm{bind}})/2} = t^{p_1} & (t < t_1) \\ t^{(3-D_{\mathrm{f}_2,\mathrm{bind}})/2} = t^{p_2} & (t_1 < t < t_2 = t_c) \\ t^{1/2} & (t > t_c). \end{cases} \tag{12.1b}$$

12.3. RESULTS

First, it is appropriate to mention that a fractal analysis will be applied to data obtained from surface plasmon resonance studies of biomolecular interactions. This is but one possible way to analyze the diffusion-limited binding kinetics assumed present in all of the systems analyzed. The parameters thus obtained would provide a useful comparision of different situations. Alternate explanations involving saturation, first-order reaction, and no diffusion limitations are possible, but they seem to be deficient in describing the heterogeneity that inherently exists on the surface. The binding reaction on the biosensor surface (SPR or other biosensor) is a complex reaction, and the fractal analysis via the fractal dimension and the binding rate coefficient provides a useful lumped-parameter analysis of the diffusion-limited situation.

Note that the SPR (for example, the BIACORE instrumentation manufactured by Pharmacia Biosensor) utilizes a carboxymethylated dextran surface which, according to Pharmacia, under appropriate and careful usage leads to diffusion-free binding kinetics. There are references to this effect available in the literature whereby first-order kinetics without heterogeneity on the surface describes the diffusion-free binding kinetics (Karlsson *et al.*, 1991; Lundstrom, 1994). Furthermore, good performance is also demonstrated for small molecules (Karlsson and Stahlberg, 1995). This is a widely used and expensive biosensor, and what we are offering here is an alternate explanation to describe the binding kinetics that includes both diffusional limitations and heterogeneity on the surface. This would be especially necessary if the SPR were not carefully utilized. Finally, our analysis would be of more value if we could offer an analysis of two different sets of experiments on the same sensing surface, one clearly diffusion-limited and one kinetically limited, to see whether a fractal analysis is really required for the second case. However, since we are analyzing the data available in the literature, we are unable to judge whether the data has been obtained under diffusion-free conditions. Thus, to be conservative, we have assumed that diffusion limitations are present and heterogeneity exists on the surface in all of the cases analyzed.

12.3.1. FRACTAL DIMENSION AND THE BINDING RATE COEFFICIENT

Wink *et al.* (1998) analyzed interferon-γ (IFN-γ) in solution using a liposome sandwich immunoassay developed specifically for SPR analysis. A 16-kDa cytokine (a capture monoclonal antibody, MD-2) was directly adsorbed onto a polystyrene layer. This polystyrene layer covers a gold surface attached to microtiter plates. These authors indicate that after the addition of the IFN-γ, a biotinylated detecting antibody is added. Avidin is the bridging molecule between the biotinylated antibody and the biotinylated liposomes. Figure 12.1 shows the curves obtained using Eq. (12.1a) for the binding of 20 ng/ml IFN-γ in solution to 16-kDa cytokine (capture antibody, MD-2) adsorbed on a polystyrene surface. Note that without the addition of the liposomes there was hardly any noticeable shift in the resonance angle.

A single-fractal analysis is adequate to describe the binding kinetics. The entire binding curve is utilized to obtain the fractal dimension and the binding rate coefficient. Table 12.1a shows the values of the binding rate coefficient, k, and the fractal dimension, D_f. The values presented in the table were obtained from a regression analysis using Sigmaplot (1993) to model the experimental data using Eq. (12.1a), wherein (analyte $\cdot$ receptor) $= kt^p$. The k and D_f values presented in the table are within 95% confidence limits. For example, for the binding of 20 ng/ml IFN-γ in solution to 16-kDa cytokine (capture monoclonal antibody, MD-2) adsorbed on a polystyrene surface, the k value is 263.76 ± 3.915. The 95% confidence limit indicates that 95% of the k values will lie between 259.845 and 267.675. This indicates that the values are precise and significant. The curves presented in the figures are theoretical

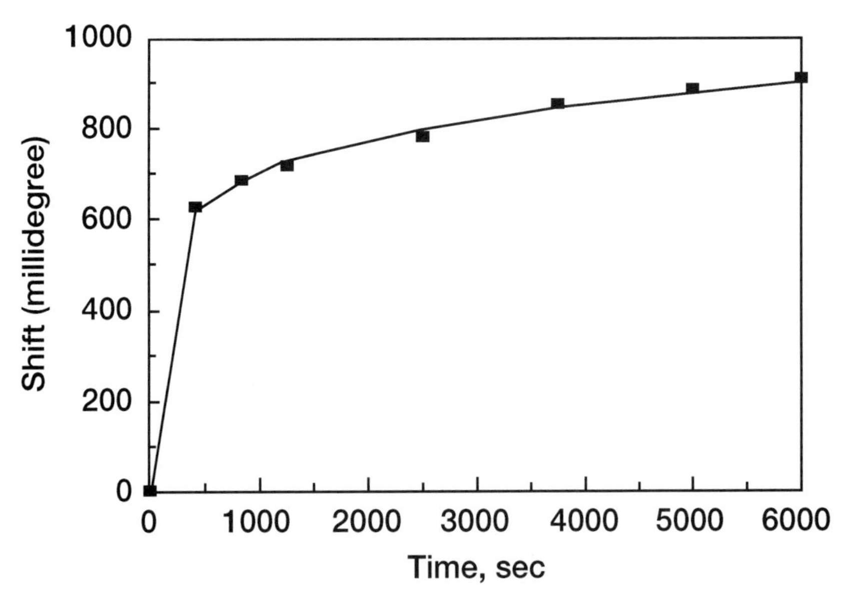

FIGURE 12.1 Binding of 20 ng/ml interferon-γ in solution to 16-kDa cytokine (capture monoclonal antibody, MD-2) adsorbed on a polystyrene surface (Wink *et al.*, 1998).

TABLE 12.1 Influence of Different Parameters on Fractal Dimensions and Binding Rate Coefficients for Different Analyte-Receptor Binding Reactions Utilizing Surface Plasmon Resonance: Single-Fractal Analysis

Analyte in solution/ receptor on surface	Binding rate coefficient, k	Fractal dimension, D_f	Reference
(a) 20 ng/ml interferon-γ/ 16-kDa cytokine (capture monoclonal antibody, MD-2) absorbed on polystyrene surface	263.76 ± 3.915	2.7134 ± 0.0120	(Wink *et al.*, 1998)
(b Protein-A-purified rabbit anti-ET-1 antibody + ET-$1_{16\text{-}21}$/ ET-$1_{15\text{-}21}$–BSA immobilized on extended carboxymethylated hydrogel matrix	3.3927 ± 0.1401	1.3658 ± 0.0544	(Laricchia-Robbio *et al.*, 1997)
Polyclonal rat anti-$1_{15\text{-}21}$/ ET-$1_{15\text{-}21}$–BSA immobilized on extended carboxymethylated hydrogel matrix	62.781 ± 0.915	2.112 ± 0.0122	(Laricchia-Robbio *et al.*, 1997)
Polyclonal rabbit anti-ET-$1_{15\text{-}21}$/ ET-$1_{15\text{-}21}$–BSA immobilized on extended carboxymethylated hydrogel matrix	39.143 ± 0.816	2.0622 ± 0.0278	(Laricchia-Robbio *et al.*, 1997)
Rabbit anti-ET-1 antibody/ immobilized ET-1	113.63 ± 4.428	2.3492 ± 0.0474	(Laricchia-Robbio *et al.*, 1997)

curves. All of the examples to be presented involve the presence of SPR. In general, the trends to be presented are similar to the ones observed when either the SPR or another type of biosensor is utilized.

Laricchia-Robbio *et al.* (1997) used a surface plasmon resonance biosensor for the detection and epitope mapping of endothelin-1. Endothelin-1 (ET-1) is a vasoconstrictor peptide that consists of 21 amino acids (Yanagisawa *et al.*, 1988). Yanagisawa *et al.* originally isolated this peptide from porcine endothelial cells. Three isoforms of human ET (ET-1, ET-2, ET-3) have been identified (Inoue *et al.*, 1989). Larrichia-Robbio *et al.* indicate that since ET-1 and its isoform exhibit high activity in hypertension and vasospasm, diagnostic tests are being developed for these compounds. Thus it is important to understand the structure–function relationships of these compounds and analyze their interactions with other receptors. Thus, these authors developed antibodies not only against ET-1, but also against the C-terminal eptapeptide, ET-1_{15-21}, due to its importance for receptor binding in ET-1.

The binding of protein-A-purified rabbit anti-ET-1 antibody incubated with the peptide ET-1_{16-21} in solution to ET-1_{15-21}–BSA coupled to an extended carboxymethylated hydrogel matrix in a BIACORE biosensor was analyzed by Laricchia-Robbio *et al.* (1997). Figure 12.2a shows the curves obtained using Eq. (12.2a) for a single-fractal analysis. Table 12.1b shows the values of k and D_f. Figure 12.2b shows the binding of polyclonal rat anti-ET-1_{15-21} in solution to ET-1_{15-21}–BSA immobilized on an extended carboxymethylated hydrogel matrix. Here too, a single-fractal analysis is sufficient to describe the binding kinetics and Table 12.1b shows the values of k and D_f.

Figure 12.2c shows the binding of polyclonal rabbit anti-ET-1_{15-21} in solution to ET-1_{15-21}–BSA immobilized on an extended carboxymethylated hydrogel matrix. Once again, a single-fractal analysis is sufficient to adequately describe the binding kinetics. Table 12.1b shows the values of k and D_f. Note that as one goes from the polyclonal rabbit anti-ET-1_{15-21} to the polyclonal rat anti-ET-1_{15-21}, D_f increases by about 2.4%—from 2.0622 to 2.1120—and k increases by about 37.4% from 39.143 to 62.781. Note also that increases in the fractal dimension and in the binding rate coefficient are in the same direction: An increase in the degree of heterogeneity on the biosensor surface (increase in D_f) leads to an increase in the binding rate coefficient. At present, no explanation is offered for this increase in the binding rate coefficient exhibited by a more heterogeneous surface, except that this is a phenomenological conclusion from the fractal analysis.

Larrichia-Robbio *et al.* (1997) performed experiments of epitope mapping to determine whether the binding of the first antibody to the immobilized antigen would affect the binding of a second antibody or vice versa. Their work also provided them with information on the number of distinct epitopes

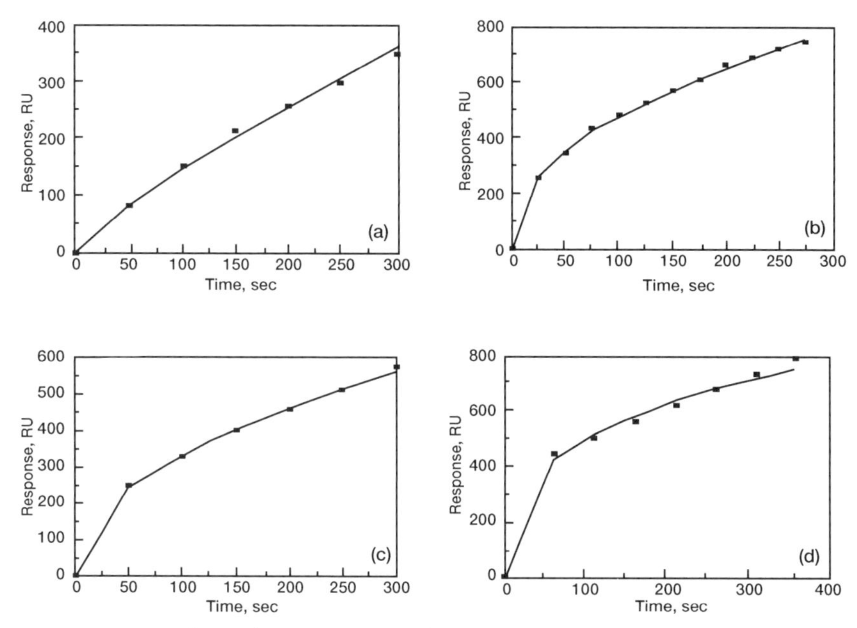

FIGURE 12.2 Binding and epitope mapping of human endothelin-1 (ET-1) by surface plasmon resonance (Laricchia-Robbio *et al.*, 1997). (a) Protein-A-purified rabbit anti-ET-1 antibody + ET-$1_{16\text{–}21}$ in solution to ET-$1_{15\text{–}21}$–BSA immobilized on extended carboxymethylated hydrogel matrix; (b) polyclonal rat anti-ET-$1_{15\text{–}21}$–BSA immobilized on extended carboxymethylated hydrogel matrix; (c) polyclonal rabbit anti-ET-$1_{15\text{–}21}$ in solution to ET-$1_{15\text{–}21}$-BSA immobilized on extended carboxymethylated hydrogel matrix; (d) rabbit anti-ET-1 antibody in solution to immobilized ET-1.

on the surface of ET-1. Figure 12.2d shows the binding of rabbit anti-ET-1 in solution to immobilized ET-1. Table 12.1b shows the values of k and D_f. Once again, a single-fractal analysis is adequate to describe the binding kinetics.

Figure 12.3 shows that the binding rate coefficient increases as the fractal dimension increases. This is in accord with the prefactor analysis of fractal aggregates (Sorenson and Roberts, 1997) and of the analyte-receptor binding kinetics observed for biosensor applications (Sadana and Sutaria, 1997; Sadana, 1998). For the data presented in Table 12.1, the binding rate coefficient, k, is given by

$$k = (0.4467 \pm 0.0799) D_f^{6.4405 \pm 0.3980}. \tag{12.2}$$

This predictive equation fits the values of k presented in Table 12.1 reasonably well. The very high exponent dependence indicates that the binding rate coefficient is very sensitive to the degree of heterogeneity that

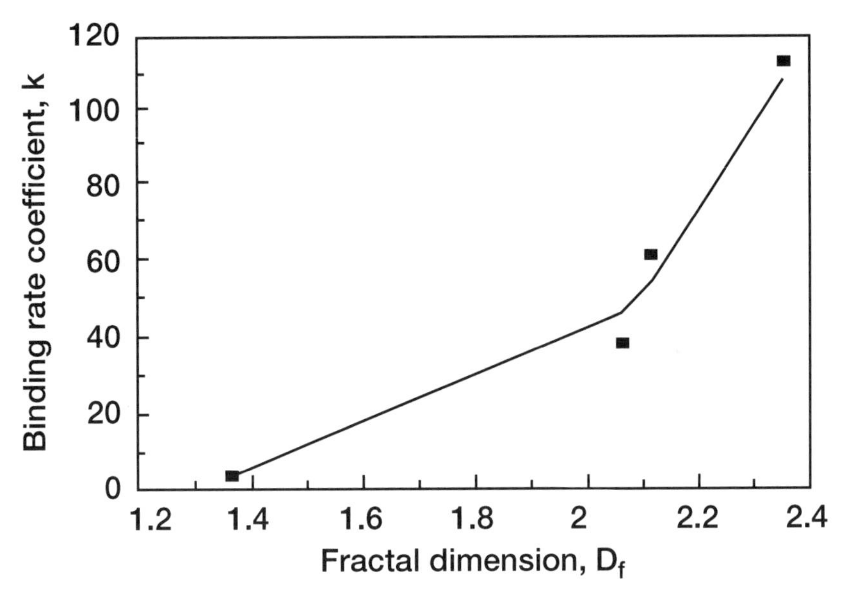

FIGURE 12.3 Influence of the fractal dimension, D_f, on the binding rate coefficient, k.

exists on the surface. More data points are required to more firmly establish this equation.

Peterlinz *et al.* (1997) analyzed the hybridization of thiol-tethered DNA on a passivated gold surface using two-color surface plasmon resonance spectroscopy. The novel two-color SPR method (Peterlinz and Georgiadis, 1996) allowed them to make quantitative the number of ss-DNA molecules per unit area for tethered DNA films. Peterlinz *et al.* tethered an immobilized oligonucleotide array that contains ss-DNA molecules with known sequence (probe) to a surface. When this surface was exposed to the target molecule (an ss-DNA molecule of unknown sequence), only those molecules whose sequence is complementary to the tethered molecule will bind (hybridize) to the probe on the surface.

Figure 12.4 shows the hybridization of 0.5 μM solution of complementary DNA fragment in 1.0 M NaCl to a DNA mercaptohexanol film. Peterlinz *et al.* indicate that in the film, the 25-base oligomer is tethered to the gold surface via an alkanethiol covalently linked to the 5′ position of the ss-DNA. The mercaptohexanol was added to minimize nonspecific adsorption. In this case, a single-fractal analysis does not provide an adequate fit, and thus a dual-fractal analysis was used. Table 12.2a shows the values of k and D_f for a single-fractal analysis and k_1, k_2, D_{f_1} and D_{f_2} for a dual-fractal analysis. Clearly, the dual-fractal analysis provides a better fit. For the dual-fractal analysis, note

TABLE 12.2 Influence of Different Parameters on Fractal Dimensions and Binding Rate Coefficients for Different Analyte-Receptor Binding Reactions Utilizing Surface Plasmon Resonance: Single and Dual-Fractal Analysis

Analyte in solution/receptor on surface	k	D_f	k_1	k_2	D_{f_1}	D_{f_2}	Reference
(a) 0.5 μM complementary DNA fragment/DNA tethered to a gold surface	2.144 ± 0.149	2.915 ± 0.030	2.141 ± 0.108	2.768 ± 0.025	2.747 ± 0.046	$3.0 - 0.014$	(Peterlinz *et al.*, 1997)
(b) 2×10^{-7} M human chorionic gonadotrophin (hCG)/α-hCG [1C]	1.230 ± 0.245	1.331 ± 0.154	0.9986 ± 0.0007	5.304 ± 0.264	0.8172 ± 0.096	2.746 ± 0.191	(Berger *et al.*, 1998)
10^{-8} M luteininzing hormone (LH)/ α-hCG [1C]	3.637 ± 1.542	1.889 ± 0.300	2.467 ± 0.601	37.37 ± 0.891	0.9282 ± 0.2940	$3.0 - 0.093$	(Berger *et al.*, 1998)
10^{-8} M luteininzing hormone (LH)/ αhCG [7B]	4.093 ± 0.229	1.881 ± 0.049	na	na	na	na	(Berger *et al.*, 1998)
2×10^{-7} M human chorionic gonadotrophin (hCG)/α-hCG [3A]	3.413 ± 0.326	2.293 ± 0.166	3.177 ± 0.247	6.779 ± 0.205	2.115 ± 0.101	$2.980 - 0.147$	(Berger *et al.*, 1998)
2×10^{-7} M human chorionic gonadotrophin (hCG)/α-hCG [3A]	5.268 ± 0.991	2.456 ± 0.156	4.342 ± 0.467	11.498 ± 0.578	1.923 ± 0.161	$3.0 - 0.122$	(Berger *et al.*, 1998)
10^{-8} M luteininzing hormone (LH)/ αhCG [3A]	2.053 ± 0.387	1.539 ± 0.169	1.726 ± 0.172	5.622 ± 0.354	1.024 ± 0.149	2.625 ± 0.262	(Berger *et al.*, 1998)
(c) 10^{-6} M bSA/plasmon carrying gold layer onto which four-channel flow cell was pressed	1.817 ± 0.288	1.572 ± 0.057	1.694 ± 0.253	3.038 ± 0.045	1.303 ± 0.081	2.098 ± 0.0358	(Berger *et al.*, 1998)
10^{-5} M bSA/plasmon carrying gold layer onto which four-channel flow cell was pressed	6.016 ± 0.047	2.053 ± 0.094	4.520 ± 0.977	11.246 ± 0.078	1.085 ± 0.216	2.688 ± 0.008	(Berger *et al.*, 1998)
10^{-4} M bSA/plasmon carrying gold layer onto which four-channel flow cell was pressed	10.632 ± 3.586	1.989 ± 0.113	9.211 ± 1.477	25.50 ± 0.128	0.6476 ± 0.145	2.902 ± 0.0047	(Berger *et al.*, 1998)

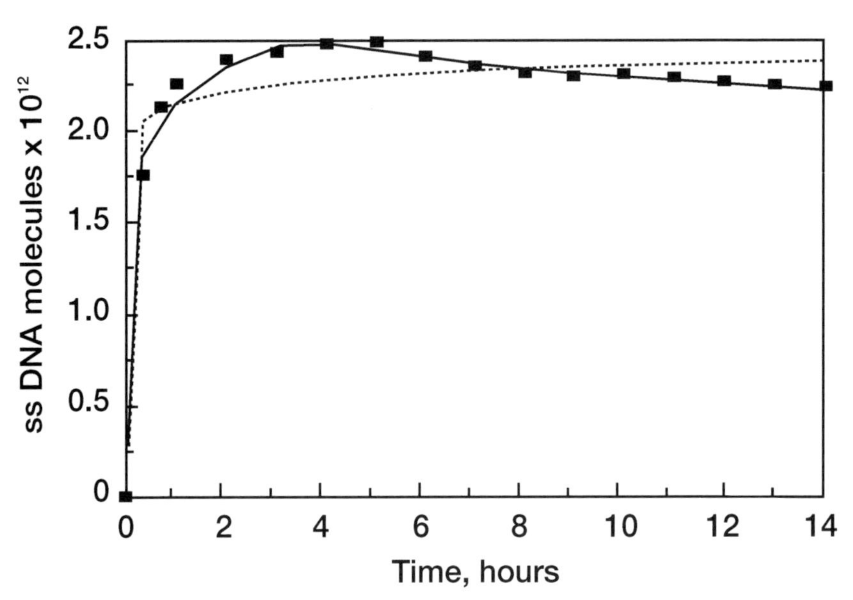

FIGURE 12.4 Hybridization (binding) of 0.5 μM complementary DNA fragment in solution to DNA tethered to a gold surface (Peterlinz *et al.*, 1997).

that as the fractal dimension increases by about 9.21%—from $D_{f_1} = 2.747$ to $D_{f_2} = 3.0$—the binding rate coefficient increases by a factor of 1.29—from $k_1 = 2.141$ to $k_2 = 2.768$. (Recall that the highest value that the fractal dimension can have is 3.0.) Also, the changes in the fractal dimension and in the binding rate coefficient are in the same direction. An increase in the degree of heterogeneity on the SPR surface leads to an increase in the binding rate coefficient. This also has been observed for analyte-receptor binding reactions occurring on other biosensor surfaces (Sadana and Sutaria, 1997; Sadana, 1998).

Berger *et al.* (1998) employed SPR multisensing to monitor four separate immunoreactions simultaneously by using a multichannel SPR instrument. These authors utilized a plasmon-carrying gold layer onto which a four-channnel cell was pressed. The gold layer was imaged at a fixed angle of incidence, which permitted the monitoring of changes in reflectance. Antibodies were coated to the surface, and antigens in solution were applied to the surface. Three different monoclonals of the α-hCG (human chorionic gonadotrophin) (1C, 7B, and 3A) were utilized. Human chorionic gonadotrophin and luteinizing hormone (LH) were used as antigens.

Figure 12.5a shows the binding curves obtained using a single-fractal analysis [Eq. (12.1a)] and a dual-fractal analysis [Eq. (12.1b)] for 2 ×

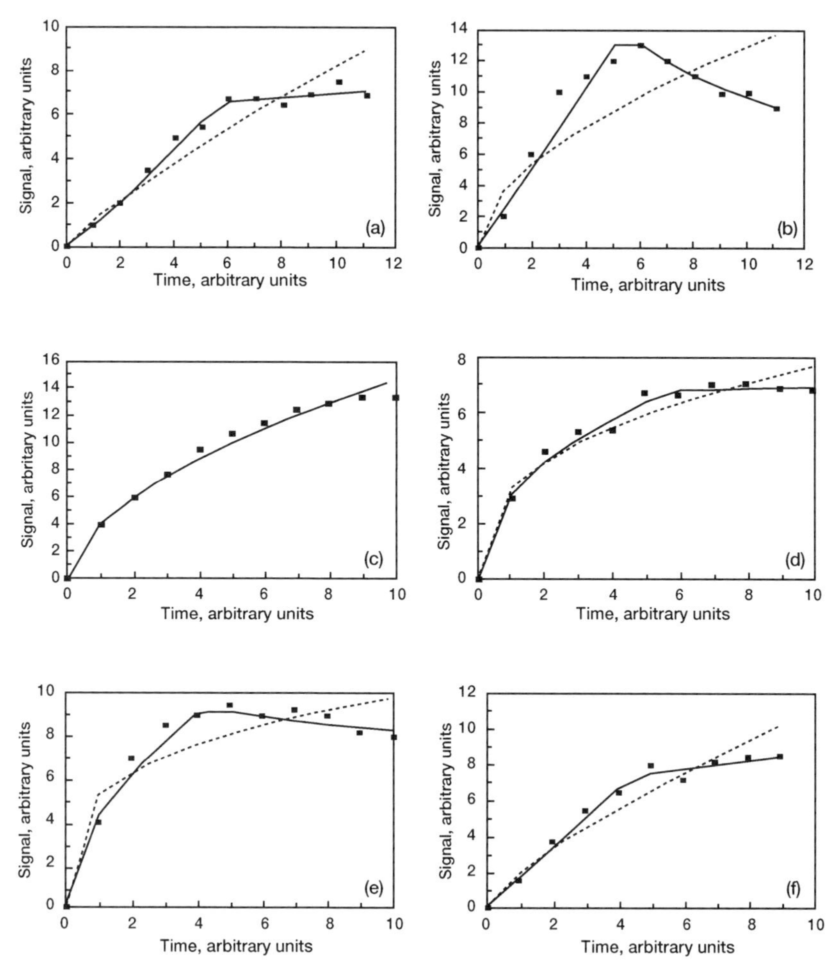

FIGURE 12.5 Binding of analyte in solution to receptor immobilized on a multichannel SPR surface (Berger *et al.*, 1998). (a) 2×10^{-7} M human chorionic gonadotrophin (hCG) in solution/α-hCG [1C] immobilized on surface (- - -, single-fractal analysis; —, dual-fractal analysis; this applies throughout the figures); (b) 10^{-8} M luteinizing hormone (LH) in solution/α-hCG [1C] immobilized on surface; (c) 10^{-8} M luteinizing hormone (LH) in solution/α-hCG [7B] immobilized on surface; (d) 2×10^{-7} M hCG in solution/α-hCG [3A] immobilized on surface; (e) 2×10^{-7} M hCG in solution/α-hCG [3A] immobilized on surface; (f) 10^{-8} M LH in solution/α-hCG [3A] immobilized on surface.

10^{-7} M hCG in solution to the α-hCG [1C] immobilized to a multisensing SPR surface. In this case, a single-fractal analysis does not provide an adequate fit, and thus a dual-fractal analysis was used. Table 12.2b shows the values of

k and D_f for a single-fractal analysis and k_1, k_2, D_{f_1} and D_{f_2} for a dual-fractal analysis. Once again, the dual-fractal anaysis clearly provides a better fit. Also for the dual-fractal analysis, note once again that as the fractal dimension increases by a factor of 3.35—from $D_{f_1} = 0.8172$ to $D_{f_2} = 2.746$—the binding rate coefficient increases by a factor of 5.31—from $k_1 = 0.9986$ to $k_2 = 5.304$. Thus, the binding rate coefficient is, once again, sensitive to the degree of heterogeneity that exists on the surface.

Figure 12.5b shows the binding of 10^{-8} M luteinizing hormone (LH) in solution to α-hCG [1C] immobilized to a multisensing SPR surface. Again, a single-fractal analysis does not provide an adequate fit so a dual-fractal analysis was utilized. Table 12.2b shows the values of k and D_f for a single-fractal analysis and k_1, k_2, D_{f_1}, and D_{f_2} for a dual-fractal analysis. For the dual-fractal analysis, as the fractal dimension increases by a factor of 3.23—from $D_{f_1} = 0.9282$ to $D_{f_2} = 3.0$—the binding rate coefficient increases by a factor of 14.1—from $k_1 = 2.467$ to $k_2 = 37.37$. In this case, the binding rate coefficient is very sensitive to the degree of heterogeneity that exists on the surface. Once again, the changes in the fractal dimension and in the binding rate coefficient are in the same direction.

The binding curve in Fig. 12.5b shows a maximum, and one might argue that a monotonically increasing function of time cannot explain the section of the curve where the signal decreases. It is for this reason that a dual-fractal analysis is required. k_1 and D_{f_1} are obtained from the increasing section of the curve, and k_2 and D_{f_2} are obtained from the decreasing section. For the decreasing section of the curve, the slope, b, of the signal versus time curve is negative. The fractal dimension, D_{f_2}, in this case is evaluated from $(3 - D_f)/2 = b$. Since b is negative, the value of D_{f_2} is set at the highest possible value—3.0. Note that the value of D_{f_2} reported in the table is $D_{f_2} = 3.0 - 0.093$. We did not use the $\pm$ sign since D_{f_2} cannot be higher than 3.0. This is just one possible explanation for the maximum exhibited in the binding curve. Another might be to resort to a model including saturation, simple first-order adsorption with normal diffusion limitation but with limited amounts of sites and/or reversibility. But these types of models seem to have a serious deficiency in that they do not incorporate the heterogeneity that exists on the surface. It is this heterogeneity on the surface under diffusion-limited conditions that we are trying to characterize using a single- or a dual-fractal analysis.

Of course, there is room to provide a better explanation for the decreasing section of the curve in a model that includes fractals, diffusion, and heterogeneity on the surface. Another possible explanation could be that desorption may lead to a decrease in the signal. (In the examples analyzed, no desorption is assumed.) Structural changes in the sensing layer may also lead to a decrease in the signal, but this is less likely. Ramsden *et al.* (1994) have also utilized random sequential adsorption to explain some of the deviations of protein adsorption kinetics from simple Langmuir first-order kinetics (and from pure diffusional limitations). Recognize that this represents "receptor-less" adsorption, wherein the protein adsorbs directly to a surface.

Figure 12.5c shows the binding of 10^{-8} M luteinizing hormone (LH) in solution to α-hCG [7B] immobilized to a SPR multisensing surface. In this case, in contrast to the curves in Fig. 12.5b for this multisensing system, a single-fractal analysis is sufficient to adequately describe the binding kinetics. Table 12.2b shows the values of k and D_f. The fact that a single-fractal analysis is adequate to describe the binding kinetics, in contrast to the previous two cases, implies that there is change in the binding mechanism between the binding of LH in solution to α-hCG [7B] immobilized to an SPR surface and the binding of both 10^{-8} M LH in solution to α-hCG [1C] immobilized to an SPR surface and 2×10^{-7} M hCG in solution to α-hCG [1C] immobilized to an SPR surface.

Figure 12.5d shows the binding of 2×10^{-7} M hCG in solution to α-hCG [3A] immobilized to a multisensing SPR surface. A single-fractal analysis does not provide an adequate fit, and a dual-fractal analysis is required. Table 12.2b shows the values of k and D_f for a single-fractal analysis and k_1, k_2, D_{f_1}, and D_{f_2} for a dual-fractal analysis. For the dual-fractal analysis, as the fractal dimension increases by a factor of about 1.41—from $D_{f_1} = 2.115$ to $D_{f_2} = 2.980$—the binding rate coefficient increases by a factor of about 2.13—from $k_1 = 3.177$ to $k_2 = 6.779$. As observed previously, the changes in the fractal dimension and the binding rate coefficient are in the same direction. It is interesting to compare the results obtained for the binding of the hCG/α-hCG [1C] and hCG/α-hCG [3A] systems. They are the same system, except for the monoclonal antibodies (α-hCG), which are type 1C and 3A. Note that the fractal dimensions (D_{f_1} and D_{f_2}) and the binding rate coefficients (k_1 and k_2) are all higher for type 3A than for type 1C. Apparently, for type 3A there is a higher degree of heterogeneity (D_{f_1} and D_{f_2}) on the surface when compared to type 1C, and this leads to higher values of the binding rate coefficients (k_1 and k_2).

Figure 12.5e shows the binding of 2×10^{-7} M hCG in solution to α-hCG [3A] immobilized to a multisensing SPR surface. Since this is exactly the same system as plotted in Fig. 12.5d, one may consider this as a repeat, or reproducibility, run. Note, however, that the values of the corresponding binding rate coefficients and fractal dimensions obtained for the single-fractal as well as the dual-fractal analysis are quite different from each other. This indicates that the results are not quite reproducible, at least for this case. Differences between similar systems with regard to signal size and kinetics, however, may depend on how many binding sites (antibodies) one has bound to the coupling matrix. Once again, the binding curve in Fig. 12.5e exhibits a maximum as in Fig. 12.5b. The same explanation applies here, so it is not repeated.

Figure 12.5f shows the binding of 10^{-8}M luteinizing hormone (LH) in solution to α-hCG [3A] immobilized to a multisensing SPR surface. Once again, a single-fractal analysis does not provide an adequate fit, and a dual-fractal analysis is required. Table 12.2b shows the values of k and D_f for a single-fractal analysis and k_1, k_2, D_{f_1} and D_{f_2} for a dual-fractal analysis. For the dual-fractal analysis, as the fractal dimension increases by a factor of about 2.56—from $D_{f_1} = 1.024$ to $D_{f_2} = 2.625$—the binding rate coefficient increases by a factor of about 3.25—from $k_1 = 1.726$ to $k_2 = 5.622$.

Compare the results obtained in Figs. 12.5c and 12.5f. They are both the same systems, LH/α-hCG, except that for Fig. 12.5c we have type 7B for α-hCG and for Fig. 12.5f we have type 3A for α-hCG. Note that in Fig. 12.5c a single-fractal analysis is sufficient to adequately describe the binding kinetics, whereas for Fig. 12.5f a dual-fractal analysis is required. This indicates that there is a difference in the binding mechanisms when these two different types of α-hCG are immobilized to the multisensing surface, everything else being the same.

It would be of interest to note the influence of the fractal dimension (or the degree of heterogeneity that exists on the surface) on the binding rate coefficient. However, not enough data is available for a particular condition. In lieu of that, we will use all the data presented in Table 12.2b for the dual-fractal analysis of the binding of analyte-receptor systems using multichannel SPR. Figure 12.6 plots the data for these binding rate coefficients as a function of the fractal dimension. Note that data for both α-hCG and LH are plotted together. Because of this, the result we present should be viewed with caution.

The binding rate coefficient, k_1, is given by

$$k_1 = (1.7647 \pm 0.6660) D_{f_1}^{1.1774 \pm 0.3105}. \tag{12.3}$$

Figure 12.6a shows that this predictive equation is very reasonable, considering we have plotted data for two different sets of systems and that in one system there are three different types of monoclonal antibodies of α-hCG (1C, 3A, and 7B) immobilized to the SPR multichannel surface. Equation (12.3) indicates that the binding rate coefficient, k_1, is only marginally sensitive to the surface roughness or the degree of heterogeneity that exists on the surface, as seen by the low exponent dependence of the binding rate coefficient on the fractal dimension (slightly more than 1).

We also made an initial attempt to analyze the influence of the fractal dimension, D_{f_2} on the binding rate coefficient, k_2. For the SPR multichannel

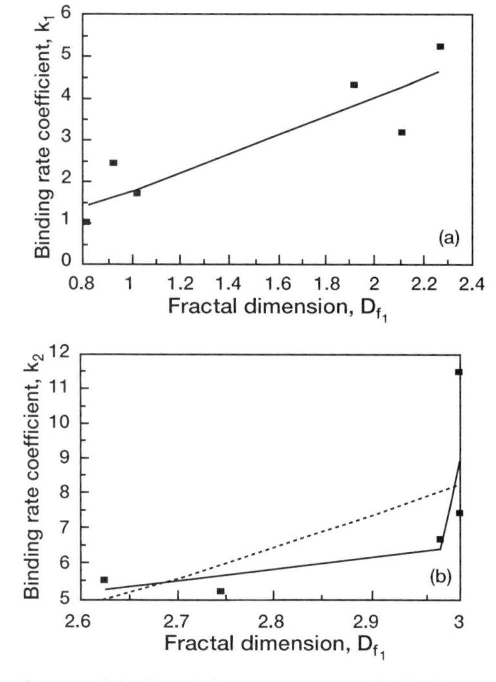

FIGURE 12.6 (a) Influence of the fractal dimension, D_{f_1}, on the binding rate coefficient, k_1. (b) Influence of the fractal dimension, D_{f_2} on the binding rate coefficient, k_2 (- - -), and of the fractal dimension, D_{f_1} and (D_{f_1} and D_{f_2}), on the binding rate coefficient, k_2 (—).

data given in Table 12.2b, the binding rate coefficient, k_2 is given by

$$k_2 = (0.1489 \pm 0.0404) D_{f_2}^{3.6440 \pm 1.9434}. \tag{12.4a}$$

Figure 12.6b shows that there is quite a bit of scatter (dotted line), due to the noticeable bend in the curve. This is indicated by the error in the exponent. A better fit could be obtained if more parameters were used. Initially, there is a degree of heterogeneity on the surface (D_{f_1}), which leads to another (higher value) degree of heterogeneity on the surface (D_{f_2}) as the reaction proceeds. Let us presume that the binding rate coefficient, k_2, at any time depends on not only the present degree of heterogeneity on the surface (D_{f_2}), but also on any previous one (D_{f_1}). In that case, the binding rate coefficient is given by

$$\begin{aligned} k_2 = {} & (1.0752 \pm 0.1136) D_{f_1}^{1.6598 \pm 1.1037} \\ & + (1.1\text{E} - 22 \pm 0.4\text{E} - 22) D_{f_2}^{48.016 \pm 55.717}. \end{aligned} \tag{12.4b}$$

This predictive equation fits the data presented in Fig. 12.6b (solid line) slightly better than does Eq. (12.4a). Recall that the highest value that the fractal dimension can have is 3, and near that value the binding rate coefficient rises rather sharply, which contributes to the high value of the exponent for D_{f_2}. More data points are required to more firmly establish the predictive equations around the fractal dimension value of 3. Note that we have used only the two lowest values of k_2 when the SPR multichannel surface exhibited a D_{f_2} value of 3. Apparently, in the region close to $D_{f_2} = 3$, the k_2 versus D_{f_2} curve would exhibit asymptotic characteristics.

It is of interest to analyze not only analyte-receptor binding kinetics using the SPR biosensor but also analyte-receptorless binding kinetics, for example, the binding of proteins in solution directly (receptorless) to an SPR surface. Figures 12.7a–12.7c show the binding of 10^{-6} to 10^{-4} M bSA in solution directly to an SPR multichannel surface (Berger *et al.*, 1998). In each case, a single-fractal analysis does not provide an adequate fit, and a dual-fractal analysis is required. Table 12.2c shows the values of k and D_f for a single-fractal analysis and k_1, k_2, D_{f_1}, and D_{f2} for a dual-fractal analysis. Note for the dual-fractal analysis that as the bSA concentration in solution increases from 10^{-6} M to 10^{-4} M both binding rate coefficients (k_1 and k_2) exhibit increases, the fractal dimension, D_{f_1}, decreases, and the fractal dimension, D_{f_2}, increases. The variation in the signal versus time is very small in Figs. 12.7b and 12.7c where one calculates the binding rate coefficient, k_2, and the fractal dimension, D_{f_2}. It is this very small variation that leads to very small values of b (the slope of the curve). The fractal dimension is evaluated from $(3 - D_f)/2 = b$. It is this very small value of b that leads to D_f values close to 3 (the maximum value). In other words, the degree of heterogeneity on the surface is now close to or at its maximum value.

In the bSA concentration range (10^{-6} to 10^{-4} M) in solution analyzed, the binding rate coefficient is given by

$$k_1 = (284.775 \pm 33.095)[\text{bSA}]^{0.3676 \pm 0.03376}. \quad (12.5a)$$

Figure 12.8a shows that only three data points are available. Nevertheless, the fit is quite reasonable. More data points would more firmly establish this equation. The low exponent dependence of k_1 on the bSA concentration indicates that the binding rate coefficient, k_1, is only mildly sensitive to the analyte concentration in solution.

Similarly, in the bSA concentration range (10^{-6} to 10^{-4} M) in solution analyzed, the binding rate coefficient, k_2, is given by

$$k_2 = (1950.72 \pm 431.98)[\text{bSA}]^{0.4630 \pm 0.0614}. \quad (12.5b)$$

Once again, Fig. 12.8b shows that only three data points are available.

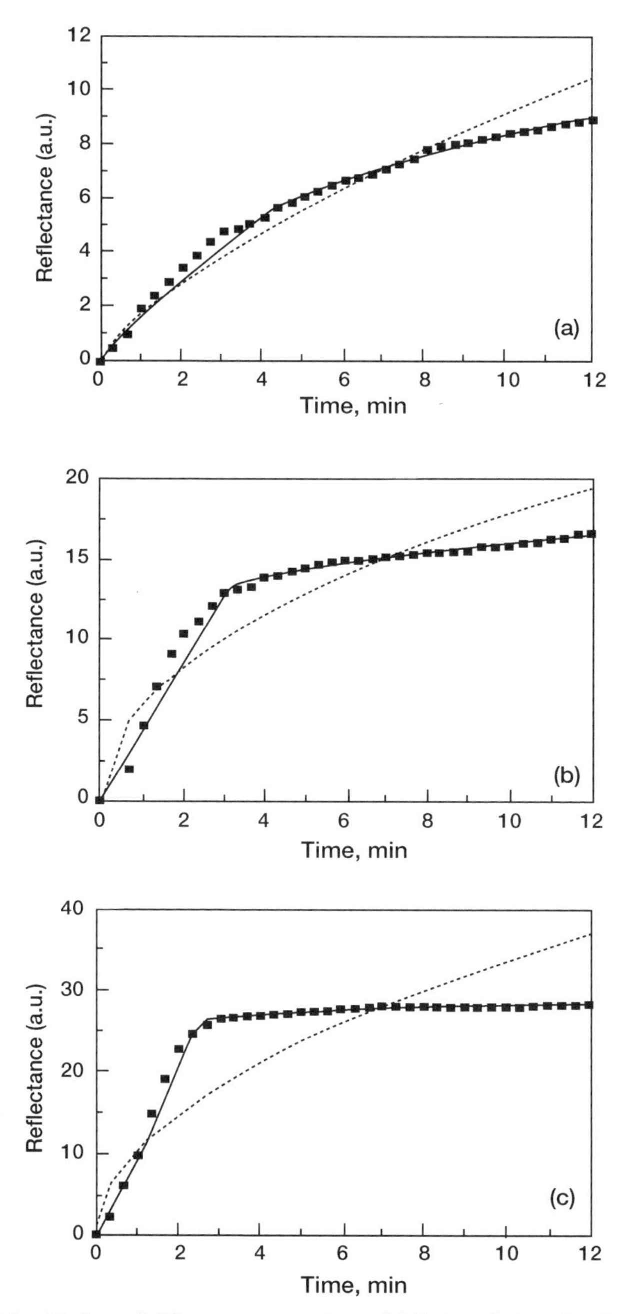

FIGURE 12.7. Binding of different concentrations of bSA in solution directly to the SPR biosensor surface (Berger *et al.*, 1998). (a) 10^{-6} M; (b) 10^{-5} M; (c) 10^{-4} M.

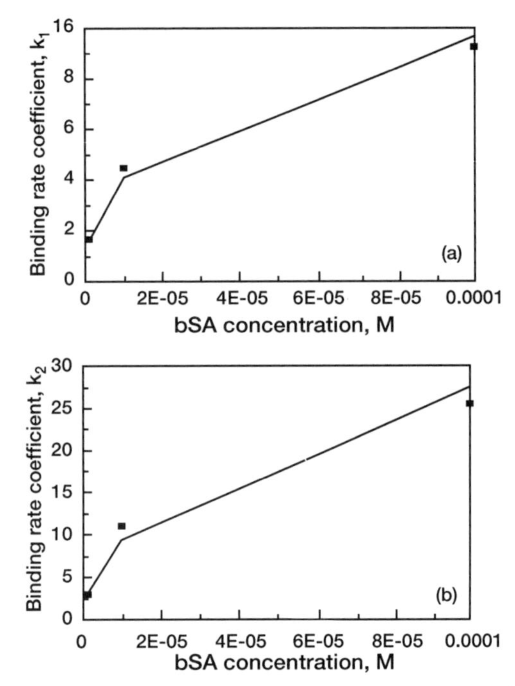

FIGURE 12.8 Influence of the bSA concentration (in M) in solution on (a) the binding rate coefficient, k_1, and (b) the binding rate coefficient, k_2.

Nevertheless, the fit is quite reasonable. More data points would more firmly establish this equation. Once again, the low exponent dependence of k_2 on the bSA concentration indicates that k_2 is only mildly sensitive to the analyte concentration in solution. The exponent dependence exhibited by both k_1 and k_2 on the bSA concentration in solution are quite close to each other (0.3676 and 0.4630, respectively), with k_2 being slightly more sensitive than k_1.

Figure 12.9a shows that D_{f_1} decreases as the bSA concentration in solution increases. In the bSA concentration range (10^{-6} to 10^{-4}M) in solution analyzed, the fractal dimension is given by

$$D_{f_1} = (0.1897 \pm 0.0241)[\text{bSA}]^{-0.1429 \pm 0.0366}. \tag{12.6a}$$

This predictive equation fits the values of D_f presented in Table 12.2c reasonably well. Since only three points are available, more data points are required to more firmly establish this equation. The fractal dimension, D_{f_1} is only mildly sensitive to the bSA concentration in solution, as noted by the low value (magnitude) of the exponent.

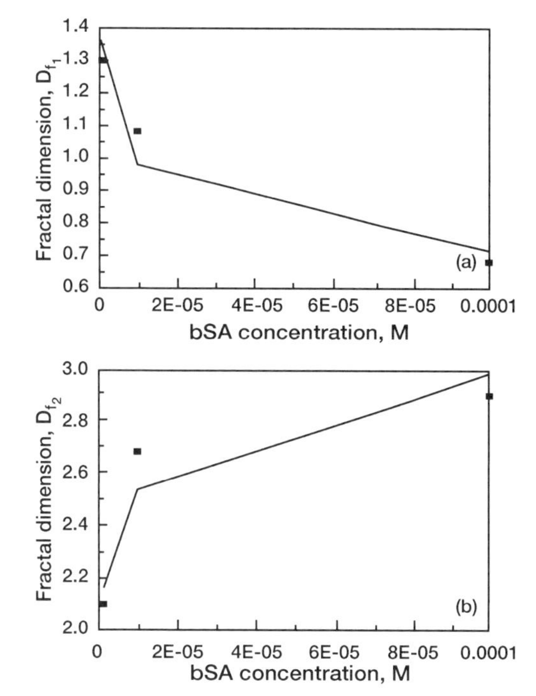

FIGURE 12.9 Influence of the bSA concentration (in M) in solution on (a) the fractal dimension, D_{f_1}, and (b) the fractal dimension, D_{f_2}.

In the bSA concentration range (10^{-6} to 10^{-4} M) in solution analyzed, the fractal dimension D_{f_2} is given by

$$D_{f_2} = (5.7125 \pm 0.4152)[\text{bSA}]^{0.0704 \pm 0.0215}. \tag{12.6b}$$

Figure 12.9b shows that D_{f_2} increases as the bSA concentration in solution increases. Once again, this predictive equation fits the values of D_{f_2} presented in Table 12.2c reasonably well. Once again, only three data points are available, so more data points are required to more firmly establish this equation. The fractal dimension, D_{f_2}, is only very slightly sensitive to the bSA concentration in solution, as noted by the very low value of the exponent. Interestingly, neither D_{f_1} nor D_{f_2} are sensitive to the bSA concentration in solution. In other words, the degree of heterogeneity on the surface is, for all practical purposes, independent of the bSA concentration in solution.

Figure 12.10a shows the decrease in k_1 with an increase in the D_{f_1}. In the 10^{-6} to 10^{-4} M bSA concentration range in solution analyzed, the binding

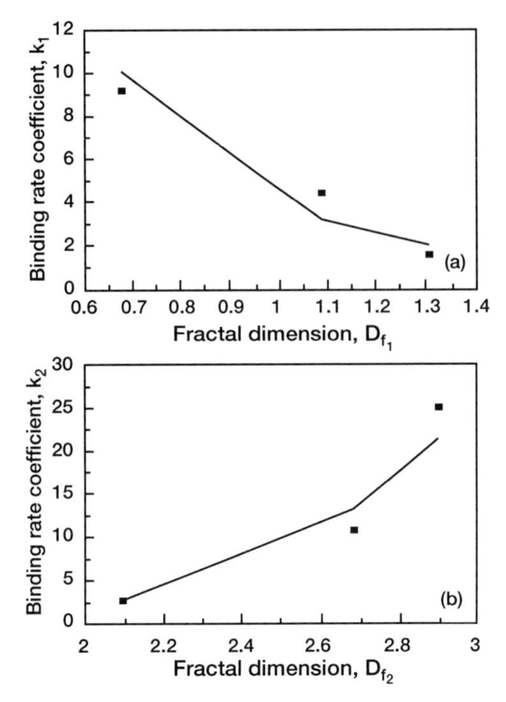

FIGURE 12.10 (a) Decrease in the binding rate coefficient, k_1, with an increase in the fractal dimension, D_{f_1}. (b) Increase in the binding rate coefficient, k_2, with an increase in the fractal dimension, D_{f_2}.

rate coefficient is given by

$$k_1 = (3.981 \pm 1.979) D_{f_1}^{-2.356 \pm 0.8394}. \tag{12.7a}$$

This predictive equation fits the values of k_1 presented in Table 12.2c reasonably well. Again, more data points are required over a wide range of D_{f_1} values to more firmly establish this equation. The fractional high exponent dependence of the binding rate coefficient, k_1, on the fractal dimension, D_{f_1} further reinforces the fractal nature of this system. The somewhat high exponent dependence of k_1 on D_{f_1} indicates that the binding rate coefficient is sensitive to the degree of heterogeneity that exists on the surface.

Figure 12.10b shows the increase in k_2 with an increase in D_{f_2}. In the 10^{-6} to 10^{-4} M bSA concentration range in solution analyzed, the binding rate coefficient, k_2, is given by

$$k_2 = (0.0284 \pm 0.0081) D_{f_2}^{6.2415 \pm 1.0371}. \tag{12.7b}$$

This predictive equation fits the values of k_2 presented in Table 12.2c reasonably well. As before, more data points are required over a wide range of D_{f_2} values to more firmly establish this equation. The very high fractional exponent dependence of the binding rate coefficient, k_2, on the fractal dimension, D_{f_2}, indicates that k_2 is very sensitive to the degree of heterogeneity that exists on the surface. Also, the fractional exponent dependence of k_2 on D_{f_2} further reinforces the fractal nature of the system.

Table 12.3 summarizes some of the binding rate and fractal dimension equations obtained.

12.3.2. Fractal Dimension and the Binding and Dissociation Rate Coefficients

Single-Fractal Analysis of the Dissociation Rate Coefficient

We propose that a similar mechanism is involved (except in reverse) for the dissociation step as for the binding rate coefficient. In this case, the dissociation takes place from a fractal surface. The diffusion of the dissociated particle (receptor [Ab] or analyte [Ag]) from the solid surface (e.g., analyte [Ag]-receptor [Ab] complex coated surface) in to solution may be given, as a first approximation, by

$$(\text{Analyte} \cdot \text{Receptor}) \sim -k' t^{(3-D_{f,\text{diss}})/2} \quad (t > t_{\text{diss}}). \qquad (12.8\text{a})$$

Here, $D_{f,\text{diss}}$ is the fractal dimension of the surface for the dissociation step and t_{diss} represents the start of the dissociation step, which corresponds to the highest concentration of the analyte-receptor on the surface. Henceforth, its concentration only decreases. $D_{f,\text{bind}}$ may or may not be equal to $D_{f,\text{diss}}$. Equation (12.8a) indicates that during the dissociation step, the concentration of the product $\text{Ab} \cdot \text{Ag}(t)$ in the reaction $\text{Ag} \cdot \text{Ab} \rightarrow \text{Ab} + \text{Ag}$ on a solid fractal surface scales at short time frames as $(\text{Ag} \cdot \text{Ab}) \sim -t^p$ with the coefficient, p, now equal to $(3 - D_{f,\text{diss}})/2$. In essence, the assumptions that are applicable for the association (or binding) step are applicable for the dissociation step. Once again, this equation is associated with the short-term diffusional properties of a random walk on a fractal surface. Note that in a perfectly stirred kinetics on a regular (nonfractal) structure (or surface), k_{diss} is a constant; that is, it is independent of time. In other words, the limit of regular structures (or surfaces) in the absence of diffusion-limited kinetics leads to k_{diss} being independent of time. In all other situations, one would expect a scaling behavior given by $k_{\text{diss}} \sim -k' t^{-b}$ with $-b = p < 0$. Once again, the appearance of p different from 0 is the consequence of two different

TABLE 12.3 Binding Rate Coefficient Equations as a Function of the Analyte Concentration and the Fractal Dimension, and Fractal Dimension Equations as a Function of the Analyte Concentration for an SPR Biosensor: Single- or Dual-Fractal Analysis

Analyte-receptor system	Binding rate coefficient or fractal dimension equation	Reference
Human endothelin-1 antibody/ ET-$1_{15\text{-}21}$–BSA or ET-1 immobilized on surface	$k = (0.4467 \pm 0.0799) D_f^{6.4405 \pm 0.3980}$	(Larichia-Robbio *et al.*, 1997)
Luteininzing hormone (LH) or human chronic gonadotrophin	$k_1 = (1.7647 \pm 0.6660) D_{f_1}^{1.1774 \pm 0.3105}$	(Berger *et al.*, 1998)
Human chorionic gonadotrophin (hCG)/different types of $\alpha - \text{hCG}$ immobilized on multisensing surface	$k_2 = (0.1489 \pm 0.0404) D_{f_2}^{3.6440 \pm 1.9434}$	(Berger *et al.*, 1998)
10^{-6} to 10^{-4}M bSA/receptorless surface	$k_1 = (3.981 \pm 1.979) D_{f_1}^{-2.356 \pm 0.8394}$	(Berger *et al.*, 1998)
	$k_2 = (0.0284 \pm 0.0081) D_{f_2}^{6.2415 \pm 1.0371}$	
	$k_1 = (284.775 \pm 33.095)[\text{bSA}]^{0.3676 \pm 0.0337}$	
	$k_2 = (1950.72 \pm 431.98)[\text{bSA}]^{0.4630 \pm 0.0614}$	
	$D_{f_1} = (0.1987 \pm 0.0241)[\text{bSA}]^{-0.1429 \pm 0.0366}$	
	$D_{f_1} = (5.7125 \pm 0.4152)[\text{bSA}]^{0.0704 \pm 0.0215}$	

phenomena: the heterogeneity (fractality) of the surface and the imperfect mixing (diffusion-limited) condition. The ratio, $K = k_{\text{diss}}/k_{\text{bind}}$, besides providing physical insights into the analyte-receptor system, is of practical importance since it may be used to help determine (and possibly enhance) the regenerability, reusability, stability, and other SPR biosensor performance parameters.

Dual-Fractal Analysis of the Dissociation Rate Coefficient

Once again, we propose that a similar mechanism is involved (except in reverse) for the dissociation step as for the binding rate coefficient(s). In this case, the dissociation takes place from a fractal surface. The diffusion of the dissociated particle (receptor [Ab] or analyte [Ag]) from the solid surface (e.g., analyte [Ag]-receptor [Ab] complex coated surface) into solution may be given, as a first approximation, by

$$(\text{Analyte} \cdot \text{Receptor}) \sim \begin{cases} -t^{(3-D_{f_1,\text{diss}})/2} & (t_{\text{diss}} < t < t_{d_1}) \\ -t^{(3-D_{f_2,\text{diss}})/2} & (t_{d_1} < t < t_{d_2}). \end{cases} \tag{12.8b}$$

Note that different combinations of the binding and dissociation steps are possible as far as the fractal analysis is concerned. Each of these steps or phases can be represented by either a single-or a dual-fractal analysis. For example, where the binding or the association step may be adequately described by a single-fractal analysis, the dissociation step may also be described by a single-fractal analysis although it is quite possible that the dissociation step may require a dual-fractal analysis. Also, where the association or the binding step may require a dual-fractal analysis, the dissociation step may require either a single- or a dual-fractal analysis. In effect, four possible combinations are possible: single fractal (association)–single fractal (dissociation); single-fractal (association)–dual-fractal (dissociation); dual-fractal (association)–single-fractal (dissociation); and dual-fractal (association)–dual-fractal (dissociation). Presumably, it is only by the analysis of a large number of association–dissociation analyte-receptor reaction systems that this point may be further clarified.

Patten *et al.* (1996) analyzed the binding of different concentrations of the Fab fragment $48G7^L48G7^H$ in solution to the p-nitrophenyl phosphonate (PNP) transition state analogue immobilized on a BIACORE (SPR) biosensor surface. These authors wanted to better understand the evolution of immune function. Figure 12.11a shows the curves obtained for the binding of $0.339\,\mu$M $48G7^L48G7^H$ in solution to the PNP immobilized on the BIACORE biosensor surface and for the dissociation of the $48G7^L48G7^H$-PNP complex from the same surface, along with its eventual diffusion in to solution. A dual-

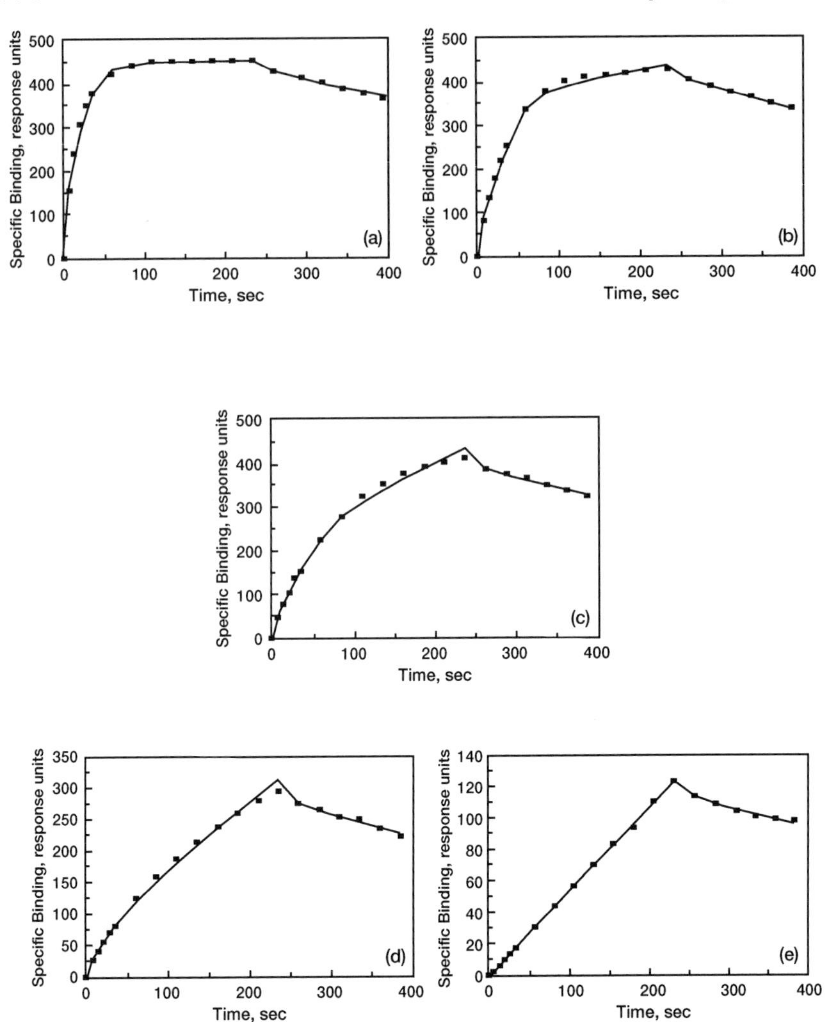

FIGURE 12.11 Binding and dissociation of different concentrations (in μM) of Fab fragment $48G7^L48G7^H$ (analyte) in solution to p-nitrophenyl phosphonate (PNP) transition state analogue (receptor) immobilized on a BIACORE biosensor surface (Patten *et al.*, 1996). (a) 0.339; (b) 0.169; (c) 0.085 (binding modeled as a dual-fractal analysis); (d) 0.042; (e) 0.021 (binding modeled as a single-fractal analysis); In all cases the dissociation is modeled as a single-fractal analysis.

fractal analysis is required to adequately describe the binding kinetics [Eq. (12.1b)], and a single-fractal analysis [eq. (12.8a)] is sufficient to describe the dissociation kinetics.

Table 12.4 shows the values of the binding rate coefficient, k_{bind}, the dissociation rate coefficient, k_{diss}, the fractal dimension for binding, $D_{f,bind}$, and the fractal dimension for dissociation, $D_{f,diss}$. The values of the binding and dissociation rate coefficient(s) and the fractal dimension(s) for association or adsorption (or binding) and dissociation presented in the table were obtained from a regression analysis using Sigmaplot (1993) to model the experimental data using Eq. (12.1b), wherein (analyte $\cdot$ receptor) $= k_{1,bind}t^p$ and $k_{2,bind}t^p$ for the binding step(s), and Eq. (12.8a), wherein (analyte $\cdot$ receptor) $= -k_{diss}t^p$ for the dissociation step. The binding and dissociation rate coefficient values presented in Table 12.4 are within 95% confidence limits. For example, for the binding of 0.339 μM $48G7^L48G7^H$ in solution to the PNP-immobilized surface, the reported value of $k_{1,bind}$ is 65.8 ± 5.91. The 95% confidence limit indicates that 95% of the $k_{1,bind}$ values will lie between 59.9 and 71.7. This indicates that the values are precise and significant. The curves presented in the figures are theoretical curves.

Figures 12.11b–12.11e show the binding of 0.169, 0.085, 0.042, and 0.021 μM $48G7^L48G7^H$ in solution to a PNP-immobilized surface as well as the dissociation of the $48G7^L48G7^H$-PNP complex from the surface. For the binding of 0.169 and the 0.085 μM $48G7^L48G7^H$ in solution to the PNP-immobilized surface, a dual-fractal analysis is required to adequately describe the binding kinetics. For the lower concentrations (0.042 and 0.021 μM $48G7^L48G7^H$ in solution to the PNP-immobilized surface), a single-fractal analysis is sufficient. This indicates that there is a change in the binding mechanism as one goes from the lower to the higher $48G7^L48G7^H$ concentration in solution. The dissociation kinetics for each of the $48G7^L48G7^H$ concentrations utilized may be adequately described by a single-fractal analysis.

The authors (Patten *et al.*, 1996) indicate that nonspecific binding has been eliminated by subtracting out sensorgrams from negative control surfaces. Our analysis here does not include this nonselective adsorption or binding. We do recognize that, in some cases, this may be a significant component of the adsorbed material and that this rate of association, which is of a temporal nature, would depend on surface availability. If we were to accommodate the nonselective adsorption into the model, there would be an increase in the degree of heterogeneity on the surface since by its very nature nonspecific adsorption is more heterogeneous than specific adsorption or binding. This would lead to higher fractal dimension values since the fractal dimension is a direct measure of the degree of heterogeneity that exists on the surface.

Table 12.4 indicates that $k_{1,bind}$ increases as the $48G7^L48G7^H$ concentration in solution increases. Figure 12.12a shows the increase in $k_{1,bind}$ (or k_1, to avoid the double subscript nomenclature) with an increase in $48G7^L48G7^H$ concentration in solution. In the $48G7^L48G7^H$ concentration range (0.021 to

TABLE 12.4 Fractal Dimensions and Binding and Dissociation Rate Coefficients Using a Single- and Dual-Fractal Analysis for the Binding of Fab $48G7^L$ $48G7^H$ (Analyte)in Solution to p-Nitrophenyl Phosphonate (PNP) Transition State Analogue (Receptor) Immobilized on a BIACORE Biosensor Surface (Patten *et al.*, 1996)

$48G7^L 48G7^H$ (in μM)/ PNP on surface	k_{bind}	$k_{1,bind}$	$k_{2,bind}$	k_{diss}	$D_{f,bind}$	$D_{f_1,bind}$	$D_{f_2,bind}$	$D_{f,diss}$
0.339	na	65.8 ± 5.91	357 ± 4.3	3.47 ± 0.13	na	2.04 ± 0.10	2.91 ± 0.02	1.70 ± 0.05
0.169	na	29.8 ± 1.8	186 ± 5.2	1.93 ± 0.14	na	1.80 ± 0.06	2.67 ± 0.04	1.44 ± 0.09
0.085	na	12.5 ± 0.4	39.7 ± 1.7	2.73 ± 0.19	na	1.58 ± 0.04	2.13 ± 0.07	1.65 ± 0.09
0.042	6.31 ± 0.24	na	na	2.10 ± 0.12	1.57 ± 0.02	na	na	1.62 ± 0.08
0.021	0.529 ± 0.00	na	na	1.73 ± 0.09	1.00 ± 0.00	na	na	1.89 ± 0.06

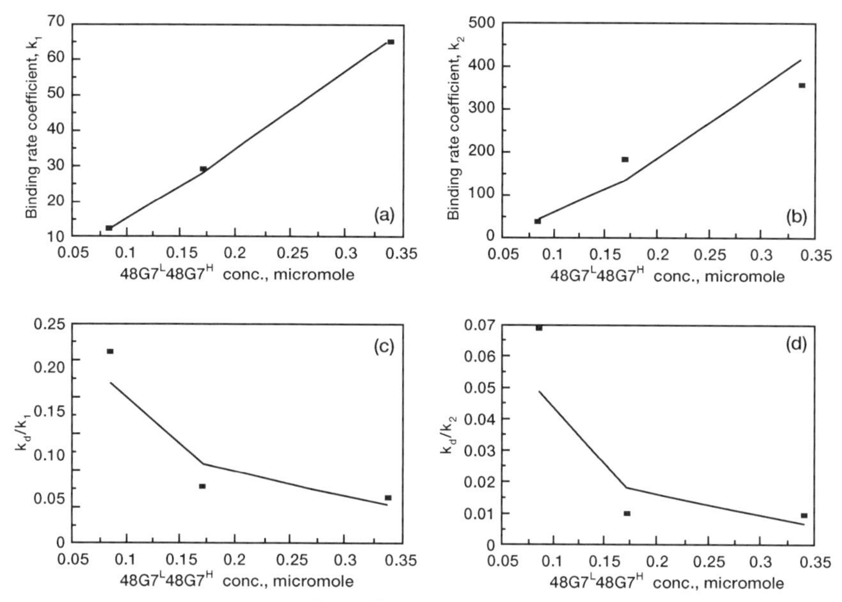

FIGURE 12.12 Influence of $48G7^L48G7^H$ (analyte) concentration in solution on (a) binding rate coefficient, k_1 (or $k_{1,\mathrm{bind}}$), (b) binding rate coefficient, k_2 (or $k_{2,\mathrm{bind}}$), and (c) affinity, k_d/k_1.

0.339 μM) analyzed, $k_{1,\mathrm{bind}}$ is given by

$$k_{1,\mathrm{bind}} = (244 \pm 8.8)[48G7^L48G7^H]^{1.20 \pm 0.04}. \tag{12.9a}$$

The fit is reasonable, although more data points are required to establish this equation more firmly. Nevertheless, Eq. (12.9a) is of value since it provides an indication of the change in $k_{1,\mathrm{bind}}$ as the $48G7^L48G7^H$ concentration in solution changes in the range analyzed. The fractional exponent dependence indicates the fractal nature of the system.

Table 12.4 indicates that $k_{2,\mathrm{bind}}$ increases as the $48G7^L48G7^H$ concentration in solution increases. Figure 12.2b shows the increase in $k_{2,\mathrm{bind}}$ (or k_2) with an increase in $48G7^L48G7^H$ concentration in solution. In the $48G7^L48G7^H$ concentration range (0.085 to 0.339 μM) analyzed, $k_{2,\mathrm{bind}}$ is given by

$$k_{2,\mathrm{bind}} = (2306 \pm 1046)[48G7^L48G7^H]^{1.59 \pm 0.38}. \tag{12.9b}$$

The fit is reasonable, but there is some scatter in the data so more data points are required to establish this equation more firmly. Nevertheless, Eq. (12.9b) is of value since it provides an indication of the change in $k_{2,\mathrm{bind}}$ as the

$48G7^L48G7^H$ concentration in solution increases. The fractional exponent dependence once again indicates the fractal nature of the system. $k_{2,bind}$ is slightly more sensitive than $k_{1,bind}$ on the $48G7^L48G7^H$ concentration in solution, as noted by their exponent values (1.59 and 1.20, respectively). The binding rate coefficients exhibit an order of dependence on the analyte concentration between 1 and 2.

The ratios $k_{diss}/k_{1,bind}$ (or k_d/k_1) and $k_{diss}/k_{2,bind}$ (or k_d/k_2) are important since they provide a measure of affinity of the receptor for the analyte—in this case, the PNP immobilized on the BIACORE biosensor surface for the $48G7^L48G7^H$ in solution. For the dual-fractal analysis examples presented, Fig. 12.12c shows the decrease in k_d/k_1 as the $48G7^L48G7^H$ concentration in solution increases. In the 0.085 to 0.339 μM $48G7^L48G7^H$ concentration in solution analyzed, the ratio k_d/k_1 is given by

$$k_d/k_1 = (0.015 \pm 0.008)[48G7^L48G7^H]^{-1.02 \pm 0.42}. \tag{12.9c}$$

Figure 12.12c shows that the ratio k_d/k_1 decreases as the $48G7^L48G7^H$ concentration in solution increases. There is scatter in the data, reflected in the error estimates for both the exponent as well as the coefficient. Although more data points are required to more firmly establish this equation, it is of value since it provides an indication of the affinity along with its quantitative change with a change in the $48G7^L48G7^H$ concentration in solution. The dependence of k_d/k_1 is clearly nonlinear even though the exponent estimate is close to 1. However, the error estimate for the exponent does have a value of ± 0.42, which underscores the nonlinearity dependence. At the lower $48G7^L48G7^H$ concentrations, the k_d/k_1 value is higher, at least in the concentration range analyzed. Thus, if affinity is a concern and one has the flexibility of selecting the analyte concentration to be analyzed, then one should utilize lower concentrations of $48G7^L48G7^H$.

For the dual-fractal analysis examples presented, Fig. 12.12d shows the decrease in k_d/k_2 as the $48G7^L48G7^H$ concentration in solution increases. In the 0.085 to 0.339 μM $48G7^L48G7^H$ concentration in solution analyzed, the ratio k_d/k_2 is given by

$$k_d/k_2 = (0.0015 \pm 0.0032)[48G7^L48G7^H]^{-1.41 \pm 0.77}. \tag{12.9d}$$

Figure 12.12d shows that the ratio k_d/k_2 decreases as the $48G7^L48G7^H$ concentration in solution increases. There is quite a bit of scatter in the data, reflected in the error estimates for both the exponent as well as (especially) the coefficient. More data points are definitely required to more firmly establish this equation. Nevertheless, Eq. (12.9d) is of value since it provides an indication of the affinity along with its quantitative change with a change in the $48G7^L48G7^H$ concentration in solution. The dependence of k_d/k_2 on

$48G7^L48G7^H$ concentration is clearly nonlinear. At the lower $48G7^L48G7^H$ concentrations, the k_d/k_2 value is higher, at least in the concentration range analyzed. Thus, if affinity is a concern and one has the flexibility of selecting the analyte concentration to be analyzed, then one should utilize lower concentrations of $48G7^L48G7^H$.

Table 12.4 and Fig. 12.13a show that the binding rate coefficient, k_1 (or $k_{1,\text{bind}}$), increases as the fractal dimension, D_{f_1} (or $D_{f_1,\text{bind}}$), increases. In the $48G7^L48G7^H$ concentration range (0.085 to 0.339 μM) analyzed, k_1 is given by

$$k_1 = (0.578 \pm 0.006) D_{f_1}^{6.69 \pm 0.06} \tag{12.10a}$$

This predictive equation fits the values of k_1 presented in Table 12.4 reasonably well. The very high exponent dependence indicates that the binding rate coefficient is very sensitive to the degree of heterogeneity that exists on the surface. More data points are required to more firmly establish this equation.

Table 12.4 and Fig. 12.13b show that the binding rate coefficient, k_2 (or $k_{2,\text{bind}}$), increases as the fractal dimension, D_{f_2} (or $D_{f_2,\text{bind}}$), increases. In the $48G7^L48G7^H$ concentration range (0.085 to 0.339μM) analyzed, k_2 is given by

$$k_2 = (0.207 \pm 0.007) D_{f_2}^{6.96 \pm 0.15} \tag{12.10b}$$

This predictive equation fits the values of k_2 presented in Table 12.4 reasonably well. The very high exponent dependence indicates that the binding rate coefficient is very sensitive to the degree of heterogeneity that exists on the surface. More data points are required to more firmly establish this equation. It is interesting that the order of dependence on the surface heterogeneity is approximately the same for both k_1 (exponent dependence 6.69) and for k_2 (exponent dependence 6.96).

Table 12.4 and Fig. 12.13c show that the dissociation rate coefficient, k_{diss} (or k_d), increases as the fractal dimension for the dissociation step, $D_{f,\text{diss}}$ (or $D_{f,\text{d}}$), increases. For the data presented in Table 12.4, k_d is given by

$$k_d = (0.574 \pm 0.06) D_{f,d}^{3.26 \pm 0.87}. \tag{12.10c}$$

This predictive equation fits the values of k_d presented in Table 12.4 reasonably well. The same fitting procedure has been applied in Figs. 12.13a–12.13c. The curves fit the data points in Figs. 12.13a and 12.13b rather well. The fit in Fig. 12.13c is not as good. This is due to the scatter in the data, reflected in the error estimates for the exponent dependence. The dissociation rate coefficient is quite sensitive to the degree of heterogeneity that exists on the surface, as indicated by the approximately third-order dependence on the fractal dimension, $D_{f,d}$. More data points are required to more firmly establish this equation.

As indicated earlier, the ratios k_d/k_1 or k_d/k_2 are important since they provide a measure of the affinity of the receptor for the analyte. It would be of interest to analyze the influence of surface heterogeneity on the affinity of the

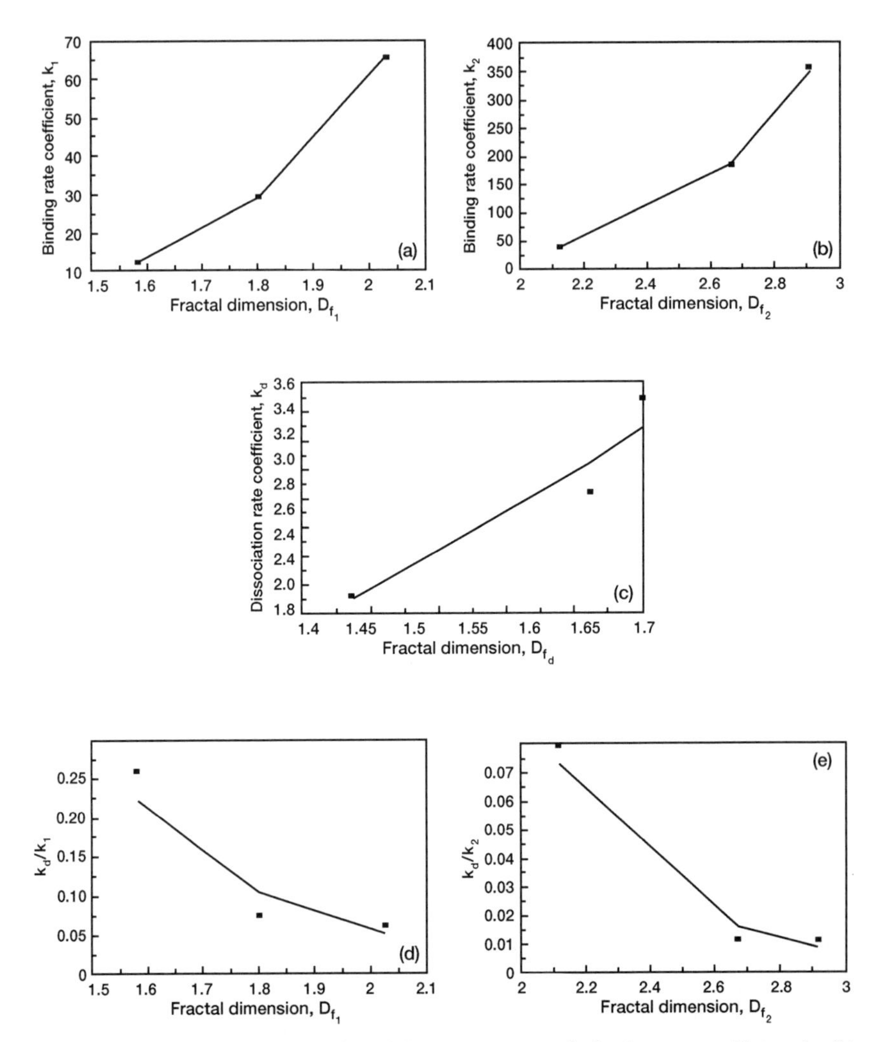

FIGURE 12.13 Influence of (a) the fractal dimension, D_{f_1}, on the binding rate coefficient, k_1, (b) the fractal dimension, D_{f_2}, on the binding rate coefficient, k_2, (c) the fractal dimension, $D_{f,d}$, on the binding rate coefficient, k_d, (d) the fractal dimension, D_{f_1}, on the affinity, k_d/k_1, and (e) the fractal dimension, D_{f_2}, on the affinity, k_d/k_2.

receptor for the analyte. Table 12.4 and Fig. 12.13d show that the ratio k_d/k_1 decreases as the fractal dimension for binding, D_{f_1}, increases. It would perhaps be more appropriate to obtain some sort of average of the fractal dimension for the binding and the dissociation steps, and then fit k_d/k_1 versus the appropriate average of the fractal dimensions. Since no appropriate average of the fractal dimension for the two steps is available, only the D_{f_1} (for the binding step) is used.

For the data presented in Table 12.4, k_d/k_1 is given by

$$k_d/k_1 = (2.62 \pm 1.26) D_{f_1}^{-5.74 \pm 2.24}. \tag{12.10d}$$

This predictive equation fits the values of the affinity presented in Table 12.4 reasonably well. There is some scatter in the data, reflected in the error estimates for the exponent dependence as well as the coefficient. Some of the error may be attributed to not using an appropriate weighted average of the fractal dimension for the binding and the dissociation phases. Nevertheless, the affinity is very sensitive to the degree of heterogeneity that exists on the surface, as noted by the high value of the exponent. Although more data points are required to more firmly establish this equation, it is of value since it provides an indication of the influence of the degree of heterogeneity on the surface on the affinity. If the affinity is a primary concern, one should apparently try to minimize the degree of heterogeneity on the surface since this is deleterious for affinity. In other words, one should keep the surface as homogeneous as possible if the affinity is a primary concern.

Table 12.4 and Fig. 12.13e shows that the ratio k_d/k_2 decreases as the fractal dimension for binding, D_{f_2}, increases. Here too, it would perhaps be more appropriate to obtain some sort of average of the fractal dimension for the binding and the dissociation steps, and then fit k_d/k_2 versus the appropriate average of the fractal dimensions. However, as indicated above, since no appropriate average of the fractal dimension for the two steps is available, just the D_{f_2} (for the binding step) is used. For the data presented in Table 12.4, k_d/k_2 is given by

$$k_d/k_2 = (9.69 \pm 4.45) D_{f_2}^{-6.67 \pm 1.64}. \tag{12.10e}$$

This predictive equation fits the values of k_d/k_2 presented in Table 12.4 reasonably well. There is some scatter in the data, reflected in the error estimates for the exponent dependence. Once again, some of the error may be attributed to not using an appropriate weighted average of the fractal dimension for the binding and the dissociation phases. Nevertheless, Eq. (12.10e) is of value since it provides an indication of the influence of the degree of heterogeneity on the surface on the affinity. Once again, if affinity is a primary concern, one should apparently minimize the degree of heterogeneity on the surface since an increasing heterogeneity is deleterious for affinity. In other words, one should keep the surface as homogeneous as is possible if affinity is a primary concern.

Cooper and Williams (1999) utilized the SPR biosensor to analyze the binding kinetics of anti-β-galactosidase in solution to β-galactosidase immobilized on a lipid monolayer. These authors emphasize that it is important to make quantitative the bivalent binding of analytes in solution to receptors on the surface for the prediction of their behavior for *in vivo* reactions. They further emphasize that their lipid monolayers closely resemble the surface of a cellular membrane and that their lipid membrane mimics the membrane surface behavior *in vivo*.

Figure 12.14a shows the curves obtained using Eq. (12.1a) for the binding of 0.63 nM anti-β-galactosidase in solution to 500 RU of β-galactosidase immobilized on a lipid monolayer and using Eq. (12.8a) for the dissociation of the anti-β-galactosidase-β-galactosidase complex from the same surface along with its eventual diffusion into solution. A single-fractal analysis is

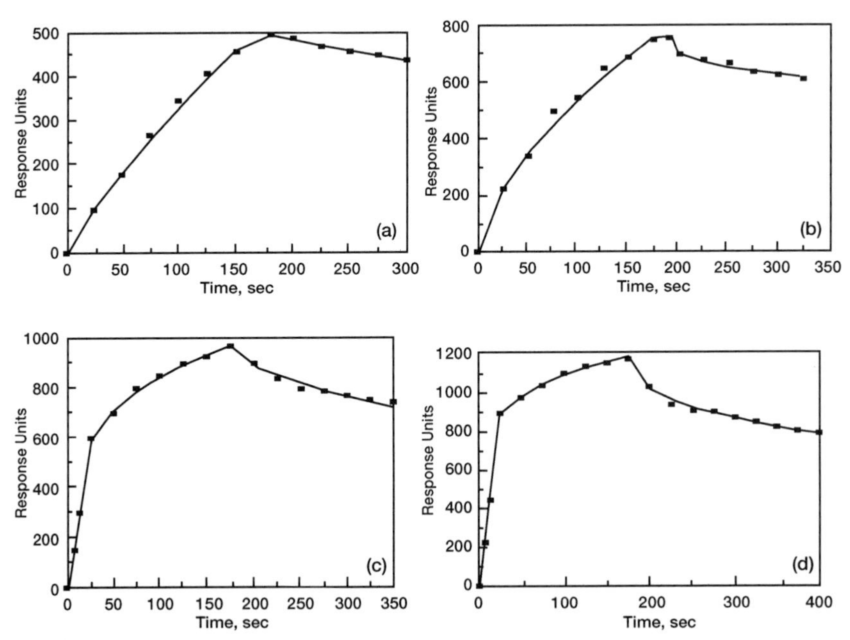

FIGURE 12.14 Binding and dissociation of different concentrations (in nM) of anti-β-galactosidase IgM in solution to β-galactosidase immobilized on a lipid monolayer using a SPR biosensor (Cooper and Williams, 1999). (a) 0.63; (b) 1.25, binding modeled as a single-fractal analysis; (c) 2.5; (d) 5.0, binding modeled as a dual-fractal analysis. In all cases the dissociation is modeled as a single-fractal analysis.

sufficient to adequately describe the binding kinetics [Eq. (12.1a)] as well as the dissociation kinetics [Eq. (12.8a)].

Table 12.5 shows the values of the binding rate coefficients, k_{bind}, the dissociation rate coefficient, k_{diss}, the fractal dimension for binding, $D_{f,\text{bind}}$, and the fractal dimension for dissociation, $D_{f,\text{diss}}$. Figure 12.14b shows the curves obtained for the binding and the dissociation of 1.25 nM anti-β-galactosidase in solution to 500 RU of β-galactosidase immobilized on a lipid monolayer. Once again, a single-fractal analysis is sufficient to adequately describe the binding as well as the dissociation kinetics.

Figures 12.14c and 12.14d show the curves obtained for the binding and the dissociation of 2.5 and 5.0 nM anti-β-galactosidase in solution, respectively, to 500 RU β-galactosidase immobilized on a lipid monolayer. In both cases, a dual-fractal analysis is required to adequately describe the binding kinetics. The dissociation kinetics in both cases is, once again, adequately described by a single-fractal analysis. The fact that the binding of the lower concentrations may be adequately described by a single-fractal analysis whereas the higher concentrations require a dual-fractal analysis indicates that there is a change in the binding mechanism as one goes from the lower (0.63 and 1.25) to the higher (2.5 and 5.0) concentrations of nM anti-β-galactosidase in solution. Also, apparently there is no significant change in the dissociation kinetics for the 0.63 to 5.0 nM anti-β-galactosidase in solution since all of the four concentrations analyzed (0.63, 1.25, 2.5, and 5.0 nM) may be adequately described by a single-fractal analysis.

12.4. CONCLUSIONS

A predictive approach using fractal analysis of the binding of analyte in solution to a receptor immobilized on the SPR biosensor surface provides a quantitative indication of the state of disorder (fractal dimension, $D_{f,\text{bind}}$) and the binding rate coefficient, k_{bind}, on the surface. In addition, fractal dimensions for the dissociation step, $D_{f,\text{diss}}$, and dissociation rate coefficients, k_{diss}, were also presented. This provides a more complete picture of the analyte-receptor reactions occurring on the SPR surface than does an analysis of the binding step alone, as done previously (Sadana, 1999). Besides, one may also use the numerical values for the rate coefficients for the binding and dissociation steps to classify the analyte-receptor biosensor system as moderate binding, extremely fast dissociation; moderate binding, fast dissociation; moderate binding, moderate dissociation; moderate binding, slow dissociation; fast binding, extremely fast dissociation; fast binding, fast dissociation; or fast binding, moderate dissociation; or fast binding, slow dissociation.

Table 12.5 Fractal Dimensions and Binding and Dissociation Rate Coefficients Using a Single- and a Dual-Fractal Analysis for the Binding of Anti-β-galactosidase IgM in Solution to β-galactosidase Immobilized on a Monolayer Using a Surface Plasmon Resonance (SPR) Biosensor (Cooper and Williams, 1999)

Anti-β-galactosidase concentration (in nM) in solution/ β-galactosidase on a lipid monolayer	k_{bind}	$k_{1,bind}$	$k_{2,bind}$	k_{diss}	$D_{f,bind}$	$D_{f_1,bind}$	$D_{f_2,bind}$	$D_{f,diss}$
0.63	6.87 ± 0.37	na	na	0.585 ± 0.087	1.32 ± 0.06	na	na	1.03 ± 0.19
1.25	33.4 ± 1.8	na	na	25.1 ± 2.7	1.79 ± 0.06	na	na	2.31 ± 0.09
2.5	na	24.4 ± 0.08	263 ± 3.3	11.7 ± 1.2	na	1.01 ± 0.00	2.49 ± 0.02	1.82 ± 0.11
5	na	36.4 ± 0.15	533 ± 5.61	39.2 ± 2.49	na	1.02 ± 0.00	1.69 ± 0.02	2.15 ± 0.06

The fractal dimension value provides a quantitative measure of the degree of heterogeneity that exists on the SPR surface for the analyte-receptor systems analyzed. The degree of heterogeneity for the binding and the dissociation phases is, in general, different for the same reaction. This indicates that the same surface exhibits different degrees of heterogeneity for the binding and the dissociation reaction. We examined examples wherein either a single-or a dual-fractal analysis was required to describe the binding kinetics. In some cases, both types of fractal analysis was required for one analyte-receptor system where an experimental condition was changed. The dual-fractal analysis was used only when the single-fractal analysis did not provide an adequate fit. This was done by the regression analysis provided by Sigmaplot (1993). The dissociation step was adequately described by a single-fractal analysis for all of the examples presented.

In accord with the prefactor analysis for fractal aggregates (Sorenson and Roberts, 1997), quantitative (predictive) expressions were developed for the binding rate coefficient, k_{bind}, the dissociation rate coefficient, k_{diss}, and the affinity as a function of the fractal dimension. Predictive equations were also developed for the binding and dissociation rate coefficients and the affinity as a function of the analyte concentration in solution. The parameter $K(=k_{\text{diss}}/k_{\text{bind}})$ values presented provide an indication of the stability, reusability, and regenerability of the SPR biosensor. Also, depending on one's final goal, it was found that a higher or a lower K value could be beneficial for a particular analyte-receptor SPR system.

The fractal dimension for the binding ($D_{f,\text{bind}}$) or the dissociation phase ($D_{f,\text{diss}}$) is not a typical independent variable, such as analyte concentration, that may be directly manipulated. It is estimated from Eqs. (12.1a) and (12.1b) [or Eqs. (12.8a) and (12.8b)], and one may consider it as a derived variable. Note that a change in the degree of heterogeneity on the surface would, in general, lead to changes in both the binding and the dissociation rate coefficients. Thus, this may require a little thought and manipulation. The binding and the dissociation rate coefficients are rather sensitive to their respective fractal dimensions or the degree of heterogeneity that exists on the biosensor surface, as may be noted by the high orders of dependence. It is suggested that the fractal surface (roughness) leads to turbulence, which enhances mixing, decreases diffusional limitations, and leads to an increase in the binding or the dissociation rate coefficient (Martin *et al.*, 1991).

More such studies are required to verify whether the binding and the dissociation rate coefficients are indeed sensitive to their respective fractal dimensions or the degree of heterogeneity that exists on the SPR biosensor surface. If this is correct, then experimentalists may find it worth their effort to pay a little more attention to the nature of the SPR surface and how it may be manipulated to control the relevant parameters and SPR (and other) biosensor performance in desired directions. In a more general sense, the treatment should also be applicable to non-biosensor applications wherein further physical insights could be obtained. The analysis, if extendable and applicable to cellular surfaces where analyte-receptor reactions occur, should be very useful. This is especially true if the cellular surface heterogeneities could be modulated in desired directions, thus effectively manipulating the reaction velocity and direction of these cellular membrane surface reactions.

REFERENCES

Anderson, J., NIH Panel Review Meeting, Case Western Reserve University, Cleveland, OH, July 1993.

Berger, C. E. H., Beumer, T. A. M., Kooyman, R. P. H, and Greve, J., *Anal. Chem.* **70**, 703 (1998).

Bluestein, R. C., Diaco, R., Hutson, D. D., Miller, W. K., Neelkantan, N. V., Pankratz, T. J., Tseng, S. Y., and Vickery, E. K., *Clin. Chem.* **33**(9), 1543 (1987).

Cooper, M. A. and Williams, D. H., *Anal. Biochem.* **276**, 36–47 (1999).

Havlin, S., in *The Fractal Approach to Heterogeneous Chemistry: and Surfaces, Colloids, Polymers* (D. Avnir, ed.), Wiley, New York, p. 251, 1989.

Inoue, A., Yanagisawa, M., Kimura, S., Kasuya, Y., Myiauchi, T., Goto, K, and Msaki, T., *Proc. Natl. Acad. Sci. USA* **86**, 2863 (1989).

Karlsson, R. *et al.*, *J. Immunol. Methods* **145**, 229 (1991).

Karlsson, R. and Stahlberg, R., *Anal. Biochem.* **228**, 274–280 (1995).

Laricchia-Robbio, L., Moscato, S., Guidi, A., Vigano, S., Rovero, P., and Revoltella, R. P., *Biosen. & Bioelectr.* **12**(8), 765 (1997).

Lofas, S., Malmqvist, M., Ronnberg, I., Stenberg, E., Liedberg, B., and Lundstrom, I., *Sensors Actuators* **5**, 79 (1991).

Lukosz, W., *Biosen. & Bioelectr.* **6**, 215 (1991).

Lundstrom, I., *Biosen. & Bioelectr.* **9**, 725 (1994).

Martin, S. J., Granstaff, V.E., and Frye, G. C., *Anal. Chem.* **65**, 2910 (1991).

Nygren, H. A. and Stenberg, M., *J. Colloid Interface Sci.* **107**, 560 (1985).

Nylander, C., Liedberg, B., and Lind, T., *Sensors Actuators* **3**, 79 (1982).

O'Shannessy, D. J., Brigham-Burke, M., Soneson, K. K., Hensley, P., and Brooks, I., in "Methods of Enzymology," (M. L. Johnson and L. Brand, eds.), Vol. 240 (Part B), pp. 323–349, Academic Press, San Diego, 1995

Patten, P. A., Gray, N. S., Yang, P. L., Marks, C. B., Wedemayer, G. J., Boniface, J. J., and Stevens, R. C., *Science* **271** (5252), 1086 (1996).

Peterlinz, K. A. and Georgiadis, R., *Opt. Commun.* **130**, 260 (1996).

Peterlinz, K. A., Georiadis, R. M., Herne, T. M., and Tarlov, M. J., *J. Am. Chem. Soc.* **119**, 3401 (1997).

Ramsden *et al.*, *Phys. Rev. E* **50**, 5072 (1994).

Sadana, A. and Madagula, A., *Biosen. & Bioelectr.* **9**, 45 (1994).

Sadana, A. and Sutaria, M., *Biophys. Chem.* **65**, 29 (1997).

Sadana, A. and Sutaria, M., *Biotech. Progr.* **13**, 464 (1997a).

Sadana, A. and Sutaria, M., *Appl. Biochem. Biotech.* **62**(2–3), 275 (1997b).

Sadana, A., *Appl. Biochem. & Biotech.* **73**(2–3), 89 (1998).

Sadana, A., *Biosen. and Bioelectr.* **14**, 515–531 (1999).

Service, R. F., *Science* **278**, 806 (1997).

Sigal, G. B., Bamdad, C., Barberis, A., Strominger, J., and Whitesides, G. M., *Anal. Chem.* **68**(3), 490 (1996).

Sigmaplot, Scientific Graphing Software, User's Manual, Jandel Scientific, San Rafael, CA, 1993.

Silin, V. and Plant, A., *Trends in Biotechnol.* **15**, 353–359 (1997).

Sorenson, C. M. and Roberts, G. C., *J. Colloid & Interface Sci.* **186**, 447 (1997).

Wink, T., van Zullen, S. J., Bult, A., and van Bennekom, W.P., *Anal. Chem.* **70**(5), 827 (1998).

Yanagisawa, M., Kurihara, H., Kimura, S., Tomobe, Y., Kobayashi, M., Mitsui, Y., Yazaki, Y., Goto, K., and Masaki, T., *Nature* **322**, 411 (1988).

CHAPTER 13

Economics and Market for Biosensors

13.1. INTRODUCTION

Since biosensors are real-time (or near real-time) measuring and detection devices, the market for them is bound to grow in the future. This is in spite of the numerous difficulties that need to be overcome to make them more efficient and cost-effective. Kilmetz and Bridge (1997) emphasize that although the biosensor market is growing, it remains immature and imbalanced. This market will be significantly influenced by global demand in the areas of environmental, health, and safety laws. The primary impetus for the development of biosensors still is in the health field. The medical market is large, and of the clinical diagnostic applications in use today, home glucose testing for diabetics claims close to 90% of the market. According to the Cranfield University Report (1997), other applications include other medical (2%), environmental (2%), and other miscellaneous applications (2%). This same report emphasizes that biosensors offer the sensitivity, specificity, and, more particularly, the convenience required by the average person. This convenience should greatly assist in expanding the market for biosensors, especially if biosensors can be developed for a wide range of applications in the health and environmental fields.

A particular advantage of real-time measurements is that the results obtained may be acted upon immediately. Furthermore, biosensors have the potential to provide precise, real-time results in a user-friendly format—essential ingredients for clinical diagnostics. Thus, it is not surprising that the major use of biosensors is in the home testing market. However, the use of biosensors must expand in other markets if these devices are to make an impact as real-time measuring instruments. Some of the issues that need to be addressed before this becomes a reality include:

(a) suitable methods for the mass production of biosensors,
(b) more compatability between the molecular recognition step and the transduction step (molecular and interface engineering),
(c) more efficient transduction of the signal that is biologically recognized (perhaps "direct" sensing),
(d) the development of nano- and microfabricated sensors including array-based sensors.

With the advent of nanotechniques and the recent emphasis on nanoengineering, the influence of these and other aspects that are constraining the further development of biosensors should gradually be minimized. However, nanotechniques and nanoengineering are not without some inherent limitations, as indicated by Wilding and Kricka (1999). These authors analyzed how small microchips can be made without exceeding the limits of detection. They also note that if more than one site is involved in the detection step, the resolution of the detection method may also become a problematic issue.

13.2. MARKET SIZE AND ECONOMICS

The market for biosensors is still a niche market, and as expected there are various estimates for their total market. Data will be presented here for the years it is available in the open literature. Since data from industrial sources is inaccessible, it of course, is not included, although it would provide a more *realistic* picture of the past, present, and future market for the overall biosensor market in different areas. However, considering the consolidation fever prevalent in the current business world, it is not surprising that any type of financial projections or information is, and will continue to be, jealously guarded. As in all businesses, but particularly in developing ones, a biosensor company needs to be nimble and be able to seize opportunities as they occur.

Before we talk about numbers and market projections, it is perhaps worthwhile to mention some of the major companies in the biosensor market. According to the Cranfield University Report (1997) the three major companies are MediSense (acquired by Abbott in 1996), Bayer, and Boehringer Mannheim. Other companies include Affinity Sensors, BIACORE, YSI, Chiron Diagnostics, Diametrics Medical, i-STAT, Molecular Devices Corporation, Nova Biomedical, Universal Sensors Incorporated, Sandia Laboratories, Texas Instruments, Eppendorf, and LifeScan. Anticipating the increasing potential in this area, some of the larger companies are repositioning at least a part of their effort. For example, Smith (2001) indicates that the production of biochip and microarray technology has become a focus for Packard Biosciences. Packard acquired GSLI Life Sciences in October 2000 to create the spin-off company, Packard Bio Chip Technologies.

Similarly, DuPont—with its expertise in polymer thick films—is positioning itself to be a strong player in the biosensor market. DuPont acquired Cyngus, Inc., which developed the GlucoWatch biographer. This is the first noninvasive biosensor based on DuPont's thick film paste, and it easily monitors glucose levels twenty-four hours a day by analyzing fluids drawn through the skin. This is definitely an improvement in the quality of life for diabetics. Assured that it can make a major impact in biosensor development (Hodgson, 1999), DuPont is well-positioned to create a plastic display that could potentially replace the most expensive parts of the biosensor. Besides, their product will be lighter and of enhanced visual quality.

For example, in 1999, researchers at DuPont were investigating the development of disposable polymer thick film (PTF) biosensors. PTF inks are paints that contain a dispersed or dissolved phase and that acquire their final properties by drying. As the paint is cured on a suitable substrate, a specific electronic or biological function is developed in the film. A very specific advantage of PTF products is that they are compact, lightweight, environmentally friendly, inexpensive, and, most important of all, they lend themselves to manufacturing techniques. They can be easily folded, twisted, or bent, all of which are required properties for the components of a biosensor. With the ever-increasing pressure to provide cost-effective biosensors, DuPont is apparently a major player in the development of biosensors.

The universities and governments are also collaborating and consolidating their strengths for biosensor development. For example, Cranfield University is a world leader in biosensor research. Another group in England is the University of Manchester Biosensor Group. The Irish government too has created a National Center for Sensor Research (NCSR) (Bradley, 2001) at Dublin City University (DCU). DCU has a track record for biomedical and environmental sensors, and the Irish government provided $13.2 million to establish the NCSR. Its role is to develop chemical and biological sensors to solve society-related problems. One of the goals of the NCSR is to relate higher-education funding with the local economy.

The British government too has provided more than $2 million in funding (with matching funds from industrial backers) to promote the lab-on-a-chip concept (Henry, 1999) to be coordinated at the University of Hull. There are of course, many other examples. The given examples simply point out the importance placed by universities, industry, and government on collaborative schemes to facilitate the development of biosensors for different applications.

The Cranfield University Report (1997) estimated that in 1996 the world biosensor market was $508 million. In that year, glucose testing (medical) was the major application, with a sales of $170 million for MediSense (Abbott) and a sales of $165 million for Kyoto Diiachi/Bayer/Menarini. These two companies accounted for about two-thirds (65.9%) of the total sales of biosensors. The Frost and Sullivan (1998) report estimated the total U.S. market for biosensors at $115 million. This study analyzed the U.S. biosensor market in detail and indicated the following four areas where biosensor applications are expected to grow: medical home diagnostics market; medical point-of-care market; medical research market; and environmental, industrial, and other markets.

Rajan (1999) indicated that the biosensor market is expected to grow at a rate of 17% from 1998 to 2003. If we were to extrapolate this author's 17% growth factor for the years 1996 to 1997 and the Cranfield University numbers, the 1997 worldwide market for biosensors should have been around $592 million. This is close to the $610 million estimate of biosensor sales provided by Theta Reports (1998). The Theta study indicates that more than 50 different biosensor systems are available worldwide. Also, out of this $610 million in sales, $500 million was generated by clinical diagnostic applications. Out of this $500 million, 90% was related in some form or the other to home glucose testing for diabetics. Thus, glucose home testing is, and will continue to be, a major driving force for biosensor sales. This is not surprising since no other disease combines the two requirements that lead to mass monitoring: a large portion of the population are afflicted with diabetes (up to 1%) and frequent recording of blood glucose levels (up to 2 to 4 times a day) is required (Medical Device Technology, 1997). In the United States alone, more than 16 million individuals (half of which are undiagnosed) are estimated to suffer from diabetes (SBI International, 1997). Also, this last report indicates that between 600,000 and 700,00 new cases are diagnosed every year.

Theta Reports (1998) indicates that from 1997 to 2000 there will be a slow period of growth for biosensors, unlike the 17% growth predicted by Rajan (1999). However, the overall biosensor market for 2000 was expected to be $2 billion. Between 2000 and 2005, Theta Reports indicates a substantial increase in biosensor sales, with the sales increasing by a factor of 4.4 from $2 billion in 2000 to $8.8 billion in 2005. Theta Reports further estimates the sales of clinical genosensors to reach $1.6 billion in 2005. Chemcor Corporation, a developer of genosensor technology, estimates the market for this technology to exceed $1 billion by 2002. Ruth (2001) further estimated the market for molecular and cytogenetic testing devices to be $66 million in 2000. This was expected to increase to $100 million in 2005.

Quantech (2000) estimates that excluding home diagnostics, the overall worldwide *in-vitro* diagnostic market is $20 billion. This number is an order of magnitude higher than that predicted ($2 billion) by Rajan (1999). Since it does not include home diagnostics, the very wide discrepancies in the estimates are to be expected. Nevertheless, the estimates do provide an order of magnitude set of numbers that also indicate the range of the market estimates. Quantech further indicates that companies and laboratories account for most of the market in this area. For example, STAT testing is an important aspect of this market. STAT tests are required by physicians and surgeons during surgery and in emergency departments because of the time-sensitive nature of the needed treatments and the rapid decisions required. Furthermore, based on the surface plasmon resonance (SPR) principles, Quantech has developed a menu that provides different tests for a physician to help make a treatment decision. Some of the tests that can be run quickly include a test for three cardiac markers (myoglobin, CK-MB, and Troponin I) for heart attacks, a quantitative test for pregnancy (to determine whether it is safe to perform some procedures), a blood count panel, a kidney panel, and a coagulation panel.

The Japanese are very practical minded, and their approach to biosensor development is no exception. They have determined that biosensor technology is and will have a significant impact on daily life. Dambrot (1999) indicates that quite a few Japanese companies are making a wide variety of biosensors. Some of these include Dainippon Printing (immune-system monitoring), INAX (*in vitro* measurement of albumen in urine), Itoh (high-sensitivity meat freshness), Nissin Seifun (fruit ripeness), and Toto (health and medical monitoring). As around the world, in Japan the main application for biosensors will be disposable biosensors in the health care field. Other applications include the determination of food quality and in telemetric biosensors (for monitoring fatigue in sports, athletics, and a driver's state of alertness). Professor Karube and others at Tokyo University are developing a "toilet sensor" to monitor various bodily functions. This will be especially helpful in managing the health of the elderly.

Dambrot provided an estimate of $16 billion for the Japanese biosensor market for the year 2000. This is seemingly much higher than the numbers provided earlier in this chapter, and this the number is for Japan only. Once again, this highlights the discrepancies from different sources, as expected, for the worldwide biosensor market.

In her report on the market for biosensors, Rajan (1999) indicates that a substantial amount of money and much effort is required to bring a biosensor to the market. However, with the ever-increasing research in this area the costs are bound to decrease, and biosensor applications should be able to expand as well. The overall biosensor market for 1998 was estimated at $765 million, with medical sales of $692 million constituting, as expected, a large percentage (90.4%). With the growing concerns of health care and personal well-being, Rajan estimates that by 2003 the medical sales share will increase to 93% of the total sales. She estimates that the medical sales of biosensors will exceed $1.5 billion.

The sales of biosensors may be estimated for 2001 to 2003 using the 17.0% average annual growth rate provided by Rajan. Starting with the estimate of $765 million for 1998, the estimates for 1999, 2000, 2001, 2002, and 2003 are (in $ million) 895, 1047, 1225, 1433, and 1677, respectively. Similarly, starting with an estimate of $692 million for the medical sales of biosensors in 1998 and using an average annual growth rate of 17.6%, the estimates for 1999, 2000, 2001, 2002, and 2003 are (in $ million) 814, 957, 1125, 1323, and 1556, respectively. Other areas where growth of biosensors is expected include industrial, environmental, government, and research. However, the combined sales of biosensors in these areas will continue to be only a small fraction of the total sales.

For 2000 and beyond estimates have been presented for worldwide sales of biosensors ranging from $2 billion (Rajan, 1999) to about $20 billion (to about $16 billion for Japan alone) (Quantech, 2000). Keeping these estimates and projections in mind, it is perhaps conservative to say that in the first 3–5 years of the new millennium, the overall worldwide sales of biosensors should be around $10–12 billion.

At present BIACORE is not one of the major players in the biosensor market. However, it does manufacture and sell the surface plasmon resonance (SPR) biosensor that is becoming increasing popular as various organizations are finding different applications for it. Although the sales figures for the BIACORE biosensor for the years 1995 to 2000 are available (BIACORE, 2001), these figures include the instrument and the reagents and other materials required for it (see Table 13.1 and Fig. 13.1). No sales figures for the instrument alone were available. A conservative estimate is that the reagents costs account for about 15%–20% of the total sales. The BIACORE 3000 that came out in 1998 is estimated to cost about $275,000. Using the figures in Figure 13.1 we can obtain a predictive equation to predict the sales figures for 2001 to about 2005. As expected, if we used more parameters, we could obtain a better predictive equation.

Table 13.1 show the sales figures in SEK (Swedish Kroner). These sales figures may be predicted using the following two-parameter equation:

$$\text{Sales SEK (1000s)} = (246333.2 \pm 18557.7)[\text{year}]^{0.21 \pm 0.06}, \tag{13.1}$$

TABLE 13.1 Total Sales Figures for the BIACORE AB (2001)

Year	Sales SEK (1000s)
1995	209012
1996	260352
1997	266523
1998	288753
1999	340414
2000	361600

where year 0 refers to 1995, year 1 refers to 1996, and so on. Using this equation, the coefficient for regression had a value of 0.816. The fit is reasonable; however, a better fit may be obtained with a four-parameter equation (along with a higher coefficient of regression).

Using the four-parameter equation, the sales figures given in Table 13.1 may be given by

$$\begin{aligned}\text{Sales SEK (1000s)} &= (255741.2 \pm 8867.4)[\text{year}]^{0.088 \pm 0.043} \\ &\quad + (178733.1 \pm 56501.7)[\text{year}]^{0.45 \pm 0.09}.\end{aligned} \tag{13.2}$$

As expected, the four-parameter equation fits the sales figures presented in the

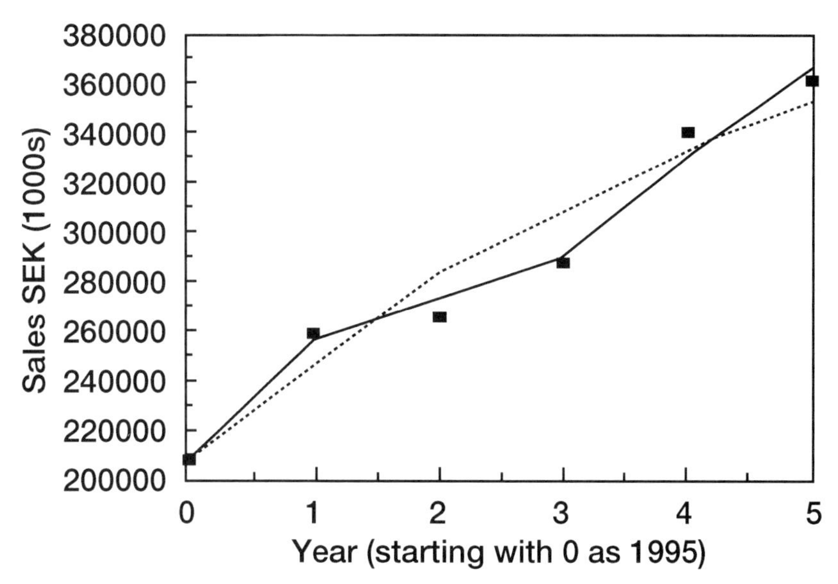

FIGURE 13.1 Sales figures (in Swedish Kroner, 1000s) for the BIACORE biosensor for 1995 to 2000. 1995 is taken arbitrarily as year zero. Approximately U.S.$ 1.0 = 10 Swedish Kroner.

table rather well. The coefficient of regression in this case is 0.964 (for the later part of the figure, which is more important in making sales projections for 2001, 2002, and beyond). It would, of course, be of tremendous interest, at least for BIACORE AB organization, to see if Eq. (13.2) does accurately predict the sales figures for the early years of the new millennium.

As to be expected during the development and bringing to market of any (profitable or potentially profitable) technology, there are claims and counterclaims. An example is the recent misunderstanding between Oxford Gene Technology (OGT), Oxford, England, and Affymetrix, Santa Clara, California, regarding claims to critical DNA microarray technology developed by OGT (Robertson, 2001). The ruling went against OGT, and that company is expected to lose millions of dollars (or British pounds) in royalties. However, Robertson adds that this decision should not hamper the development of the chip industry, especially since new entrants in this area are bound to move to alternative microarray systems. Thus (biosensor) companies need to be nimble, as mentioned earlier, and they need par excellence research and manufacturing abilities. Robertson very nicely points out the "intellectual minefield" that is prevalent in this area. Perhaps this is also true of areas other than biosensor development.

It would also be appropriate to provide an example of some emerging trends in biosensor development that seem to exhibit market and/or application potential. Piletsky *et al.* (2001) indicate that molecularly imprinted polymers (MIP) are particularly suitable for use in biosensor development. MIPs exhibit high affinity and selectivity (similar to natural receptors), are more stable than their natural counterparts, are easy to prepare, and are easily adaptable to different applications. Also, MIPs can be synthesized as receptors for analytes for which no enzyme or receptor is available. Besides, they are cheaper than natural receptors. Also, an important aspect that is often neglected when analyzing newer techniques is the manufacturing and fabrication capabilities of the technique. Fortunately, as Piletsky *et al.* indicate, MIPs (or the polymerization step) are easily amenable to microfabrication steps involved in biosensor technology. These authors note that MIP sensors have been developed for herbicides, sugars, nucleic and amino acid derivatives, drugs, toxins, etc. The authors suggest that the market for multisensors (electronic noses and tongues) could be worth as much as $4 billion. In their opinion MIP sensors are well suited to make an impact in the area of testing product quality as well as in the perfumes and wine industries.

13.3. DEVELOPMENT COST OF A BIOSENSOR

In our attempt to provide some cost estimates for the development of a biosensor for different applications, we found that such data is difficult to obtain in the open literature. However, we provide two examples.

In one instance, the Swedish Rescue Service wanted an "artificial dog nose" to help localize minefields and free other areas of land. (Incidentally, this is also a high priority for the United Nations. For example, the estimated cost to clear Angola of minefields is about $40 billion or SEK 400 billion). Biosensor Applications (1999) in Sweden intended to deliver a prototype in August 1999 that would have cost SEK 1.1 million (roughly $100,000). Field testing in a central military field in Sweden was planned and an actual application in Angola was also being considered.

In another instance, a multianalyte biosensor instrument was proposed for the Environmental Protection Agency (EPA) for detecting phenolic compounds and for pathogens (Schmidt, 1997). The intent was to reduce the analysis turn-around time—from approximately 19 days to 15 minutes—and to reduce the annual cost of environmental analyses in the United States by more than $20 million. Schmidt emphasizes that the major barrier to the development of a biosensor for environmental measurement is the diversity of the environmental market. Thus, no one analyte application is large enough to justify the development cost of a specific biosensor instrument. This points out the need for a disposable biosensor that is capable of being adapted to dozens of analytes. Schmidt proposed that he could develop the required biosensor in 6 to 7 months at a cost of approximately $70,000.

Sensitive detection and rapid screening systems are critically needed to detect environmental chemicals that exhibit estrogenic and androgenic agonist and antagonist activities. For example, the pesticide methoprene binds to retinoic acid (vitamin A) receptors and results in developmental abnormalities. Similarly, the phytochemical diethylstilbestrol and the pesticide DDT interact with estrogen receptors and have been implicated in developmental and reproductive disorders as well as hormone-dependent cancers. Also, dioxin or TCDD, a by-product of manufacturing processes using trichlorophenols, also has been implicated in developmental defects and in tumor formation. Another example is the organophosphate insectide methyl parathion that disrupts neuronal networks in the central and peripheral nervous systems, resulting acutely in tremor, seizures, coma, or death and chronically in behavior abnormalities or sustained neuropathy. Thus, inexpensive *in-vitro* assays for the detection of steriodiogenic activity in environmental chemicals in the Mississippi River basin has been proposed (University of Mississippi, 2000). An initial estimate of $250,000–300,000 was obtained for the development of this biosensor with the time period of 1–2 years (Vo-Dinh, 2000). Sometimes, the high cost of development includes the high overhead charges of a particular institution or organization.

However, there is a critical need for such biosensors since, as Wiseman *et al.* (1999) indicate, there is increasing public concern about the gender hazard caused by environmental and dietary phytoestrogens and xenoestrogens. This leads to increased risk of damage to species ecowebs (Polis, 1988). An early assessment of risk is required for the developing ecological disruption that includes *in vitro* assays for compounds known as estrogen mimics (Lynch and Wiseman, 1988) and *in vivo* bioindicators. Furthermore, Wiseman *et al.* suggest another approach that involves analyzing the ecoweb breakdown in five or more species (ecotranslators).

In their review of environmental biosensors, Rogers and Gerlach (1996) indicate that the new instruments and methods being developed do exhibit promise for the continuous *in situ* monitoring of toxic compounds. These authors compared two different detection systems (an immunoassay kit and a biosensor system) for the monitoring of groundwater pump-and-treat systems. Their projections were derived from cost-per-sample versus initial investment cost. They estimated the cost per assay for the immunoassay kits ranged from $50 to $75, and the cost for the biosensor was $8. The biosensor cost was slightly offset by the start-up costs for the immunoassay kits, which were considerably cheaper: The start-up costs for the immunoassay kits and the biosensor system were $3000 and $20,000, respectively.

Bradley (1998) indicates that a new class of sensors for environmental monitoring of gases such as CO_2, SO_2, O_3 and nitrogen oxide has been described by Dasgupta *et al.* (1998). This sensor is based on an amorphous Teflon polymer. A 20-μm-thick tube filled with liquid acts as a liquid-core optical fiber. This is highly permeable to various gases, including those of environmental interest. Dasgupta *et al.* indicate that in its simplest form the sensor can be fabricated for less than $100, and it can be used to monitor pollutant gases at ambient temperature. Finally, the response times are less than 1 sec because the design facilitates diffusion, which is often a hindrance in biosensor development and in analyzing analyte-receptor binding kinetics.

Georgia Institute of Technology (1999) in Atlanta, Georgia, has developed a biosensor that is apparently expected to improve food safety. It identifies and determines concentrations of multiple pathogens such as *E. coli* 0157:H7 and *Salmonella*. This biosensor can detect these pathogens in food products in less than 2 hours while in operation on a processing plant floor. Of course, its most important contribution is the considerable reduction in time required to assess the presence of contamination. The biosensor designed costs $1000 to $5000 and can detect cells at levels of 500 cells per ml. Current laboratory techniques cost around $12,000 to $20,000 and can only detect cell levels of

5000 cells per ml. In addition, the laboratory techniques take from 8 to 24 hours to yield results.

We now discuss the products for the clinical glucose market. Weetall (1996) indicates that this market is unique in that it is large enough to encourage stand-alone products. In other words, it does not have to provide a menu of tests, to be competitive. In 1996, Weetall indicated that the products for this market range from several cents for paper strips to around a few dollars for disposable electrodes used in commercially available electrochemical devices. Present-day costs include, for example, around $60–$70 for the device that produces reliable results from a small blood spot (which is less traumatic than having to prick and press your finger to get a reasonable size spot as with some less expensive devices). The strips that are used to get a quantitative result cost around a dollar.

Finally, as a last example, we discuss the development of a biosensor that closely mimics biological sensory functions (Downard, 1998). No costs were available, but it took 10 years and a team of 60 scientists and engineers to convert this ion-channel switch (ICS) biosensor from a concept to a practical device (Cornell, 1997). The advantage of this sensor is that it directly provides a functional test of the interaction between a potential drug and an artificial cell membrane. According to Cornell, changes in the ion flux across the membrane may be detected as a change in the membrane's electrical conductance. This author indicates that in essence they have designed a tethered membrane that permits one to quickly screen drugs of importance very efficiently.

13.4. COST REDUCTION METHODS

The biological receptor used in biosensors may itself account for quite a high fraction of the entire cost of the biosensor, depending on the receptor used. Table 13.2 shows the approximate costs of making antibody fragments for possible immobilization on a biosensor surface (Harris, 1999). The table indicates that to keep costs low, one should use microorganisms or plants to make the required antibodies, if possible.

As indicated by Turner (2000), inexpensive biological receptors is a critical issue in the high development costs of biosensors. There is a lack of suitable biological recognition molecules that are inexpensive to manufacture and also stable during storage. Turner further indicates that this problem is exacerbated when one is manufacturing high-density arrays to be used in medical diagnostics, functional genomics, proteomics, environmental monitoring, food preservation and safety, etc.

TABLE 13.2 Approximate Costs for Making Antibody Fragments (Harris, 1999)

Source	Yield	Cost ($ per gram)
Mammalian cells	0.5–1.0 g/l	450.00
Transgenic milk	10 g/l	90.00
Bacteria	3 g/l	1.00
Transgenic plants	2 g/kg	0.30
Viral vector	10 g/kg	0.06

Nova Biomedical (1997) indicates that their analyzers are cost-effective because their biosensors are long-lived and reusable. Furthermore, their Nova Stat Profile analyzers consolidate up to five different analyzers. They indicate that their reusable biosensor technology is about five to ten times less expensive than handheld disposable devices. Furthermore, the consumable(s) cost to perform a Chem 6 plus hemacrit on a typical disposable cartridge-based device is $5.00 per sample, or roughly $0.83 per test. This same test on a Nova Stat Profile analyzer is only $0.80 per sample, or roughly $0.11 per test. For 25 samples per day, this leads to a cost savings of about $105 per day. Assuming that tests on samples are performed 300 days per year, this is a saving of about $31,500 per year. Another advantage of the Nova Stat profile Analyzer is that the tests results are given in about 5 min. Thus, the turn-around time is very low as compared to routine testing in a central laboratory.

13.5. CONCLUSIONS

Biosensors seem to offer tremendous potential in their application to different and ever-expanding areas. Both the academic and industrial sectors have realized this and are gradually putting more and more resources in this area. As in all research and development areas one has to keep the economic aspects in focus. Universities, with their culture of openness and their emphasis on research, publish findings quickly in the open literature, but tend to pay scant attention to economic aspects.

Industrial organizations may be expected to undertake a careful and thorough evaluation of economic and other aspects before investing in biosensors. Such information is jealously guarded. In particular, economic forecasting organizations and trade groups which collect such information consider it a source of income. Thus, very little reliable economic information on biosensors is available in the open literature. More information needs to be made available, so that it can be used in the future development of biosensors.

Perhaps universities should encourage their students to write theses on economic aspects of biosensor development, though the research component of this type of endeavor will be understandably questioned. This chapter attempts to make a start on this critical need. As the field matures, collaborations between universities and the industrial sector will be strengthened if there is a freer flow of economic information.

REFERENCES

BIACORE AB Company, Five year financial overview, 2001.

Biosensor Applications, "Swedish Rescue Service orders first 'Artificial Dog Nose," March 1999.

Bradley, D., *Anal. Chem. (News and Features)* December, 771A (1998).

Bradley, D., *Anal. Chem.* February, 70A–71A (2001).

Cornell, B., *Nature* **387**, 580 (1997).

Cranfield University Report on Biosensors, 1997.

Dambrot, S. M., *Bio/Technology*, **129** (1999).

Dasgupta, P., *et al.*, *Anal. Chem.*, 4661 (1998).

Downard, A., *Anal. Chem., (News and Features)* 774A, December (1998).

DuPont Microcircuit Materials, Disposable Polymer Thick Film Biosensors, 1999.

Frost and Sullivan, "Emerging Applications Attract Large Healthcare Companies to Biosensors Market," Report 5459-32-U.S. Biosensors Markets, November 1998.

Georgia Institute of Technology, "Biosensor Expected to Improve Food Safety," Atlanta, GA, September, 1999.

Harris, W. J., *TIBTECH* **17**, 290–294 (1999).

Henry, C., *Anal. Chem. (News and Features)* February 86A (1999).

Hodgson, J., *DuPont Photopolymer & Electronic Materials Targets Information Technologies for the 21st Century*, December 16, 1999.

Kilmetz, S. D. and Bridge, R. S., *Med. Electr.* **28**, 56–58 (1997).

Lynch, J. M. and Wiseman, A., eds., *Environmental Biomonitoring: The Biotechnology–Ecotoxicology Interface*, Cambridge University Press, Cambridge, 1988.

Medical Device Technology, "Biosensors Markets—Opportunities and Obstacles," 1994.

Nova Biomedical, "Critical Care Testing Solutions," 1997.

Piletsky, S. A., Alcock, S., and Turner, A. P. F., *TIBTECH* **19**(1), 9–13 (2001).

Polis, G. A., *Nature* **395**, 744–745 (1998).

Quantech, Quantech Home, Eagan, MN, 2000.

Rajan, M., "Biosensors Market to Grow at 17% Annually from 1998–2003," Study RC-053X, Business Communications Co., Norwalk, CT, March 8, 1999.

Robertson, D., *Nature Biotech.* **19**, 13 (2001).

Rogers, K. R. and Gerlach, C. L., *Envir. Sci. & Technol.*, November (1996).

Ruth, L., *Anal. Chem.*, January, 16A (2001).

SBI, Inc., "Glucose Biosensor," Kettering, OH, 1997.

Schmidt, J. C., EPA Contract 68D70028, September 1997–March 1998.

Smith, W., *Anal. Chem.*, January, 16A (2001).

Theta Reports, Report 501, New York, NY, October 1998.

Turner, A. P. F., *Science* **290**, 1315–1316 (2000).

University of Mississippi, "UM Environmental Signals and Sensors," University of Mississippi Medical Center/Center for Disease Control (CDC), Atlanta, October 2000.

Vo-Dinh, T., Personal communication, 2000.
Weetall, H. H., *Biosen. & Bioelectr.* **11**(1/2), Guest Editorial, i–iv (1996).
Wilding, P. and Kricka, L. J., *TIBTECH* **17**, 465 (1999).
Wiseman, A., Goldfarb, P. S., Ridgway, T. J., and Wiseman, H., *Trends in Biotech.* **17**(2), 43 (1999).
Wiseman, A., Goldfarb, P. S., Ridgway, T. J., and Wiseman, H., *Trends in Biotech.* **18**(1), 1 (2000).

INDEX

E

F

S

T

U

Y

RECEIVED
JAN 28 2002